Student Solutions Manual for Charles P. McKeague's Essentials of Elementary & Intermediate Algebra
Second Edition

Prepared by

Ross Rueger
Department of Mathematics
College of the Sequoias
Visalia, California

**Student Solutions Manual for
Charles P. McKeague's Essentials of Elementary
& Intermediate Algebra, Second Edition**

Ross Rueger

Publisher: XYZ Textbooks

Sales: Amy Jacobs, Richard Jones, Bruce Spears, Rachael Hillman

Cover Design: Kyle Schoenberger

© 2018 by McKeague Textbooks LLC

ALL RIGHTS RESERVED. No part of this work covered by the copyright herein may be reproduced, transmitted, stored or used in any form or by any means graphic, electronic, or mechanical, including but not limited to photocopying, recording, scanning, digitizing, taping, Web distribution, information networks, or information storage and retrieval systems, except as permitted under Section 107 or 108 of the 1976 United States Copyright Act, without the prior written permission of the publisher.

ISBN-13: 978-1-63098-156-3 / ISBN-10: 1-63098-156-7

> For product information and technology assistance, contact us at
> **XYZ Textbooks, 1-877-745-3499**
> For permission to use material from this text or product,
> e-mail: **info@mathtv.com**

XYZ Textbooks
1339 Marsh Street
San Luis Obispo, CA 93401
USA

Printed in the United States of America

For your course and learning solutions, visit **www.xyztextbooks.com**

Student Solutions Manual to Accompany

Essentials of Elementary & Intermediate Algebra: A Combined Course

by
Charles P. McKeague

Prepared by
Ross Rueger
Department of Mathematics
College of the Sequoias
Visalia, California

XYZ Textbooks

Contents

Preface			v
Chapter 0	**The Basics**		**1**
	0.1	Notation and Symbols	1
	0.2	Real Numbers	2
	0.3	Addition of Real Numbers	4
	0.4	Subtraction of Real Numbers	5
	0.5	Properties of Real Numbers	6
	0.6	Multiplication of Real Numbers	8
	0.7	Division of Real Numbers	9
	0.8	Subsets of the Real Numbers	11
	0.9	Addition and Subtraction with Fractions	12
		Chapter 0 Test	15
Chapter 1	**Linear Equations and Inequalities**		**17**
	1.1	Simplifying Expressions	17
	1.2	Addition Property of Equality	19
	1.3	Multiplication Property of Equality	23
	1.4	Solving Linear Equations	28
	1.5	Formulas	34
	1.6	Applications	38
	1.7	More Applications	42
	1.8	Linear Inequalities	47
	1.9	Compound Inequalities	52
		Chapter 1 Test	55
Chapter 2	**Linear Equations and Inequalities in Two Variables**		**59**
	2.1	Paired Data and Graphing Ordered Pairs	59
	2.2	Solutions to Linear Equations in Two Variables	63
	2.3	Graphing Linear Equations in Two Variables	65
	2.4	More on Graphing: Intercepts	71
	2.5	The Slope of a Line	77
	2.6	Finding the Equation of a Line	81
	2.7	Linear Inequalities in Two Variables	84
		Chapter 2 Test	88

Chapter 3	Systems of Linear Equations		**91**
	3.1	Solving Linear Systems by Graphing	91
	3.2	The Elimination Method	95
	3.3	The Substitution Method	98
	3.4	Systems of Equations in Three Variables	102
	3.5	Matrix Solutions to Linear Systems	110
	3.6	Determinants and Cramer's Rule	123
	3.7	Applications of Systems of Equations	137
		Chapter 3 Test	144

Chapter 4	Exponents and Polynomials		**151**
	4.1	Multiplication with Exponents	151
	4.2	Division with Exponents	153
	4.3	Operations with Monomials	156
	4.4	Addition and Subtraction of Polynomials	158
	4.5	Multiplication with Polynomials	159
	4.6	Binomial Squares and Other Special Products	161
	4.7	Dividing a Polynomial by a Monomial	163
	4.8	Dividing a Polynomial by a Polynomial	165
		Chapter 4 Test	167

Chapter 5	Factoring		**169**
	5.1	The Greatest Common Factor and Factoring by Grouping	169
	5.2	Factoring Trinomials	170
	5.3	More Trinomials to Factor	172
	5.4	The Difference of Two Squares	173
	5.5	The Sum and Difference of Two Cubes	175
	5.6	Factoring: A General Review	176
	5.7	Solving Equations by Factoring	177
	5.8	Applications	181
		Chapter 5 Test	185

Chapter 6	Functions and Function Notation		**189**
	6.1	Introduction to Functions	189
	6.2	Function Notation	191
	6.3	Variation	194
	6.4	Algebra and Composition with Functions	198
	6.5	Regression Analysis and the Coefficient of Correlation	200
		Chapter 6 Test	204

Chapter 7	**Rational Expressions and Rational Functions**	**207**
	7.1 Basic Properties and Reducing to Lowest Terms	207
	7.2 Multiplication and Division of Rational Expressions	210
	7.3 Addition and Subtraction of Rational Expressions	213
	7.4 Complex Fractions	220
	7.5 Equations With Rational Expressions	225
	7.6 Applications	232
	7.7 Division of Polynomials	237
	Chapter 7 Test	241

Chapter 8	**Rational Exponents and Roots**	**245**
	8.1 Rational Exponents	245
	8.2 Simplified Form for Radicals	247
	8.3 Addition and Subtraction of Radical Expressions	249
	8.4 Multiplication and Division of Radical Expressions	251
	8.5 Equations Involving Radicals	253
	8.6 Complex Numbers	258
	Chapter 8 Test	260

Chapter 9	**Quadratic Equations**	**263**
	9.1 Completing the Square	263
	9.2 The Quadratic Formula	267
	9.3 Additional Items Involving Solutions to Equations	271
	9.4 More Equations	275
	9.5 Graphing Parabolas	278
	9.6 Quadratic Inequalities	284
	Chapter 9 Test	291

Chapter 10	**Exponential and Logarithmic Functions**	**295**
	10.1 Exponential Functions	295
	10.2 The Inverse of a Function	298
	10.3 Logarithms are Exponents	303
	10.4 Properties of Logarithms	306
	10.5 Common Logarithms and Natural Logarithms	309
	10.6 Exponential Equations and Change of Base	311
	Chapter 10 Test	316

Appendix A	**Piecewise Functions**	**319**

Preface

This *Student Solutions Manual* contains complete solutions to all odd-numbered exercises, and complete solutions to all chapter test exercises, of *Essentials of Elementary and Intermediate Algebra: A Combined Course* by Charles P. McKeague. I have attempted to format solutions for readability and accuracy, and apologize to you for any errors that you may encounter. If you have any comments, suggestions, error corrections, or alternative solutions please feel free to send me an email (address below).

Please use this manual with some degree of caution. Be sure that you have attempted a solution, and re-attempted it, before you look it up in this manual. Mathematics can only be learned by *doing*, and not by observing! As you use this manual, do not just read the solution but work it along with the manual, using my solution to check your work. If you use this manual in that fashion then it should be helpful to you in your studying.

I would like to thank Matthew Hoy and Amy Jacobs at XYZ Textbooks for their help with this project and for getting back to me with corrections so quickly. Producing a manual such as this is a team effort, and this is an excellent team to work with.

I wish to express my appreciation to Pat McKeague for asking me to be involved with this textbook. His books continue to refine the subject of elementary and intermediate algebra, and you will find the text very easy to read and understand. I especially appreciate his efforts through XYZ Textbooks to make textbooks affordable for students to purchase.

Good luck!

Ross Rueger
College of the Sequoias
rossrueger@gmail.com

July, 2017

Chapter 0
The Basics

0.1 Notation and Symbols

1. The equivalent expression is $x+5=14$.

3. The equivalent expression is $5y<30$.

5. The equivalent expression is $5y \geq y-16$.

7. The equivalent expression is $\dfrac{x}{3}=x+2$.

9. Evaluating the exponent: $3^2 = 3 \cdot 3 = 9$

11. Evaluating the exponent: $7^2 = 7 \cdot 7 = 49$

13. Evaluating the exponent: $2^3 = 2 \cdot 2 \cdot 2 = 8$

15. Evaluating the exponent: $4^3 = 4 \cdot 4 \cdot 4 = 64$

17. Evaluating the exponent: $2^4 = 2 \cdot 2 \cdot 2 \cdot 2 = 16$

19. Evaluating the exponent: $10^2 = 10 \cdot 10 = 100$

21. Evaluating the exponent: $11^2 = 11 \cdot 11 = 121$

23. Using the order of operations: $2 \cdot 3 + 5 = 6 + 5 = 11$

25. Using the order of operations: $2(3+5) = 2(8) = 16$

27. Using the order of operations: $5 + 2 \cdot 6 = 5 + 12 = 17$

29. Using the order of operations: $(5+2) \cdot 6 = 7 \cdot 6 = 42$

31. Using the order of operations: $5 \cdot 4 + 5 \cdot 2 = 20 + 10 = 30$

33. Using the order of operations: $5(4+2) = 5(6) = 30$

35. Using the order of operations: $8 + 2(5+3) = 8 + 2(8) = 8 + 16 = 24$

37. Using the order of operations: $(8+2)(5+3) = (10)(8) = 80$

39. Using the order of operations: $20 + 2(8-5) + 1 = 20 + 2(3) + 1 = 20 + 6 + 1 = 27$

41. Using the order of operations: $5 + 2(3 \cdot 4 - 1) + 8 = 5 + 2(12-1) + 8 = 5 + 2(11) + 8 = 5 + 22 + 8 = 35$

43. Using the order of operations: $8 + 10 \div 2 = 8 + 5 = 13$

45. Using the order of operations: $4 + 8 \div 4 - 2 = 4 + 2 - 2 = 4$

47. Using the order of operations: $3 + 12 \div 3 + 6 \cdot 5 = 3 + 4 + 30 = 37$

49. Using the order of operations: $3 \cdot 8 + 10 \div 2 + 4 \cdot 2 = 24 + 5 + 8 = 37$

51. Using the order of operations: $(5+3)(5-3) = (8)(2) = 16$

53. Using the order of operations: $5^2 - 3^2 = 5 \cdot 5 - 3 \cdot 3 = 25 - 9 = 16$

55. Using the order of operations: $(4+5)^2 = 9^2 = 9 \cdot 9 = 81$

57. Using the order of operations: $4^2 + 5^2 = 4 \cdot 4 + 5 \cdot 5 = 16 + 25 = 41$

59. Using the order of operations: $3 \cdot 10^2 + 4 \cdot 10 + 5 = 300 + 40 + 5 = 345$

61. Using the order of operations: $2 \cdot 10^3 + 3 \cdot 10^2 + 4 \cdot 10 + 5 = 2{,}000 + 300 + 40 + 5 = 2{,}345$

63. Using the order of operations: $10 - 2(4 \cdot 5 - 16) = 10 - 2(20-16) = 10 - 2(4) = 10 - 8 = 2$

65. Using the order of operations: $4[7 + 3(2 \cdot 9 - 8)] = 4[7 + 3(18-8)] = 4[7 + 3(10)] = 4(7+30) = 4(37) = 148$

67. Using the order of operations: $5(7-3) + 8(6-4) = 5(4) + 8(2) = 20 + 16 = 36$

69. Using the order of operations: $3(4 \cdot 5 - 12) + 6(7 \cdot 6 - 40) = 3(20-12) + 6(42-40) = 3(8) + 6(2) = 24 + 12 = 36$

71. Using the order of operations: $3^4 + 4^2 \div 2^3 - 5^2 = 81 + 16 \div 8 - 25 = 81 + 2 - 25 = 58$

73. Using the order of operations: $5^2 + 3^4 \div 9^2 + 6^2 = 25 + 81 \div 81 + 36 = 25 + 1 + 36 = 62$

75. Using the order of operations: $20 \div 2 \cdot 10 = 10 \cdot 10 = 100$

77. Using the order of operations: $24 \div 8 \cdot 3 = 3 \cdot 3 = 9$
79. Using the order of operations: $36 \div 6 \cdot 3 = 6 \cdot 3 = 18$
81. Using the order of operations: $48 \div 12 \cdot 2 = 4 \cdot 2 = 8$
83. Using the order of operations: $16 - 8 + 4 = 8 + 4 = 12$
85. Using the order of operations: $24 - 14 + 8 = 10 + 8 = 18$
87. Using the order of operations: $36 - 6 + 12 = 30 + 12 = 42$
89. Using the order of operations: $48 - 12 + 17 = 36 + 17 = 53$
91. There are $5 \cdot 2 = 10$ cookies in the package.
93. The total number of calories is $210 \cdot 2 = 420$ calories.
95. There are $7 \cdot 32 = 224$ chips in the bag.
97. For a person eating 3,000 calories per day, the recommended amount of fat would be $80 + 15 = 95$ grams.
99.   a.   The amount of caffeine is: $6(100) = 600$ mg
      b.   The amount of caffeine is: $2(45) + 3(47) = 90 + 141 = 231$ mg

101. Completing the table:

Activity	Calories burned in 1 hour
Bicycling	374
Bowling	265
Handball	680
Jogging	680
Skiing	544

103. The sequence is counting numbers, so the next number is 5.
105. The sequence is even numbers, so the next number is 10.
107. The sequence is squares, so the next number is $5^2 = 25$.
109. The sequence is to add the previous two terms, so the next number is $4 + 6 = 10$.

0.2 Real Numbers

1. Labeling the point: (point at 4 on number line)
3. Labeling the point: (point at −4 on number line)
5. Labeling the point: 1.5 (point between 1 and 2)
7. Labeling the point: 9/4 (point between 2 and 3)
9. Building the fraction: $\dfrac{3}{4} = \dfrac{3}{4} \cdot \dfrac{6}{6} = \dfrac{18}{24}$
11. Building the fraction: $\dfrac{1}{2} = \dfrac{1}{2} \cdot \dfrac{12}{12} = \dfrac{12}{24}$
13. Building the fraction: $\dfrac{5}{8} = \dfrac{5}{8} \cdot \dfrac{3}{3} = \dfrac{15}{24}$
15. Building the fraction: $\dfrac{3}{5} = \dfrac{3}{5} \cdot \dfrac{12}{12} = \dfrac{36}{60}$
17. Building the fraction: $\dfrac{11}{30} = \dfrac{11}{30} \cdot \dfrac{2}{2} = \dfrac{22}{60}$
19. The opposite of 10 is −10, the reciprocal is $\dfrac{1}{10}$, and the absolute value is $|10| = 10$.
21. The opposite of $\dfrac{3}{4}$ is $-\dfrac{3}{4}$, the reciprocal is $\dfrac{4}{3}$, and the absolute value is $\left|\dfrac{3}{4}\right| = \dfrac{3}{4}$.
23. The opposite of $\dfrac{11}{2}$ is $-\dfrac{11}{2}$, the reciprocal is $\dfrac{2}{11}$, and the absolute value is $\left|\dfrac{11}{2}\right| = \dfrac{11}{2}$.
25. The opposite of −3 is 3, the reciprocal is $-\dfrac{1}{3}$, and the absolute value is $|-3| = 3$.

27. The opposite of $-\frac{2}{5}$ is $\frac{2}{5}$, the reciprocal is $-\frac{5}{2}$, and the absolute value is $\left|-\frac{2}{5}\right|=\frac{2}{5}$.

29. The opposite of x is $-x$, the reciprocal is $\frac{1}{x}$, and the absolute value is $|x|$.

31. The correct symbol is $<$: $-5<-3$
33. The correct symbol is $>$: $-3>-7$
35. Since $|-4|=4$ and $-|-4|=-4$, the correct symbol is $>$: $|-4|>-|-4|$
37. Since $-|-7|=-7$, the correct symbol is $>$: $7>-|-7|$
39. The correct symbol is $<$: $-\frac{3}{4}<-\frac{1}{4}$
41. The correct symbol is $<$: $-\frac{3}{2}<-\frac{3}{4}$
43. Simplifying the expression: $|8-2|=|6|=6$
45. Simplifying the expression: $|5 \cdot 2^3 - 2 \cdot 3^2| = |5 \cdot 8 - 2 \cdot 9| = |40-18| = |22| = 22$
47. Simplifying the expression: $|7-2|-|4-2| = |5|-|2| = 5-2 = 3$
49. Simplifying the expression: $10-|7-2(5-3)| = 10-|7-2(2)| = 10-|7-4| = 10-|3| = 10-3 = 7$
51. Simplifying the expression:
$$15-|8-2(3 \cdot 4-9)|-10 = 15-|8-2(12-9)|-10$$
$$= 15-|8-2(3)|-10$$
$$= 15-|8-6|-10$$
$$= 15-|2|-10$$
$$= 15-2-10$$
$$= 3$$

53. Multiplying the fractions: $\frac{2}{3} \cdot \frac{4}{5} = \frac{8}{15}$

55. Multiplying the fractions: $\frac{1}{2}(3) = \frac{1}{2} \cdot \frac{3}{1} = \frac{3}{2}$

57. Multiplying the fractions: $\frac{1}{4}(5) = \frac{1}{4} \cdot \frac{5}{1} = \frac{5}{4}$

59. Multiplying the fractions: $\frac{4}{3} \cdot \frac{3}{4} = \frac{12}{12} = 1$

61. Multiplying the fractions: $6\left(\frac{1}{6}\right) = \frac{6}{1} \cdot \frac{1}{6} = \frac{6}{6} = 1$

63. Multiplying the fractions: $3 \cdot \frac{1}{3} = \frac{3}{1} \cdot \frac{1}{3} = \frac{3}{3} = 1$

65. Expanding the exponent: $\left(\frac{3}{4}\right)^2 = \frac{3}{4} \cdot \frac{3}{4} = \frac{9}{16}$

67. Expanding the exponent: $\left(\frac{2}{3}\right)^3 = \frac{2}{3} \cdot \frac{2}{3} \cdot \frac{2}{3} = \frac{8}{27}$

69. Expanding the exponent: $\left(\frac{1}{10}\right)^4 = \frac{1}{10} \cdot \frac{1}{10} \cdot \frac{1}{10} \cdot \frac{1}{10} = \frac{1}{10,000}$

71. The sequence is reciprocals of odd numbers, so the next number is $\frac{1}{9}$.

73. The sequence is reciprocals of squares, so the next number is $\frac{1}{5^2} = \frac{1}{25}$.

75. The perimeter is $4(1 \text{ in.}) = 4 \text{ in.}$, and the area is $(1 \text{ in.})^2 = 1 \text{ in.}^2$.

77. The perimeter is $2(1.5 \text{ in.}) + 2(0.75 \text{ in.}) = 3.0 \text{ in.} + 1.5 \text{ in.} = 4.5 \text{ in.}$, and the area is $(1.5 \text{ in.})(0.75 \text{ in.}) = 1.125 \text{ in.}^2$.

79. The perimeter is $2.75 \text{ cm} + 4 \text{ cm} + 3.5 \text{ cm} = 10.25 \text{ cm}$, and the area is $\frac{1}{2}(4 \text{ cm})(2.5 \text{ cm}) = 5.0 \text{ cm}^2$.

81. A loss of 8 yards corresponds to -8 on a number line. The total yards gained corresponds to -2 yards.
83. The temperature can be represented as $-64°$. The new (warmer) temperature corresponds to $-54°$.
85. The wind chill temperature is $-15°$.

87. His depth can be represented by –100 feet. After he descends another 5 feet, his depth is –105 feet.

89. The area is given by: $(8.5)(11) = 93.5$ square inches

The perimeter is given by: $2(8.5) + 2(11) = 17 + 22 = 39$ inches

91. The calories burned is: $2(544) + 1(299) = 1{,}088 + 299 = 1{,}387$ calories

93. The difference in calories burned is: $3(653) - 3(435) = 1{,}959 - 1{,}305 = 654$ more calories

95.
  a. This is $3 \cdot 31 = 93$ million phones.
  b. The chart shows $4.5 \cdot 31 = 139.5$ million phones, so this statement is false.
  c. The chart shows $9.5 \cdot 31 = 294.5$ million phones, so this statement is true.

0.3 Addition of Real Numbers

1. Adding all positive and negative combinations of 3 and 5:
$3 + 5 = 8$ $3 + (-5) = -2$
$-3 + 5 = 2$ $(-3) + (-5) = -8$

3. Adding all positive and negative combinations of 15 and 20:
$15 + 20 = 35$ $15 + (-20) = -5$
$-15 + 20 = 5$ $(-15) + (-20) = -35$

5. Adding the numbers: $6 + (-3) = 3$

7. Adding the numbers: $13 + (-20) = -7$

9. Adding the numbers: $18 + (-32) = -14$

11. Adding the numbers: $-6 + 3 = -3$

13. Adding the numbers: $-30 + 5 = -25$

15. Adding the numbers: $-6 + (-6) = -12$

17. Adding the numbers: $-9 + (-10) = -19$

19. Adding the numbers: $-10 + (-15) = -25$

21. Performing the additions: $5 + (-6) + (-7) = 5 + (-13) = -8$

23. Performing the additions: $-7 + 8 + (-5) = -12 + 8 = -4$

25. Performing the additions: $5 + [6 + (-2)] + (-3) = 5 + 4 + (-3) = 9 + (-3) = 6$

27. Performing the additions: $[6 + (-2)] + [3 + (-1)] = 4 + 2 = 6$

29. Performing the additions: $20 + (-6) + [3 + (-9)] = 14 + (-6) = 8$

31. Performing the additions: $-3 + (-2) + [5 + (-4)] = -3 + (-2) + 1 = -5 + 1 = -4$

33. Performing the additions: $(-9 + 2) + [5 + (-8)] + (-4) = -7 + (-3) + (-4) = -14$

35. Performing the additions: $[-6 + (-4)] + [7 + (-5)] + (-9) = -10 + 2 + (-9) = -19 + 2 = -17$

37. Performing the additions: $(-6 + 9) + (-5) + (-4 + 3) + 7 = 3 + (-5) + (-1) + 7 = 10 + (-6) = 4$

39. Using order of operations: $-5 + 2(-3 + 7) = -5 + 2(4) = -5 + 8 = 3$

41. Using order of operations: $9 + 3(-8 + 10) = 9 + 3(2) = 9 + 6 = 15$

43. Using order of operations: $-10 + 2(-6 + 8) + (-2) = -10 + 2(2) + (-2) = -10 + 4 + (-2) = -12 + 4 = -8$

45. Using order of operations: $2(-4 + 7) + 3(-6 + 8) = 2(3) + 3(2) = 6 + 6 = 12$

47. The pattern is to add 5, so the next two terms are $18 + 5 = 23$ and $23 + 5 = 28$.

49. The pattern is to add 5, so the next two terms are $25 + 5 = 30$ and $30 + 5 = 35$.

51. The pattern is to add –5, so the next two terms are $5 + (-5) = 0$ and $0 + (-5) = -5$.

53. The pattern is to add –6, so the next two terms are $-6 + (-6) = -12$ and $-12 + (-6) = -18$.

55. The pattern is to add –4, so the next two terms are $0 + (-4) = -4$ and $-4 + (-4) = -8$.

57. Yes, since each successive odd number is 2 added to the previous one.

59. Writing the expression and simplifying: $5+9=14$
61. Writing the expression and simplifying: $[-7+(-5)]+4=(-12)+4=-8$
63. The expression is: $[-2+(-3)]+10=-5+10=5$
65. The number is 3, since $-8+3=-5$.
67. The number is -3, since $-3+(-6)=-9$.
69. The expression is $-12°+4°=-8°$.
71. The expression is $\$10+(-\$6)+(-\$8)=\$10+(-\$14)=-\$4$.
73. The new balance is $-\$30+\$40=\$10$.
75. The profit was: $\$11{,}500-\$9{,}500=\$2{,}000$.
77. There was a loss in the year 2013: $\$7{,}750-\$8{,}250=-\$500$

0.4 Subtraction of Real Numbers

1. Subtracting the numbers: $5-8=5+(-8)=-3$
3. Subtracting the numbers: $3-9=3+(-9)=-6$
5. Subtracting the numbers: $5-5=5+(-5)=0$
7. Subtracting the numbers: $-8-2=-8+(-2)=-10$
9. Subtracting the numbers: $-4-12=-4+(-12)=-16$
11. Subtracting the numbers: $-6-6=-6+(-6)=-12$
13. Subtracting the numbers: $-8-(-1)=-8+1=-7$
15. Subtracting the numbers: $15-(-20)=15+20=35$
17. Subtracting the numbers: $-4-(-4)=-4+4=0$
19. Using order of operations: $3-2-5=3+(-2)+(-5)=3+(-7)=-4$
21. Using order of operations: $9-2-3=9+(-2)+(-3)=9+(-5)=4$
23. Using order of operations: $-6-8-10=-6+(-8)+(-10)=-24$
25. Performing the additions: $-22+4-10=-22+4+(-10)=-32+4=-28$
27. Using order of operations: $10-(-20)-5=10+20+(-5)=30+(-5)=25$
29. Using order of operations: $8-(2-3)-5=8-(-1)-5=8+1+(-5)=9+(-5)=4$
31. Using order of operations: $7-(3-9)-6=7-(-6)-6=7+6+(-6)=13+(-6)=7$
33. Using order of operations: $5-(-8-6)-2=5-(-14)-2=5+14+(-2)=19+(-2)=17$
35. Using order of operations: $-(5-7)-(2-8)=-(-2)-(-6)=2+6=8$
37. Using order of operations: $-(3-10)-(6-3)=-(-7)-3=7+(-3)=4$
39. Performing the additions: $16-[(4-5)-1]=16-(-1-1)=16-(-2)=16+2=18$
41. Using order of operations: $5-[(2-3)-4]=5-(-1-4)=5-(-5)=5+5=10$
43. Using order of operations:
   $21-[-(3-4)-2]-5=21-[-(-1)-2]-5=21-(1-2)-5=21-(-1)-5=21+1+(-5)=22+(-5)=17$
45. Using order of operations: $2 \cdot 8-3 \cdot 5=16-15=1$
47. Using order of operations: $3 \cdot 5-2 \cdot 7=15-14=15+(-14)=1$
49. Using order of operations: $5 \cdot 9-2 \cdot 3-6 \cdot 2=45-6-12=45+(-6)+(-12)=45+(-18)=27$
51. Using order of operations: $3 \cdot 8-2 \cdot 4-6 \cdot 7=24-8-42=24+(-8)+(-42)=24+(-50)=-26$
53. Using order of operations: $2 \cdot 3^2-5 \cdot 2^2=2 \cdot 9-5 \cdot 4=18-20=18+(-20)=-2$
55. Using order of operations: $4 \cdot 3^3-5 \cdot 2^3=4 \cdot 27-5 \cdot 8=108-40=108+(-40)=68$
57. Writing the expression and simplifying: $-7-4=-7+(-4)=-11$
59. Writing the expression and simplifying: $12-(-8)=12+8=20$

61. Writing the expression and simplifying: $-5-(-7) = -5+7 = 2$
63. Writing the expression and simplifying: $[4+(-5)]-17 = (-1)+(-17) = -18$
65. Writing the expression and simplifying: $8-5 = 3$
67. Writing the expression and simplifying: $-8-5 = -8+(-5) = -13$
69. Writing the expression and simplifying: $8-(-5) = 8+5 = 13$
71. The number is 10, since $8-10 = -2$.
73. The number is –2, since $8-(-2) = 8+2 = 10$.
75. The expression is $\$1{,}500 - \$730 = \$770$.
77. The expression is $-\$35+\$15-\$20 = -\$35+(-\$20)+\$15 = -\$55+\$15 = -\$40$.
79. The expression is $73°+10°-8° = 83°-8° = 75°\,\text{F}$.
81. The sequence of values is $4,500, $3,950, $3,400, $2,850, and $2,300. This is an arithmetic sequence, since –$550 is added to each value to obtain the new value.
83. Since the angles add up to 90°: $x = 90°-55° = 35°$
85. Since the angles add up to 180°: $x = 180°-120° = 60°$
87. a. Completing the table:

Day	0	2	4	6	8	10
Plant Height (inches)	0	1	3	6	13	23

   b. Subtracting: $13-1 = 12$. The grass is 12 inches higher after 8 days than after 2 days.
89. a. Yes, the numbers appear to be rounded to the hundreds place.
   b. Subtracting: $21{,}400-13{,}000 = 8{,}400$. There were 8,400 more participants in 2004 than in 2000.
   c. If the amount increases by another 8,400, there will be $21{,}400 + 8{,}400 = 29{,}800$ participants in 2008.

0.5 Properties of Real Numbers

1. commutative property (of addition)
3. multiplicative inverse property
5. commutative property (of addition)
7. distributive property
9. commutative and associative properties (of addition)
11. commutative and associative properties (of addition)
13. commutative property (of addition)
15. commutative and associative properties (of multiplication)
17. commutative property (of multiplication)
19. additive inverse property
21. The expression should read $3(x+2) = 3x+6$.
23. The expression should read $9(a+b) = 9a+9b$.
25. The expression should read $3(0) = 0$.
27. The expression should read $3+(-3) = 0$.
29. The expression should read $10(1) = 10$.
31. Simplifying the expression: $4+(2+x) = (4+2)+x = 6+x$
33. Simplifying the expression: $(x+2)+7 = x+(2+7) = x+9$
35. Simplifying the expression: $3(5x) = (3 \cdot 5)x = 15x$
37. Simplifying the expression: $9(6y) = (9 \cdot 6)y = 54y$
39. Simplifying the expression: $\frac{1}{2}(3a) = \left(\frac{1}{2} \cdot 3\right)a = \frac{3}{2}a$
41. Simplifying the expression: $\frac{1}{3}(3x) = \left(\frac{1}{3} \cdot 3\right)x = 1x = x$
43. Simplifying the expression: $\frac{1}{2}(2y) = \left(\frac{1}{2} \cdot 2\right)y = 1y = y$
45. Simplifying the expression: $\frac{3}{4}\left(\frac{4}{3}x\right) = \left(\frac{3}{4} \cdot \frac{4}{3}\right)x = 1x = x$
47. Simplifying the expression: $\frac{6}{5}\left(\frac{5}{6}a\right) = \left(\frac{6}{5} \cdot \frac{5}{6}\right)a = 1a = a$
49. Applying the distributive property: $8(x+2) = 8 \cdot x + 8 \cdot 2 = 8x+16$
51. Applying the distributive property: $8(x-2) = 8 \cdot x - 8 \cdot 2 = 8x-16$

53. Applying the distributive property: $4(y+1) = 4 \cdot y + 4 \cdot 1 = 4y + 4$

55. Applying the distributive property: $3(6x+5) = 3 \cdot 6x + 3 \cdot 5 = 18x + 15$

57. Applying the distributive property: $2(3a+7) = 2 \cdot 3a + 2 \cdot 7 = 6a + 14$

59. Applying the distributive property: $9(6y-8) = 9 \cdot 6y - 9 \cdot 8 = 54y - 72$

61. Applying the distributive property: $\frac{1}{3}(3x+6) = \frac{1}{3} \cdot 3x + \frac{1}{3} \cdot 6 = x + 2$

63. Applying the distributive property: $6(2x+3y) = 6 \cdot 2x + 6 \cdot 3y = 12x + 18y$

65. Applying the distributive property: $4(3a-2b) = 4 \cdot 3a - 4 \cdot 2b = 12a - 8b$

67. Applying the distributive property: $\frac{1}{2}(6x+4y) = \frac{1}{2} \cdot 6x + \frac{1}{2} \cdot 4y = 3x + 2y$

69. Applying the distributive property: $4(a+4)+9 = 4a + 16 + 9 = 4a + 25$

71. Applying the distributive property: $2(3x+5)+2 = 6x + 10 + 2 = 6x + 12$

73. Applying the distributive property: $7(2x+4)+10 = 14x + 28 + 10 = 14x + 38$

75. Applying the distributive property: $\frac{1}{2}(4x+2) = \frac{1}{2} \cdot 4x + \frac{1}{2} \cdot 2 = 2x + 1$

77. Applying the distributive property: $\frac{3}{4}(8x-4) = \frac{3}{4} \cdot 8x - \frac{3}{4} \cdot 4 = 6x - 3$

79. Applying the distributive property: $\frac{5}{6}(6x+12) = \frac{5}{6} \cdot 6x + \frac{5}{6} \cdot 12 = 5x + 10$

81. Applying the distributive property: $10\left(\frac{3}{5}x + \frac{1}{2}\right) = 10 \cdot \frac{3}{5}x + 10 \cdot \frac{1}{2} = 6x + 5$

83. Applying the distributive property: $15\left(\frac{1}{3}x + \frac{2}{5}\right) = 15 \cdot \frac{1}{3}x + 15 \cdot \frac{2}{5} = 5x + 6$

85. Applying the distributive property: $12\left(\frac{1}{2}m - \frac{5}{12}\right) = 12 \cdot \frac{1}{2}m - 12 \cdot \frac{5}{12} = 6m - 5$

87. Applying the distributive property: $21\left(\frac{1}{3} + \frac{1}{7}x\right) = 21 \cdot \frac{1}{3} + 21 \cdot \frac{1}{7}x = 7 + 3x$

89. Applying the distributive property: $6\left(\frac{1}{2}x - \frac{1}{3}y\right) = 6 \cdot \frac{1}{2}x - 6 \cdot \frac{1}{3}y = 3x - 2y$

91. Applying the distributive property: $0.09(x+2,000) = 0.09x + 180$

93. Applying the distributive property: $0.12(x+500) = 0.12x + 60$

95. Applying the distributive property: $a\left(1 + \frac{1}{a}\right) = a \cdot 1 + a \cdot \frac{1}{a} = a + 1$

97. Applying the distributive property: $a\left(\frac{1}{a} - 1\right) = a \cdot \frac{1}{a} - a \cdot 1 = 1 - a$

99. No. The man cannot reverse the order of putting on his socks and putting on his shoes.

101. Division is not a commutative operation. For example, $8 \div 4 = 2$ while $4 \div 8 = \frac{1}{2}$.

103. Computing his hours worked:
$4(2+3) = 4(5) = 20$ hours $\qquad 4 \cdot 2 + 4 \cdot 3 = 8 + 12 = 20$ hours

0.6 Multiplication of Real Numbers

1. Finding the product: $7(-6) = -42$
3. Finding the product: $-8(2) = -16$
5. Finding the product: $-3(-1) = 3$
7. Finding the product: $-11(-11) = 121$
9. Using order of operations: $-3(2)(-1) = 6$
11. Using order of operations: $-3(-4)(-5) = -60$
13. Using order of operations: $-2(-4)(-3)(-1) = 24$
15. Using order of operations: $(-7)^2 = (-7)(-7) = 49$
17. Using order of operations: $(-3)^3 = (-3)(-3)(-3) = -27$
19. Using order of operations: $-2(2-5) = -2(-3) = 6$
21. Using order of operations: $-5(8-10) = -5(-2) = 10$
23. Using order of operations: $(4-7)(6-9) = (-3)(-3) = 9$
25. Using order of operations: $(-3-2)(-5-4) = (-5)(-9) = 45$
27. Using order of operations: $-3(-6) + 4(-1) = 18 + (-4) = 14$
29. Using order of operations: $2(3) - 3(-4) + 4(-5) = 6 + 12 + (-20) = 18 + (-20) = -2$
31. Using order of operations: $4(-3)^2 + 5(-6)^2 = 4(9) + 5(36) = 36 + 180 = 216$
33. Using order of operations: $7(-2)^3 - 2(-3)^3 = 7(-8) - 2(-27) = -56 + 54 = -2$
35. Using order of operations: $6 - 4(8-2) = 6 - 4(6) = 6 - 24 = 6 + (-24) = -18$
37. Using order of operations: $9 - 4(3-8) = 9 - 4(-5) = 9 + 20 = 29$
39. Using order of operations: $-4(3-8) - 6(2-5) = -4(-5) - 6(-3) = 20 + 18 = 38$
41. Using order of operations: $7 - 2[-6 - 4(-3)] = 7 - 2(-6 + 12) = 7 - 2(6) = 7 - 12 = 7 + (-12) = -5$
43. Using order of operations:
$7 - 3[2(-4-4) - 3(-1-1)] = 7 - 3[2(-8) - 3(-2)] = 7 - 3(-16 + 6) = 7 - 3(-10) = 7 + 30 = 37$
45. Using order of operations:
$8 - 6[-2(-3-1) + 4(-2-3)] = 8 - 6[-2(-4) + 4(-5)] = 8 - 6(8 - 20) = 8 - 6(-12) = 8 + 72 = 80$
47. Multiplying the fractions: $-\dfrac{2}{3} \cdot \dfrac{5}{7} = -\dfrac{2 \cdot 5}{3 \cdot 7} = -\dfrac{10}{21}$
49. Multiplying the fractions: $-8\left(\dfrac{1}{2}\right) = -\dfrac{8}{1} \cdot \dfrac{1}{2} = -\dfrac{8}{2} = -4$
51. Multiplying the fractions: $-\dfrac{3}{4}\left(-\dfrac{4}{3}\right) = \dfrac{12}{12} = 1$
53. Multiplying the fractions: $\left(-\dfrac{3}{4}\right)^2 = \left(-\dfrac{3}{4}\right)\left(-\dfrac{3}{4}\right) = \dfrac{9}{16}$
55. Multiplying the expressions: $-2(4x) = [-2 \cdot 4]x = -8x$
57. Multiplying the expressions: $-7(-6x) = [-7 \cdot (-6)]x = 42x$
59. Multiplying the expressions: $-\dfrac{1}{3}(-3x) = \left[-\dfrac{1}{3} \cdot (-3)\right]x = 1x = x$
61. Applying the distributive property: $-4(a+2) = -4a + (-4)(2) = -4a - 8$
63. Applying the distributive property: $-\dfrac{1}{2}(3x - 6) = -\dfrac{3}{2}x - \dfrac{1}{2}(-6) = -\dfrac{3}{2}x + 3$
65. Applying the distributive property: $-3(2x - 5) - 7 = -6x + 15 - 7 = -6x + 8$
67. Applying the distributive property: $-5(3x + 4) - 10 = -15x - 20 - 10 = -15x - 30$
69. Writing the expression: $3(-10) + 5 = -30 + 5 = -25$
71. Writing the expression: $2(-4x) = -8x$
73. Evaluating: $(-9)(2) - 8 = -18 - 8 = -18 + (-8) = -26$
75. The pattern is to multiply by 2, so the next number is $4 \cdot 2 = 8$.

77. The pattern is to multiply by –2, so the next number is $40 \cdot (-2) = -80$.

79. The pattern is to multiply by $\frac{1}{2}$, so the next number is: $\frac{1}{4} \cdot \frac{1}{2} = \frac{1}{8}$

81. The pattern is to multiply by –2, so the next number is $12 \cdot (-2) = -24$.

83. Simplifying the expression: $3(x-5)+4 = 3x-15+4 = 3x-11$

85. Simplifying the expression: $2(3)-4-3(-4) = 6-4+12 = 2+12 = 14$

87. Simplifying: $\left(\frac{1}{2} \cdot 18\right)^2 = (9)^2 = (9)(9) = 81$

89. Simplifying: $\left(\frac{1}{2} \cdot 3\right)^2 = \left(\frac{3}{2}\right)^2 = \left(\frac{3}{2}\right)\left(\frac{3}{2}\right) = \frac{9}{4}$

91. Simplifying: $-\frac{1}{3}(-2x+6) = -\frac{1}{3} \cdot (-2x) - \frac{1}{3} \cdot 6 = \frac{2}{3}x - 2$

93. Simplifying: $8\left(-\frac{1}{4}x + \frac{1}{8}y\right) = 8 \cdot \left(-\frac{1}{4}x\right) + 8 \cdot \left(\frac{1}{8}y\right) = -2x + y$

95. The temperature is: $25° - 4(6°) = 25° - 24° = 1°$ F.

97. The net change in calories is: $2(630) - 3(265) = 1260 - 795 = 465$ calories

0.7 Division of Real Numbers

1. Finding the quotient: $\frac{8}{-4} = -2$

3. Finding the quotient: $\frac{-48}{16} = -3$

5. Finding the quotient: $\frac{-7}{21} = -\frac{1}{3}$

7. Finding the quotient: $\frac{-39}{-13} = 3$

9. Finding the quotient: $\frac{-6}{-42} = \frac{1}{7}$

11. Finding the quotient: $\frac{0}{-32} = 0$

13. Performing the operations: $-3 + 12 = 9$

15. Performing the operations: $-3 - 12 = -3 + (-12) = -15$

17. Performing the operations: $-3(12) = -36$

19. Performing the operations: $-3 \div 12 = \frac{-3}{12} = -\frac{1}{4}$

21. Dividing and reducing: $\frac{4}{5} \div \frac{3}{4} = \frac{4}{5} \cdot \frac{4}{3} = \frac{16}{15}$

23. Dividing and reducing: $-\frac{5}{6} \div \left(-\frac{5}{8}\right) = -\frac{5}{6} \cdot \left(-\frac{8}{5}\right) = \frac{40}{30} = \frac{4}{3}$

25. Dividing and reducing: $\frac{10}{13} \div \left(-\frac{5}{4}\right) = \frac{10}{13} \cdot \left(-\frac{4}{5}\right) = -\frac{40}{65} = -\frac{8}{13}$

27. Dividing and reducing: $-\frac{5}{6} \div \frac{5}{6} = -\frac{5}{6} \cdot \frac{6}{5} = -\frac{30}{30} = -1$

29. Dividing and reducing: $-\frac{3}{4} \div \left(-\frac{3}{4}\right) = -\frac{3}{4} \cdot \left(-\frac{4}{3}\right) = \frac{12}{12} = 1$

31. Using order of operations: $\frac{3(-2)}{-10} = \frac{-6}{-10} = \frac{3}{5}$

33. Using order of operations: $\frac{-5(-5)}{-15} = \frac{25}{-15} = -\frac{5}{3}$

35. Using order of operations: $\frac{-8(-7)}{-28} = \frac{56}{-28} = -2$

37. Using order of operations: $\frac{27}{4-13} = \frac{27}{-9} = -3$

39. Using order of operations: $\frac{20-6}{5-5} = \frac{14}{0}$, which is undefined

41. Using order of operations: $\dfrac{-3+9}{2\cdot 5-10}=\dfrac{6}{10-10}=\dfrac{6}{0}$, which is undefined

43. Using order of operations: $\dfrac{15(-5)-25}{2(-10)}=\dfrac{-75-25}{-20}=\dfrac{-100}{-20}=5$

45. Using order of operations: $\dfrac{27-2(-4)}{-3(5)}=\dfrac{27+8}{-15}=\dfrac{35}{-15}=-\dfrac{7}{3}$

47. Using order of operations: $\dfrac{12-6(-2)}{12(-2)}=\dfrac{12+12}{-24}=\dfrac{24}{-24}=-1$

49. Using order of operations: $\dfrac{5^2-2^2}{-5+2}=\dfrac{25-4}{-3}=\dfrac{21}{-3}=-7$

51. Using order of operations: $\dfrac{8^2-2^2}{8^2+2^2}=\dfrac{64-4}{64+4}=\dfrac{60}{68}=\dfrac{15}{17}$

53. Using order of operations: $\dfrac{(5+3)^2}{-5^2-3^2}=\dfrac{8^2}{-25-9}=\dfrac{64}{-34}=-\dfrac{32}{17}$

55. Using order of operations: $\dfrac{(8-4)^2}{8^2-4^2}=\dfrac{4^2}{64-16}=\dfrac{16}{48}=\dfrac{1}{3}$

57. Using order of operations: $\dfrac{-4\cdot 3^2-5\cdot 2^2}{-8(7)}=\dfrac{-4\cdot 9-5\cdot 4}{-56}=\dfrac{-36-20}{-56}=\dfrac{-56}{-56}=1$

59. Using order of operations: $\dfrac{3\cdot 10^2+4\cdot 10+5}{345}=\dfrac{300+40+5}{345}=\dfrac{345}{345}=1$

61. Using order of operations: $\dfrac{7-[(2-3)-4]}{-1-2-3}=\dfrac{7-(-1-4)}{-6}=\dfrac{7-(-5)}{-6}=\dfrac{7+5}{-6}=\dfrac{12}{-6}=-2$

63. Using order of operations: $\dfrac{6(-4)-2(5-8)}{-6-3-5}=\dfrac{-24-2(-3)}{-14}=\dfrac{-24+6}{-14}=\dfrac{-18}{-14}=\dfrac{9}{7}$

65. Using order of operations: $\dfrac{3(-5-3)+4(7-9)}{5(-2)+3(-4)}=\dfrac{3(-8)+4(-2)}{-10+(-12)}=\dfrac{-24+(-8)}{-22}=\dfrac{-32}{-22}=\dfrac{16}{11}$

67. Using order of operations: $\dfrac{|3-9|}{3-9}=\dfrac{|-6|}{-6}=\dfrac{6}{-6}=-1$

69. a. Simplifying: $20\div 4\cdot 5=5\cdot 5=25$ b. Simplifying: $-20\div 4\cdot 5=-5\cdot 5=-25$
    c. Simplifying: $20\div(-4)\cdot 5=-5\cdot 5=-25$ d. Simplifying: $20\div 4(-5)=5(-5)=-25$
    e. Simplifying: $-20\div 4(-5)=-5(-5)=25$

71. a. Simplifying: $8\div\dfrac{4}{5}=8\cdot\dfrac{5}{4}=10$ b. Simplifying: $8\div\dfrac{4}{5}-10=8\cdot\dfrac{5}{4}-10=10-10=0$
    c. Simplifying: $8\div\dfrac{4}{5}(-10)=8\cdot\dfrac{5}{4}(-10)=10(-10)=-100$
    d. Simplifying: $8\div\left(-\dfrac{4}{5}\right)-10=8\cdot\left(-\dfrac{5}{4}\right)-10=-10-10=-20$

73. Applying the distributive property: $10\left(\dfrac{x}{2}+\dfrac{3}{5}\right)=10\cdot\dfrac{x}{2}+10\cdot\dfrac{3}{5}=5x+6$

75. Applying the distributive property: $15\left(\dfrac{x}{5}+\dfrac{4}{3}\right)=15\cdot\dfrac{x}{5}+15\cdot\dfrac{4}{3}=3x+20$

77. Applying the distributive property: $x\left(\dfrac{3}{x}+1\right) = x \cdot \dfrac{3}{x} + x \cdot 1 = 3 + x$

79. Applying the distributive property: $21\left(\dfrac{x}{7} - \dfrac{y}{3}\right) = 21 \cdot \dfrac{x}{7} - 21 \cdot \dfrac{y}{3} = 3x - 7y$

81. Applying the distributive property: $a\left(\dfrac{3}{a} - \dfrac{2}{a}\right) = a \cdot \dfrac{3}{a} - a \cdot \dfrac{2}{a} = 3 - 2 = 1$

83. The quotient is $\dfrac{-12}{-4} = 3$.

85. The number is -10, since $\dfrac{-10}{-5} = 2$.

87. The number is -3, since $\dfrac{27}{-3} = -9$.

89. The expression is: $\dfrac{-20}{4} - 3 = -5 - 3 = -8$

91. Each person would lose: $\dfrac{13{,}600 - 15{,}000}{4} = \dfrac{-1{,}400}{4} = -350 = \$350$ loss

93. The change per hour is: $\dfrac{61° - 75°}{4} = \dfrac{-14°}{4} = -3.5°$ per hour

95. The number of blankets is: $12 \div \dfrac{6}{7} = 12 \cdot \dfrac{7}{6} = \dfrac{84}{6} = 14$ blankets

97. The number of bags is: $12 \div \dfrac{1}{4} = 12 \cdot \dfrac{4}{1} = 48$ bags

99. The number of spoons is: $\dfrac{3}{4} \div \dfrac{1}{8} = \dfrac{3}{4} \cdot \dfrac{8}{1} = \dfrac{24}{4} = 6$ spoons

101. The number of cartons is: $14 \div \dfrac{1}{2} = 14 \cdot \dfrac{2}{1} = 28$ cartons

103. **a.** Since they predict $50 revenue for every 25 people, their projected revenue is:
$\dfrac{\$50}{25 \text{ people}} \cdot 10{,}000 \text{ people} = \$20{,}000$

  b. Since they predict $50 revenue for every 25 people, their projected revenue is:
$\dfrac{\$50}{25 \text{ people}} \cdot 25{,}000 \text{ people} = \$50{,}000$

  c. Since they predict $50 revenue for every 25 people, their projected revenue is:
$\dfrac{\$50}{25 \text{ people}} \cdot 5{,}000 \text{ people} = \$10{,}000$

  Since this projected revenue is more than the $5,000 cost for the list, it is a wise purchase.

0.8 Subsets of the Real Numbers

1. The whole numbers are: 0, 1

3. The rational numbers are: $-3, -2.5, 0, 1, \dfrac{3}{2}$

5. The real numbers are: $-3, -2.5, 0, 1, \dfrac{3}{2}, \sqrt{15}$

7. The integers are: $-10, -8, -2, 9$

9. The irrational numbers are: π

11. This statement is true.

13. This statement is false.

15. This statement is false.

17. This statement is true.

19. This number is composite: $48 = 6 \cdot 8 = (2 \cdot 3) \cdot (2 \cdot 2 \cdot 2) = 2^4 \cdot 3$

21. This number is prime.

23. This number is composite: $1{,}023 = 3 \cdot 341 = 3 \cdot 11 \cdot 31$

25. Factoring the number: $144 = 12 \cdot 12 = (3 \cdot 4) \cdot (3 \cdot 4) = (3 \cdot 2 \cdot 2) \cdot (3 \cdot 2 \cdot 2) = 2^4 \cdot 3^2$

27. Factoring the number: $38 = 2 \cdot 19$

29. Factoring the number: $105 = 5 \cdot 21 = 5 \cdot (3 \cdot 7) = 3 \cdot 5 \cdot 7$

31. Factoring the number: $180 = 10 \cdot 18 = (2 \cdot 5) \cdot (3 \cdot 6) = (2 \cdot 5) \cdot (3 \cdot 2 \cdot 3) = 2^2 \cdot 3^2 \cdot 5$

33. Factoring the number: $385 = 5 \cdot 77 = 5 \cdot (7 \cdot 11) = 5 \cdot 7 \cdot 11$

35. Factoring the number: $121 = 11 \cdot 11 = 11^2$

37. Factoring the number: $420 = 10 \cdot 42 = (2 \cdot 5) \cdot (7 \cdot 6) = (2 \cdot 5) \cdot (7 \cdot 2 \cdot 3) = 2^2 \cdot 3 \cdot 5 \cdot 7$

39. Factoring the number: $620 = 10 \cdot 62 = (2 \cdot 5) \cdot (2 \cdot 31) = 2^2 \cdot 5 \cdot 31$

41. Reducing the fraction: $\dfrac{105}{165} = \dfrac{3 \cdot 5 \cdot 7}{3 \cdot 5 \cdot 11} = \dfrac{7}{11}$

43. Reducing the fraction: $\dfrac{525}{735} = \dfrac{3 \cdot 5 \cdot 5 \cdot 7}{3 \cdot 5 \cdot 7 \cdot 7} = \dfrac{5}{7}$

45. Reducing the fraction: $\dfrac{385}{455} = \dfrac{5 \cdot 7 \cdot 11}{5 \cdot 7 \cdot 13} = \dfrac{11}{13}$

47. Reducing the fraction: $\dfrac{322}{345} = \dfrac{2 \cdot 7 \cdot 23}{3 \cdot 5 \cdot 23} = \dfrac{2 \cdot 7}{3 \cdot 5} = \dfrac{14}{15}$

49. Reducing the fraction: $\dfrac{205}{369} = \dfrac{5 \cdot 41}{3 \cdot 3 \cdot 41} = \dfrac{5}{3 \cdot 3} = \dfrac{5}{9}$

51. Reducing the fraction: $\dfrac{215}{344} = \dfrac{5 \cdot 43}{2 \cdot 2 \cdot 2 \cdot 43} = \dfrac{5}{2 \cdot 2 \cdot 2} = \dfrac{5}{8}$

53. Factoring into prime numbers: $6^3 = (2 \cdot 3)^3 = 2^3 \cdot 3^3$

55. Factoring into prime numbers: $9^4 \cdot 16^2 = (3 \cdot 3)^4 \cdot (2 \cdot 2 \cdot 2 \cdot 2)^2 = 3^4 \cdot 3^4 \cdot 2^2 \cdot 2^2 \cdot 2^2 \cdot 2^2 = 2^8 \cdot 3^8$

57. Simplifying and factoring: $3 \cdot 8 + 3 \cdot 7 + 3 \cdot 5 = 24 + 21 + 15 = 60 = 6 \cdot 10 = (2 \cdot 3) \cdot (2 \cdot 5) = 2^2 \cdot 3 \cdot 5$

59. They are not a subset of the irrational numbers.

61. 8, 21, and 34 are Fibonacci numbers that are composite numbers.

0.9 Addition and Subtraction with Fractions

1. Combining the fractions: $\dfrac{3}{6} + \dfrac{1}{6} = \dfrac{4}{6} = \dfrac{2}{3}$

3. Combining the fractions: $\dfrac{3}{8} - \dfrac{5}{8} = -\dfrac{2}{8} = -\dfrac{1}{4}$

5. Combining the fractions: $-\dfrac{1}{4} + \dfrac{3}{4} = \dfrac{2}{4} = \dfrac{1}{2}$

7. Combining the fractions: $\dfrac{x}{3} - \dfrac{1}{3} = \dfrac{x-1}{3}$

9. Combining the fractions: $\dfrac{1}{4} + \dfrac{2}{4} + \dfrac{3}{4} = \dfrac{6}{4} = \dfrac{3}{2}$

11. Combining the fractions: $\dfrac{x+7}{2} - \dfrac{1}{2} = \dfrac{x+7-1}{2} = \dfrac{x+6}{2}$

13. Combining the fractions: $\dfrac{1}{10} - \dfrac{3}{10} - \dfrac{4}{10} = -\dfrac{6}{10} = -\dfrac{3}{5}$

15. Combining the fractions: $\dfrac{1}{a} + \dfrac{4}{a} + \dfrac{5}{a} = \dfrac{10}{a}$

17. Completing the table:

First Number a	Second Number b	The Sum of a and b $a+b$
$\dfrac{1}{2}$	$\dfrac{1}{3}$	$\dfrac{5}{6}$
$\dfrac{1}{3}$	$\dfrac{1}{4}$	$\dfrac{7}{12}$
$\dfrac{1}{4}$	$\dfrac{1}{5}$	$\dfrac{9}{20}$
$\dfrac{1}{5}$	$\dfrac{1}{6}$	$\dfrac{11}{30}$

19. Completing the table:

First Number a	Second Number b	The Sum of a and b $a+b$
$\frac{1}{12}$	$\frac{1}{2}$	$\frac{7}{12}$
$\frac{1}{12}$	$\frac{1}{3}$	$\frac{5}{12}$
$\frac{1}{12}$	$\frac{1}{4}$	$\frac{1}{3}$
$\frac{1}{12}$	$\frac{1}{6}$	$\frac{1}{4}$

21. Combining the fractions: $\frac{4}{9}+\frac{1}{3}=\frac{4}{9}+\frac{1\cdot 3}{3\cdot 3}=\frac{4}{9}+\frac{3}{9}=\frac{7}{9}$

23. Combining the fractions: $2+\frac{1}{3}=\frac{2\cdot 3}{1\cdot 3}+\frac{1}{3}=\frac{6}{3}+\frac{1}{3}=\frac{7}{3}$

25. Combining the fractions: $-\frac{3}{4}+1=-\frac{3}{4}+\frac{1\cdot 4}{1\cdot 4}=-\frac{3}{4}+\frac{4}{4}=\frac{1}{4}$

27. Combining the fractions: $\frac{1}{2}+\frac{2}{3}=\frac{1\cdot 3}{2\cdot 3}+\frac{2\cdot 2}{3\cdot 2}=\frac{3}{6}+\frac{4}{6}=\frac{7}{6}$

29. Combining the fractions: $\frac{5}{12}-\left(-\frac{3}{8}\right)=\frac{5}{12}+\frac{3}{8}=\frac{5\cdot 2}{12\cdot 2}+\frac{3\cdot 3}{8\cdot 3}=\frac{10}{24}+\frac{9}{24}=\frac{19}{24}$

31. Combining the fractions: $-\frac{1}{20}+\frac{8}{30}=-\frac{1\cdot 3}{20\cdot 3}+\frac{8\cdot 2}{30\cdot 2}=-\frac{3}{60}+\frac{16}{60}=\frac{13}{60}$

33. First factor the denominators to find the LCM:
    $30 = 2\cdot 3\cdot 5$
    $42 = 2\cdot 3\cdot 7$
    $\text{LCM} = 2\cdot 3\cdot 5\cdot 7 = 210$
    Combining the fractions: $\frac{17}{30}+\frac{11}{42}=\frac{17\cdot 7}{30\cdot 7}+\frac{11\cdot 5}{42\cdot 5}=\frac{119}{210}+\frac{55}{210}=\frac{174}{210}=\frac{2\cdot 3\cdot 29}{2\cdot 3\cdot 5\cdot 7}=\frac{29}{5\cdot 7}=\frac{29}{35}$

35. First factor the denominators to find the LCM:
    $84 = 2\cdot 2\cdot 3\cdot 7$
    $90 = 2\cdot 3\cdot 3\cdot 5$
    $\text{LCM} = 2\cdot 2\cdot 3\cdot 3\cdot 5\cdot 7 = 1{,}260$
    Combining the fractions: $\frac{25}{84}+\frac{41}{90}=\frac{25\cdot 15}{84\cdot 15}+\frac{41\cdot 14}{90\cdot 14}=\frac{375}{1{,}260}+\frac{574}{1{,}260}=\frac{949}{1{,}260}$

37. First factor the denominators to find the LCM:
    $126 = 2\cdot 3\cdot 3\cdot 7$
    $180 = 2\cdot 2\cdot 3\cdot 3\cdot 5$
    $\text{LCM} = 2\cdot 2\cdot 3\cdot 3\cdot 5\cdot 7 = 1{,}260$
    Combining the fractions:
    $\frac{13}{126}-\frac{13}{180}=\frac{13\cdot 10}{126\cdot 10}-\frac{13\cdot 7}{180\cdot 7}=\frac{130}{1{,}260}-\frac{91}{1{,}260}=\frac{39}{1{,}260}=\frac{3\cdot 13}{2\cdot 2\cdot 3\cdot 3\cdot 5\cdot 7}=\frac{13}{2\cdot 2\cdot 3\cdot 5\cdot 7}=\frac{13}{420}$

39. Combining the fractions: $\frac{3}{4}+\frac{1}{8}+\frac{5}{6}=\frac{3\cdot 6}{4\cdot 6}+\frac{1\cdot 3}{8\cdot 3}+\frac{5\cdot 4}{6\cdot 4}=\frac{18}{24}+\frac{3}{24}+\frac{20}{24}=\frac{41}{24}$

41. Combining the fractions: $\frac{1}{2}+\frac{1}{3}+\frac{1}{4}+\frac{1}{6}=\frac{1\cdot 6}{2\cdot 6}+\frac{1\cdot 4}{3\cdot 4}+\frac{1\cdot 3}{4\cdot 3}+\frac{1\cdot 2}{6\cdot 2}=\frac{6}{12}+\frac{4}{12}+\frac{3}{12}+\frac{2}{12}=\frac{15}{12}=\frac{5}{4}$

43. Combining the fractions: $1-\frac{5}{2}=1\cdot\frac{2}{2}-\frac{5}{2}=\frac{2}{2}-\frac{5}{2}=-\frac{3}{2}$

45. Combining the fractions: $1 + \frac{1}{2} = 1 \cdot \frac{2}{2} + \frac{1}{2} = \frac{2}{2} + \frac{1}{2} = \frac{3}{2}$

47. The sum is given by: $\frac{3}{7} + 2 + \frac{1}{9} = \frac{3 \cdot 9}{7 \cdot 9} + \frac{2 \cdot 63}{1 \cdot 63} + \frac{1 \cdot 7}{9 \cdot 7} = \frac{27}{63} + \frac{126}{63} + \frac{7}{63} = \frac{160}{63}$

49. The difference is given by: $\frac{7}{8} - \frac{1}{4} = \frac{7}{8} - \frac{1 \cdot 2}{4 \cdot 2} = \frac{7}{8} - \frac{2}{8} = \frac{5}{8}$

51. The pattern is to add $-\frac{1}{3}$, so the fourth term is: $-\frac{1}{3} + \left(-\frac{1}{3}\right) = -\frac{2}{3}$

53. The pattern is to add $\frac{2}{3}$, so the fourth term is: $\frac{5}{3} + \frac{2}{3} = \frac{7}{3}$

55. The pattern is to multiply by $\frac{1}{5}$, so the fourth term is: $\frac{1}{25} \cdot \frac{1}{5} = \frac{1}{125}$

57. The perimeter is: $4\left(\frac{3}{8}\right) = \frac{12}{8} = \frac{3}{2} = 1\frac{1}{2}$ feet

59. The perimeter is: $2\left(\frac{4}{5}\right) + 2\left(\frac{3}{10}\right) = \frac{8}{5} + \frac{3}{5} = \frac{11}{5} = 2\frac{1}{5}$ centimeters (cm)

61. Adding: $\frac{1}{2} + 4 = \frac{1}{2} + \frac{8}{2} = \frac{9}{2} = 4\frac{1}{2}$ pints

63. Multiplying: $\frac{5}{8} \cdot 2{,}120 = \frac{5}{8} \cdot \frac{2{,}120}{1} = \$1{,}325$

65. Adding: $\frac{1}{4} + \frac{3}{20} = \frac{5}{20} + \frac{3}{20} = \frac{8}{20} = \frac{2}{5}$ of the students

67. Completing the table:

Grade	Number of Students	Fraction of Students
A	5	$\frac{1}{8}$
B	8	$\frac{1}{5}$
C	20	$\frac{1}{2}$
below C	7	$\frac{7}{40}$
Total	40	1

69. Dividing: $\dfrac{6 \text{ acres}}{\frac{3}{5} \text{ acres/lot}} = \frac{6}{1} \cdot \frac{5}{3} = 10$ lots

Chapter 0 Test

1. Translating into symbols: $15 - x = 12$
2. Translating into symbols: $6a = 30$
3. Simplifying using order of operations: $10 + 2(7-3) - 4^2 = 10 + 2(4) - 4^2 = 10 + 8 - 16 = 2$
4. Simplifying using order of operations: $15 + 24 \div 6 - 3^2 = 15 + 24 \div 6 - 9 = 15 + 4 - 9 = 10$
5. associative property of multiplication (**d**)
6. distributive property (**e**)
7. commutative property of addition (**a**)
8. associative property of addition (**c**)
9. Simplifying: $-2(3) - 7 = -6 - 7 = -6 + (-7) = -13$
10. Simplifying: $2(3)^3 - 4(-2)^4 = 2 \cdot 27 - 4 \cdot 16 = 54 - 64 = -10$
11. Simplifying: $9 + 4(2-6) = 9 + 4(-4) = 9 - 16 = -7$
12. Simplifying: $5 - 3[-2(1+4) + 3(-3)] = 5 - 3[-2(5) + (-9)] = 5 - 3(-10 - 9) = 5 - 3(-19) = 5 + 57 = 62$
13. Simplifying: $\dfrac{-4(3) + 5(-2)}{-5 - 6} = \dfrac{-12 - 10}{-11} = \dfrac{-22}{-11} = 2$
14. Simplifying: $\dfrac{4(3-5) - 2(-6+8)}{4(-2) + 10} = \dfrac{4(-2) - 2(2)}{-8 + 10} = \dfrac{-8 - 4}{2} = \dfrac{-12}{2} = -6$
15. Simplifying: $5 + (7 + 3x) = 3x + 5 + 7 = 3x + 12$
16. Simplifying: $3(-5y) = -15y$
17. Simplifying: $-5(2x - 3) = -5(2x) + 5(3) = -10x + 15$
18. Simplifying: $\dfrac{1}{3}(6x + 12) = \dfrac{1}{3}(6x) + \dfrac{1}{3}(12) = 2x + 4$
19. The integers are: $-3, 2$
20. The rational numbers are: $-3, -\dfrac{1}{2}, 2$
21. Factoring the number: $660 = 10 \cdot 66 = (2 \cdot 5) \cdot (6 \cdot 11) = (2 \cdot 5) \cdot (2 \cdot 3 \cdot 11) = 2^2 \cdot 3 \cdot 5 \cdot 11$
22. Factoring the number: $4{,}725 = 25 \cdot 189 = (5 \cdot 5) \cdot (9 \cdot 21) = (5 \cdot 5) \cdot (3 \cdot 3 \cdot 3 \cdot 7) = 3^3 \cdot 5^2 \cdot 7$
23. Combining the fractions: $\dfrac{5}{24} + \dfrac{9}{36} = \dfrac{5 \cdot 3}{24 \cdot 3} + \dfrac{9 \cdot 2}{36 \cdot 2} = \dfrac{15}{72} + \dfrac{18}{72} = \dfrac{33}{72} = \dfrac{11}{24}$
24. Combining the fractions: $\dfrac{5}{y} + \dfrac{6}{y} = \dfrac{11}{y}$
25. Translating into symbols: $6 + (-9) = -3$
26. Translating into symbols: $-5 - (-12) = -5 + 12 = 7$
27. Translating into symbols: $6(-7) = -42$
28. Translating into symbols: $\dfrac{32}{-8} = -4$
29. The pattern is to add 4, so the next number is: $9 + 4 = 13$
30. The pattern is to multiply by $-\dfrac{1}{3}$, so the next number is: $-3 \cdot \left(-\dfrac{1}{3}\right) = 1$

Chapter 1
Linear Equations and Inequalities

1.1 Simplifying Expressions

1. Simplifying the expression: $3x - 6x = (3-6)x = -3x$
3. Simplifying the expression: $-2a + a = (-2+1)a = -a$
5. Simplifying the expression: $7x + 3x + 2x = (7+3+2)x = 12x$
7. Simplifying the expression: $3a - 2a + 5a = (3-2+5)a = 6a$
9. Simplifying the expression: $4x - 3 + 2x = 4x + 2x - 3 = 6x - 3$
11. Simplifying the expression: $3a + 4a + 5 = (3+4)a + 5 = 7a + 5$
13. Simplifying the expression: $2x - 3 + 3x - 2 = 2x + 3x - 3 - 2 = 5x - 5$
15. Simplifying the expression: $3a - 1 + a + 3 = 3a + a - 1 + 3 = 4a + 2$
17. Simplifying the expression: $-4x + 8 - 5x - 10 = -4x - 5x + 8 - 10 = -9x - 2$
19. Simplifying the expression: $7a + 3 + 2a + 3a = 7a + 2a + 3a + 3 = 12a + 3$
21. Simplifying the expression: $5(2x-1) + 4 = 10x - 5 + 4 = 10x - 1$
23. Simplifying the expression: $7(3y+2) - 8 = 21y + 14 - 8 = 21y + 6$
25. Simplifying the expression: $-3(2x-1) + 5 = -6x + 3 + 5 = -6x + 8$
27. Simplifying the expression: $5 - 2(a+1) = 5 - 2a - 2 = -2a + 3$
29. Simplifying the expression: $6 - 4(x-5) = 6 - 4x + 20 = -4x + 20 + 6 = -4x + 26$
31. Simplifying the expression: $-9 - 4(2-y) + 1 = -9 - 8 + 4y + 1 = 4y + 1 - 9 - 8 = 4y - 16$
33. Simplifying the expression: $-6 + 2(2 - 3x) + 1 = -6 + 4 - 6x + 1 = -6x - 6 + 4 + 1 = -6x - 1$
35. Simplifying the expression: $(4x-7) - (2x+5) = 4x - 7 - 2x - 5 = 4x - 2x - 7 - 5 = 2x - 12$
37. Simplifying the expression: $8(2a+4) - (6a-1) = 16a + 32 - 6a + 1 = 16a - 6a + 32 + 1 = 10a + 33$
39. Simplifying the expression: $3(x-2) + (x-3) = 3x - 6 + x - 3 = 3x + x - 6 - 3 = 4x - 9$
41. Simplifying the expression: $4(2y-8) - (y+7) = 8y - 32 - y - 7 = 8y - y - 32 - 7 = 7y - 39$
43. Simplifying the expression: $-9(2x+1) - (x+5) = -18x - 9 - x - 5 = -18x - x - 9 - 5 = -19x - 14$
45. Evaluating when $x = 2$: $3x - 1 = 3(2) - 1 = 6 - 1 = 5$
47. Evaluating when $x = 2$: $-2x - 5 = -2(2) - 5 = -4 - 5 = -9$
49. Evaluating when $x = 2$: $x^2 - 8x + 16 = (2)^2 - 8(2) + 16 = 4 - 16 + 16 = 4$
51. Evaluating when $x = 2$: $(x-4)^2 = (2-4)^2 = (-2)^2 = 4$

53. Evaluating when $x = -5$: $7x - 4 - x - 3 = 7(-5) - 4 - (-5) - 3 = -35 - 4 + 5 - 3 = -42 + 5 = -37$
    Now simplifying the expression: $7x - 4 - x - 3 = 7x - x - 4 - 3 = 6x - 7$
    Evaluating when $x = -5$: $6x - 7 = 6(-5) - 7 = -30 - 7 = -37$
    Note that the two values are the same.

55. Evaluating when $x = -5$: $5(2x+1) + 4 = 5[2(-5)+1] + 4 = 5(-10+1) + 4 = 5(-9) + 4 = -45 + 4 = -41$
    Now simplifying the expression: $5(2x+1) + 4 = 10x + 5 + 4 = 10x + 9$
    Evaluating when $x = -5$: $10x + 9 = 10(-5) + 9 = -50 + 9 = -41$
    Note that the two values are the same.

57. Evaluating when $x = -3$ and $y = 5$: $x^2 - 2xy + y^2 = (-3)^2 - 2(-3)(5) + (5)^2 = 9 + 30 + 25 = 64$

59. Evaluating when $x = -3$ and $y = 5$: $(x-y)^2 = (-3-5)^2 = (-8)^2 = 64$

61. Evaluating when $x = -3$ and $y = 5$: $x^2 + 6xy + 9y^2 = (-3)^2 + 6(-3)(5) + 9(5)^2 = 9 - 90 + 225 = 144$

63. Evaluating when $x = -3$ and $y = 5$: $(x+3y)^2 = [-3+3(5)]^2 = (-3+15)^2 = (12)^2 = 144$

65. Evaluating when $x = \frac{1}{2}$: $12x - 3 = 12\left(\frac{1}{2}\right) - 3 = 6 - 3 = 3$

67. Evaluating when $x = \frac{1}{4}$: $12x - 3 = 12\left(\frac{1}{4}\right) - 3 = 3 - 3 = 0$

69. Evaluating when $x = \frac{3}{2}$: $12x - 3 = 12\left(\frac{3}{2}\right) - 3 = 18 - 3 = 15$

71. Evaluating when $x = \frac{3}{4}$: $12x - 3 = 12\left(\frac{3}{4}\right) - 3 = 9 - 3 = 6$

73. **a.** Substituting the values for n:

n	1	2	3	4
$3n$	3	6	9	12

    b. Substituting the values for n:

n	1	2	3	4
n^3	1	8	27	64

75. Substituting $n = 1, 2, 3, 4$:
    $n = 1$: $3(1) - 2 = 3 - 2 = 1$
    $n = 2$: $3(2) - 2 = 6 - 2 = 4$
    $n = 3$: $3(3) - 2 = 9 - 2 = 7$
    $n = 4$: $3(4) - 2 = 12 - 2 = 10$
    The sequence is 1, 4, 7, 10, ..., which is an arithmetic sequence.

77. Substituting $n = 1, 2, 3, 4$:
    $n = 1$: $(1)^2 - 2(1) + 1 = 1 - 2 + 1 = 0$
    $n = 2$: $(2)^2 - 2(2) + 1 = 4 - 4 + 1 = 1$
    $n = 3$: $(3)^2 - 2(3) + 1 = 9 - 6 + 1 = 4$
    $n = 4$: $(4)^2 - 2(4) + 1 = 16 - 8 + 1 = 9$
    The sequence is 0, 1, 4, 9, ..., which is a sequence of squares.

79. Simplifying the expression: $7 - 3(2y+1) = 7 - 6y - 3 = -6y + 4$

81. Simplifying the expression: $0.08x + 0.09x = 0.17x$

83. Simplifying the expression: $(x+y) + (x-y) = x + x + y - y = 2x$

85. Simplifying the expression: $3x + 2(x-2) = 3x + 2x - 4 = 5x - 4$

87. Simplifying the expression: $4(x+1)+3(x-3) = 4x+4+3x-9 = 7x-5$

89. Simplifying the expression: $x+(x+3)(-3) = x-3x-9 = -2x-9$

91. Simplifying the expression: $3(4x-2)-(5x-8) = 12x-6-5x+8 = 7x+2$

93. Simplifying the expression: $-(3x+1)-(4x-7) = -3x-1-4x+7 = -7x+6$

95. Simplifying the expression: $(x+3y)+3(2x-y) = x+3y+6x-3y = 7x$

97. Simplifying the expression: $3(2x+3y)-2(3x+5y) = 6x+9y-6x-10y = -y$

99. Simplifying the expression: $-6\left(\frac{1}{2}x-\frac{1}{3}y\right)+12\left(\frac{1}{4}x+\frac{2}{3}y\right) = -3x+2y+3x+8y = 10y$

101. Simplifying the expression: $0.08x+0.09(x+2{,}000) = 0.08x+0.09x+180 = 0.17x+180$

103. Simplifying the expression: $0.10x+0.12(x+500) = 0.10x+0.12x+60 = 0.22x+60$

105. Evaluating the expression: $b^2-4ac = (-5)^2-4(1)(-6) = 25-(-24) = 25+24 = 49$

107. Evaluating the expression: $b^2-4ac = (4)^2-4(2)(-3) = 16-(-24) = 16+24 = 40$

109. a. Substituting $x = 8{,}000$: $-0.0035(8000)+70 = 42°F$

   b. Substituting $x = 12{,}000$: $-0.0035(12000)+70 = 28°F$

   c. Substituting $x = 24{,}000$: $-0.0035(24000)+70 = -14°F$

111. a. Substituting $d = 2$: $35+5(2) = \$45$

   b. Substituting $d = 4$: $35+5(4) = \$55$

   c. Substituting $d = 8$: $35+5(8) = \$75$

113. Simplifying: $17-5 = 12$

115. Simplifying: $2-5 = -3$

117. Simplifying: $-2.4+(-7.3) = -9.7$

119. Simplifying: $-\frac{1}{2}+\left(-\frac{3}{4}\right) = -\frac{1}{2}\cdot\frac{2}{2}-\frac{3}{4} = -\frac{2}{4}-\frac{3}{4} = -\frac{5}{4}$

121. Simplifying: $4(2\cdot 9-3)-7 = 4(18-3)-7 = 4(15)-7 = 60-7 = 53$

123. Simplifying: $4(2a-3)-7a = 8a-12-7a = a-12$

125. Evaluating when $x = 5$: $2(5)-3 = 10-3 = 7$

1.2 Addition Property of Equality

1. Solving the equation:
$$x-3 = 8$$
$$x-3+3 = 8+3$$
$$x = 11$$

3. Solving the equation:
$$x+2 = 6$$
$$x+2+(-2) = 6+(-2)$$
$$x = 4$$

5. Solving the equation:
$$a+\frac{1}{2} = -\frac{1}{4}$$
$$a+\frac{1}{2}+\left(-\frac{1}{2}\right) = -\frac{1}{4}+\left(-\frac{1}{2}\right)$$
$$a = -\frac{1}{4}+\left(-\frac{2}{4}\right)$$
$$a = -\frac{3}{4}$$

7. Solving the equation:
$$x+2.3 = -3.5$$
$$x+2.3+(-2.3) = -3.5+(-2.3)$$
$$x = -5.8$$

9. Solving the equation:
$$y+11=-6$$
$$y+11+(-11)=-6+(-11)$$
$$y=-17$$

11. Solving the equation:
$$x-\frac{5}{8}=-\frac{3}{4}$$
$$x-\frac{5}{8}+\frac{5}{8}=-\frac{3}{4}+\frac{5}{8}$$
$$x=-\frac{6}{8}+\frac{5}{8}$$
$$x=-\frac{1}{8}$$

13. Solving the equation:
$$m-6=-10$$
$$m-6+6=-10+6$$
$$m=-4$$

15. Solving the equation:
$$6.9+x=3.3$$
$$-6.9+6.9+x=-6.9+3.3$$
$$x=-3.6$$

17. Solving the equation:
$$5=a+4$$
$$5-4=a+4-4$$
$$a=1$$

19. Solving the equation:
$$-\frac{5}{9}=x-\frac{2}{5}$$
$$-\frac{5}{9}+\frac{2}{5}=x-\frac{2}{5}+\frac{2}{5}$$
$$-\frac{25}{45}+\frac{18}{45}=x$$
$$x=-\frac{7}{45}$$

21. Solving the equation:
$$4x+2-3x=4+1$$
$$x+2=5$$
$$x+2+(-2)=5+(-2)$$
$$x=3$$

23. Solving the equation:
$$8a-\frac{1}{2}-7a=\frac{3}{4}+\frac{1}{8}$$
$$a-\frac{1}{2}=\frac{6}{8}+\frac{1}{8}$$
$$a-\frac{1}{2}=\frac{7}{8}$$
$$a-\frac{1}{2}+\frac{1}{2}=\frac{7}{8}+\frac{1}{2}$$
$$a=\frac{7}{8}+\frac{4}{8}$$
$$a=\frac{11}{8}$$

25. Solving the equation:
$$-3-4x+5x=18$$
$$-3+x=18$$
$$3-3+x=3+18$$
$$x=21$$

27. Solving the equation:
$$-11x+2+10x+2x=9$$
$$x+2=9$$
$$x+2+(-2)=9+(-2)$$
$$x=7$$

29. Solving the equation:
$$-2.5+4.8=8x-1.2-7x$$
$$2.3=x-1.2$$
$$2.3+1.2=x-1.2+1.2$$
$$x=3.5$$

31. Solving the equation:
$$2y-10+3y-4y=18-6$$
$$y-10=12$$
$$y-10+10=12+10$$
$$y=22$$

33. Solving the equation:
$$2(x+3)-x=4$$
$$2x+6-x=4$$
$$x+6=4$$
$$x+6+(-6)=4+(-6)$$
$$x=-2$$

35. Solving the equation:
$$-3(x-4)+4x=3-7$$
$$-3x+12+4x=-4$$
$$x+12=-4$$
$$x+12+(-12)=-4+(-12)$$
$$x=-16$$

37. Solving the equation:
$$5(2a+1)-9a=8-6$$
$$10a+5-9a=2$$
$$a+5=2$$
$$a+5+(-5)=2+(-5)$$
$$a=-3$$

39. Solving the equation:
$$-(x+3)+2x-1=6$$
$$-x-3+2x-1=6$$
$$x-4=6$$
$$x-4+4=6+4$$
$$x=10$$

41. Solving the equation:
$$4y-3(y-6)+2=8$$
$$4y-3y+18+2=8$$
$$y+20=8$$
$$y+20+(-20)=8+(-20)$$
$$y=-12$$

43. Solving the equation:
$$-3(2m-9)+7(m-4)=12-9$$
$$-6m+27+7m-28=3$$
$$m-1=3$$
$$m-1+1=3+1$$
$$m=4$$

45. Solving the equation:
$$4x=3x+2$$
$$4x+(-3x)=3x+(-3x)+2$$
$$x=2$$

47. Solving the equation:
$$8a=7a-5$$
$$8a+(-7a)=7a+(-7a)-5$$
$$a=-5$$

49. Solving the equation:
$$2x=3x+1$$
$$(-2x)+2x=(-2x)+3x+1$$
$$0=x+1$$
$$0+(-1)=x+1+(-1)$$
$$x=-1$$

51. Solving the equation:
$$3y+4=2y+1$$
$$3y+(-2y)+4=2y+(-2y)+1$$
$$y+4=1$$
$$y+4+(-4)=1+(-4)$$
$$y=-3$$

53. Solving the equation:
$$2m-3=m+5$$
$$2m+(-m)-3=m+(-m)+5$$
$$m-3=5$$
$$m-3+3=5+3$$
$$m=8$$

55. Solving the equation:
$$4x-7=5x+1$$
$$4x+(-4x)-7=5x+(-4x)+1$$
$$-7=x+1$$
$$-7+(-1)=x+1+(-1)$$
$$x=-8$$

57. Solving the equation:
$$5x - \frac{2}{3} = 4x + \frac{4}{3}$$
$$5x + (-4x) - \frac{2}{3} = 4x + (-4x) + \frac{4}{3}$$
$$x - \frac{2}{3} = \frac{4}{3}$$
$$x - \frac{2}{3} + \frac{2}{3} = \frac{4}{3} + \frac{2}{3}$$
$$x = \frac{6}{3} = 2$$

59. Solving the equation:
$$8a - 7.1 = 7a + 3.9$$
$$8a + (-7a) - 7.1 = 7a + (-7a) + 3.9$$
$$a - 7.1 = 3.9$$
$$a - 7.1 + 7.1 = 3.9 + 7.1$$
$$a = 11$$

61. Solving the equation:
$$11y - 2.9 = 12y + 2.9$$
$$11y + (-11y) - 2.9 = 12y + (-11y) + 2.9$$
$$-2.9 = y + 2.9$$
$$-2.9 - 2.9 = y + 2.9 - 2.9$$
$$y = -5.8$$

63. **a.** Solving for R:
$$T + R + A = 100$$
$$88 + R + 6 = 100$$
$$94 + R = 100$$
$$R = 6\%$$

b. Solving for R:
$$T + R + A = 100$$
$$0 + R + 95 = 100$$
$$95 + R = 100$$
$$R = 5\%$$

c. Solving for A:
$$T + R + A = 100$$
$$0 + 98 + A = 100$$
$$98 + A = 100$$
$$A = 2\%$$

d. Solving for R:
$$T + R + A = 100$$
$$0 + R + 25 = 100$$
$$25 + R = 100$$
$$R = 75\%$$

65. Simplifying: $\frac{3}{2}\left(\frac{2}{3}y\right) = y$

67. Simplifying: $\frac{1}{5}(5x) = x$

69. Simplifying: $\frac{1}{5}(30) = 6$

71. Simplifying: $\frac{3}{2}(4) = \frac{12}{2} = 6$

73. Simplifying: $12\left(-\frac{3}{4}\right) = -\frac{36}{4} = -9$

75. Simplifying: $\frac{3}{2}\left(-\frac{5}{4}\right) = -\frac{15}{8}$

77. Simplifying: $13 + (-5) = 8$

79. Simplifying: $-\frac{3}{4} + \left(-\frac{1}{2}\right) = -\frac{3}{4} - \frac{1}{2} \cdot \frac{2}{2} = -\frac{3}{4} - \frac{2}{4} = -\frac{5}{4}$

81. Simplifying: $7x + (-4x) = 3x$

1.3 Multiplication Property of Equality

1. Solving the equation:
$$5x = 10$$
$$\frac{1}{5}(5x) = \frac{1}{5}(10)$$
$$x = 2$$

3. Solving the equation:
$$7a = 28$$
$$\frac{1}{7}(7a) = \frac{1}{7}(28)$$
$$a = 4$$

5. Solving the equation:
$$-8x = 4$$
$$-\frac{1}{8}(-8x) = -\frac{1}{8}(4)$$
$$x = -\frac{1}{2}$$

7. Solving the equation:
$$8m = -16$$
$$\frac{1}{8}(8m) = \frac{1}{8}(-16)$$
$$m = -2$$

9. Solving the equation:
$$-3x = -9$$
$$-\frac{1}{3}(-3x) = -\frac{1}{3}(-9)$$
$$x = 3$$

11. Solving the equation:
$$-7y = -28$$
$$-\frac{1}{7}(-7y) = -\frac{1}{7}(-28)$$
$$y = 4$$

13. Solving the equation:
$$2x = 0$$
$$\frac{1}{2}(2x) = \frac{1}{2}(0)$$
$$x = 0$$

15. Solving the equation:
$$-5x = 0$$
$$-\frac{1}{5}(-5x) = -\frac{1}{5}(0)$$
$$x = 0$$

17. Solving the equation:
$$\frac{x}{3} = 2$$
$$3\left(\frac{x}{3}\right) = 3(2)$$
$$x = 6$$

19. Solving the equation:
$$-\frac{m}{5} = 10$$
$$-5\left(-\frac{m}{5}\right) = -5(10)$$
$$m = -50$$

21. Solving the equation:
$$-\frac{x}{2} = -\frac{3}{4}$$
$$-2\left(-\frac{x}{2}\right) = -2\left(-\frac{3}{4}\right)$$
$$x = \frac{3}{2}$$

23. Solving the equation:
$$\frac{2}{3}a = 8$$
$$\frac{3}{2}\left(\frac{2}{3}a\right) = \frac{3}{2}(8)$$
$$a = 12$$

25. Solving the equation:
$$-\frac{3}{5}x = \frac{9}{5}$$
$$-\frac{5}{3}\left(-\frac{3}{5}x\right) = -\frac{5}{3}\left(\frac{9}{5}\right)$$
$$x = -3$$

27. Solving the equation:
$$-\frac{5}{8}y = -20$$
$$-\frac{8}{5}\left(-\frac{5}{8}y\right) = -\frac{8}{5}(-20)$$
$$y = 32$$

29. Simplifying and then solving the equation:
$$-4x - 2x + 3x = 24$$
$$-3x = 24$$
$$-\frac{1}{3}(-3x) = -\frac{1}{3}(24)$$
$$x = -8$$

31. Simplifying and then solving the equation:
$$4x + 8x - 2x = 15 - 10$$
$$10x = 5$$
$$\frac{1}{10}(10x) = \frac{1}{10}(5)$$
$$x = \frac{1}{2}$$

33. Simplifying and then solving the equation:
$$-3 - 5 = 3x + 5x - 10x$$
$$-8 = -2x$$
$$-\frac{1}{2}(-8) = -\frac{1}{2}(-2x)$$
$$x = 4$$

35. Multiplying by 8 to eliminate fractions:
$$18 - 13 = \frac{1}{2}a + \frac{3}{4}a - \frac{5}{8}a$$
$$8(5) = 8\left(\frac{1}{2}a + \frac{3}{4}a - \frac{5}{8}a\right)$$
$$40 = 4a + 6a - 5a$$
$$40 = 5a$$
$$\frac{1}{5}(40) = \frac{1}{5}(5a)$$
$$a = 8$$

37. Solving by multiplying both sides of the equation by -1:
$$-x = 4$$
$$-1(-x) = -1(4)$$
$$x = -4$$

39. Solving by multiplying both sides of the equation by -1:
$$-x = -4$$
$$-1(-x) = -1(-4)$$
$$x = 4$$

41. Solving by multiplying both sides of the equation by -1:
$$15 = -a$$
$$-1(15) = -1(-a)$$
$$a = -15$$

43. Solving by multiplying both sides of the equation by -1:
$$-y = \frac{1}{2}$$
$$-1(-y) = -1\left(\frac{1}{2}\right)$$
$$y = -\frac{1}{2}$$

45. Solving the equation:
$$3x - 2 = 7$$
$$3x - 2 + 2 = 7 + 2$$
$$3x = 9$$
$$\frac{1}{3}(3x) = \frac{1}{3}(9)$$
$$x = 3$$

47. Solving the equation:
$$2a + 1 = 3$$
$$2a + 1 + (-1) = 3 + (-1)$$
$$2a = 2$$
$$\frac{1}{2}(2a) = \frac{1}{2}(2)$$
$$a = 1$$

49. Multiplying by 8 to eliminate fractions:
$$\frac{1}{8}+\frac{1}{2}x=\frac{1}{4}$$
$$8\left(\frac{1}{8}+\frac{1}{2}x\right)=8\left(\frac{1}{4}\right)$$
$$1+4x=2$$
$$(-1)+1+4x=(-1)+2$$
$$4x=1$$
$$\frac{1}{4}(4x)=\frac{1}{4}(1)$$
$$x=\frac{1}{4}$$

51. Solving the equation:
$$6x=2x-12$$
$$6x+(-2x)=2x+(-2x)-12$$
$$4x=-12$$
$$\frac{1}{4}(4x)=\frac{1}{4}(-12)$$
$$x=-3$$

53. Solving the equation:
$$2y=-4y+18$$
$$2y+4y=-4y+4y+18$$
$$6y=18$$
$$\frac{1}{6}(6y)=\frac{1}{6}(18)$$
$$y=3$$

55. Solving the equation:
$$-7x=-3x-8$$
$$-7x+3x=-3x+3x-8$$
$$-4x=-8$$
$$-\frac{1}{4}(-4x)=-\frac{1}{4}(-8)$$
$$x=2$$

57. Solving the equation:
$$8x+4=2x-5$$
$$8x+(-8x)+4=2x+(-8x)-5$$
$$-6x-5=4$$
$$-6x-5+5=4+5$$
$$-6x=9$$
$$-\frac{1}{6}(-6x)=-\frac{1}{6}(9)$$
$$x=-\frac{3}{2}$$

59. Multiplying by 8 to eliminate fractions:
$$x+\frac{1}{2}=\frac{1}{4}x-\frac{5}{8}$$
$$8\left(x+\frac{1}{2}\right)=8\left(\frac{1}{4}x-\frac{5}{8}\right)$$
$$8x+4=2x-5$$
$$8x+(-2x)+4=2x+(-2x)-5$$
$$6x+4=-5$$
$$6x+4+(-4)=-5+(-4)$$
$$6x=-9$$
$$\frac{1}{6}(6x)=\frac{1}{6}(-9)$$
$$x=-\frac{3}{2}$$

61. Solving the equation:
$$6m - 3 = m + 2$$
$$6m + (-6m) - 3 = m + (-6m) + 2$$
$$-5m + 2 = -3$$
$$-5m + 2 + (-2) = -3 + (-2)$$
$$-5m = -5$$
$$-\frac{1}{5}(-5m) = -\frac{1}{5}(-5)$$
$$m = 1$$

63. Multiplying by 12 to eliminate fractions:
$$\frac{1}{2}m - \frac{1}{4} = \frac{1}{12}m + \frac{1}{6}$$
$$12\left(\frac{1}{2}m - \frac{1}{4}\right) = 12\left(\frac{1}{12}m + \frac{1}{6}\right)$$
$$6m - 3 = m + 2$$
$$6m + (-m) - 3 = m + (-m) + 2$$
$$5m - 3 = 2$$
$$5m - 3 + 3 = 2 + 3$$
$$5m = 5$$
$$\frac{1}{5}(5m) = \frac{1}{5}(5)$$
$$m = 1$$

65. Solving the equation:
$$9y + 2 = 6y - 4$$
$$9y + (-9y) + 2 = 6y + (-9y) - 4$$
$$-3y - 4 = 2$$
$$-3y - 4 + 4 = 2 + 4$$
$$-3y = 6$$
$$-\frac{1}{3}(-3y) = -\frac{1}{3}(6)$$
$$y = -2$$

67. a. Solving the equation:
$$2x = 3$$
$$\frac{1}{2}(2x) = \frac{1}{2}(3)$$
$$x = \frac{3}{2}$$

b. Solving the equation:
$$2 + x = 3$$
$$2 + (-2) + x = 3 + (-2)$$
$$x = 1$$

c. Solving the equation:
$$2x + 3 = 0$$
$$2x + 3 + (-3) = 0 + (-3)$$
$$2x = -3$$
$$\frac{1}{2}(2x) = \frac{1}{2}(-3)$$
$$x = -\frac{3}{2}$$

d. Solving the equation:
$$2x + 3 = -5$$
$$2x + 3 + (-3) = -5 + (-3)$$
$$2x = -8$$
$$\frac{1}{2}(2x) = \frac{1}{2}(-8)$$
$$x = -4$$

e. Solving the equation:
$$2x+3=7x-5$$
$$2x+(-7x)+3=7x+(-7x)-5$$
$$-5x+3=-5$$
$$-5x+3+(-3)=-5+(-3)$$
$$-5x=-8$$
$$-\frac{1}{5}(-5x)=-\frac{1}{5}(-8)$$
$$x=\frac{8}{5}$$

69. Solving the equation:
$$7.5x=1500$$
$$\frac{7.5x}{7.5}=\frac{1500}{7.5}$$
$$x=200$$
The break-even point is 200 tickets.

71. Solving the equation:
$$G-0.21G-0.08G=987.5$$
$$0.71G=987.5$$
$$G\approx 1{,}390.85$$
Your gross monthly income is approximately $1,390.85.

73. Solving the equation:
$$2x=4$$
$$\frac{1}{2}(2x)=\frac{1}{2}(4)$$
$$x=2$$

75. Solving the equation:
$$30=5x$$
$$5x=30$$
$$\frac{1}{5}(5x)=\frac{1}{5}(30)$$
$$x=6$$

77. Solving the equation:
$$0.17x=510$$
$$x=\frac{510}{0.17}=3{,}000$$

79. Simplifying: $3(x-5)+4=3x-15+4=3x-11$

81. Simplifying: $0.09(x+2{,}000)=0.09x+180$

83. Simplifying: $7-3(2y+1)=7-6y-3=4-6y=-6y+4$

85. Simplifying: $3(2x-5)-(2x-4)=6x-15-2x+4=4x-11$

87. Simplifying: $10x+(-5x)=5x$

89. Simplifying: $0.08x+0.09x=0.17x$

1.4 Solving Linear Equations

1. Solving the equation:
$$2(x+3) = 12$$
$$2x+6 = 12$$
$$2x+6+(-6) = 12+(-6)$$
$$2x = 6$$
$$\frac{1}{2}(2x) = \frac{1}{2}(6)$$
$$x = 3$$

3. Solving the equation:
$$6(x-1) = -18$$
$$6x-6 = -18$$
$$6x-6+6 = -18+6$$
$$6x = -12$$
$$\frac{1}{6}(6x) = \frac{1}{6}(-12)$$
$$x = -2$$

5. Solving the equation:
$$2(4a+1) = -6$$
$$8a+2 = -6$$
$$8a+2+(-2) = -6+(-2)$$
$$8a = -8$$
$$\frac{1}{8}(8a) = \frac{1}{8}(-8)$$
$$a = -1$$

7. Solving the equation:
$$14 = 2(5x-3)$$
$$14 = 10x-6$$
$$14+6 = 10x-6+6$$
$$20 = 10x$$
$$\frac{1}{10}(20) = \frac{1}{10}(10x)$$
$$x = 2$$

9. Solving the equation:
$$-2(3y+5) = 14$$
$$-6y-10 = 14$$
$$-6y-10+10 = 14+10$$
$$-6y = 24$$
$$-\frac{1}{6}(-6y) = -\frac{1}{6}(24)$$
$$y = -4$$

11. Solving the equation:
$$-5(2a+4) = 0$$
$$-10a-20 = 0$$
$$-10a-20+20 = 0+20$$
$$-10a = 20$$
$$-\frac{1}{10}(-10a) = -\frac{1}{10}(20)$$
$$a = -2$$

13. Solving the equation:
$$1 = \frac{1}{2}(4x+2)$$
$$1 = 2x+1$$
$$1+(-1) = 2x+1+(-1)$$
$$0 = 2x$$
$$\frac{1}{2}(0) = \frac{1}{2}(2x)$$
$$x = 0$$

15. Solving the equation:
$$3(t-4)+5 = -4$$
$$3t-12+5 = -4$$
$$3t-7 = -4$$
$$3t-7+7 = -4+7$$
$$3t = 3$$
$$\frac{1}{3}(3t) = \frac{1}{3}(3)$$
$$t = 1$$

17. Solving the equation:

$$4(2x+1)-7=1$$
$$8x+4-7=1$$
$$8x-3=1$$
$$8x-3+3=1+3$$
$$8x=4$$
$$\frac{1}{8}(8x)=\frac{1}{8}(4)$$
$$x=\frac{1}{2}$$

19. Solving the equation:

$$\frac{1}{2}(x-3)=\frac{1}{4}(x+1)$$
$$\frac{1}{2}x-\frac{3}{2}=\frac{1}{4}x+\frac{1}{4}$$
$$4\left(\frac{1}{2}x-\frac{3}{2}\right)=4\left(\frac{1}{4}x+\frac{1}{4}\right)$$
$$2x-6=x+1$$
$$2x+(-x)-6=x+(-x)+1$$
$$x-6=1$$
$$x-6+6=1+6$$
$$x=7$$

21. Solving the equation:

$$-0.7(2x-7)=0.3(11-4x)$$
$$-1.4x+4.9=3.3-1.2x$$
$$-1.4x+1.2x+4.9=3.3-1.2x+1.2x$$
$$-0.2x+4.9=3.3$$
$$-0.2x+4.9+(-4.9)=3.3+(-4.9)$$
$$-0.2x=-1.6$$
$$\frac{-0.2x}{-0.2}=\frac{-1.6}{-0.2}$$
$$x=8$$

23. Solving the equation:

$$-2(3y+1)=3(1-6y)-9$$
$$-6y-2=3-18y-9$$
$$-6y-2=-18y-6$$
$$-6y+18y-2=-18y+18y-6$$
$$12y-2=-6$$
$$12y-2+2=-6+2$$
$$12y=-4$$
$$\frac{1}{12}(12y)=\frac{1}{12}(-4)$$
$$y=-\frac{1}{3}$$

25. Solving the equation:

$$\frac{3}{4}(8x-4)+3=\frac{2}{5}(5x+10)-1$$
$$6x-3+3=2x+4-1$$
$$6x=2x+3$$
$$6x+(-2x)=2x+(-2x)+3$$
$$4x=3$$
$$\frac{1}{4}(4x)=\frac{1}{4}(3)$$
$$x=\frac{3}{4}$$

27. Solving the equation:

$$0.06x+0.08(100-x)=6.5$$
$$0.06x+8-0.08x=6.5$$
$$-0.02x+8=6.5$$
$$-0.02x+8+(-8)=6.5+(-8)$$
$$-0.02x=-1.5$$
$$\frac{-0.02x}{-0.02}=\frac{-1.5}{-0.02}$$
$$x=75$$

29. Solving the equation:

$$6-5(2a-3)=1$$
$$6-10a+15=1$$
$$-10a+21=1$$
$$-10a+21+(-21)=1+(-21)$$
$$-10a=-20$$
$$-\frac{1}{10}(-10a)=-\frac{1}{10}(-20)$$
$$a=2$$

31. Solving the equation:

$$0.2x-0.5=0.5-0.2(2x-13)$$
$$0.2x-0.5=0.5-0.4x+2.6$$
$$0.2x-0.5=-0.4x+3.1$$
$$0.2x+0.4x-0.5=-0.4x+0.4x+3.1$$
$$0.6x-0.5=3.1$$
$$0.6x-0.5+0.5=3.1+0.5$$
$$0.6x=3.6$$
$$\frac{0.6x}{0.6}=\frac{3.6}{0.6}$$
$$x=6$$

Essentials of Elementary and Intermediate Algebra: A Combined Course
Student Solutions Manual, Problem Set 1.4

33. Solving the equation:
$$2(t-3)+3(t-2)=28$$
$$2t-6+3t-6=28$$
$$5t-12=28$$
$$5t-12+12=28+12$$
$$5t=40$$
$$\frac{1}{5}(5t)=\frac{1}{5}(40)$$
$$t=8$$

35. Solving the equation:
$$5(x-2)-(3x+4)=3(6x-8)+10$$
$$5x-10-3x-4=18x-24+10$$
$$2x-14=18x-14$$
$$2x+(-18x)-14=18x+(-18x)-14$$
$$-16x-14=-14$$
$$-16x-14+14=-14+14$$
$$-16x=0$$
$$-\frac{1}{16}(-16x)=-\frac{1}{16}(0)$$
$$x=0$$

37. Solving the equation:
$$2(5x-3)-(2x-4)=5-(6x+1)$$
$$10x-6-2x+4=5-6x-1$$
$$8x-2=-6x+4$$
$$8x+6x-2=-6x+6x+4$$
$$14x-2=4$$
$$14x-2+2=4+2$$
$$14x=6$$
$$\frac{1}{14}(14x)=\frac{1}{14}(6)$$
$$x=\frac{3}{7}$$

39. Solving the equation:
$$-(3x+1)-(4x-7)=4-(3x+2)$$
$$-3x-1-4x+7=4-3x-2$$
$$-7x+6=-3x+2$$
$$-7x+3x+6=-3x+3x+2$$
$$-4x+6=2$$
$$-4x+6+(-6)=2+(-6)$$
$$-4x=-4$$
$$-\frac{1}{4}(-4x)=-\frac{1}{4}(-4)$$
$$x=1$$

41. Solving the equation:
$$x+(2x-1)=2$$
$$3x-1=2$$
$$3x-1+1=2+1$$
$$3x=3$$
$$\frac{1}{3}(3x)=\frac{1}{3}(3)$$
$$x=1$$

43. Solving the equation:
$$x-(3x+5)=-3$$
$$x-3x-5=-3$$
$$-2x-5=-3$$
$$-2x-5+5=-3+5$$
$$-2x=2$$
$$-\frac{1}{2}(-2x)=-\frac{1}{2}(2)$$
$$x=-1$$

45. Solving the equation:
$$15=3(x-1)$$
$$15=3x-3$$
$$15+3=3x-3+3$$
$$18=3x$$
$$\frac{1}{3}(18)=\frac{1}{3}(3x)$$
$$x=6$$

47. Solving the equation:
$$4x-(-4x+1)=5$$
$$4x+4x-1=5$$
$$8x-1=5$$
$$8x-1+1=5+1$$
$$8x=6$$
$$\frac{1}{8}(8x)=\frac{1}{8}(6)$$
$$x=\frac{3}{4}$$

49. Solving the equation:
$$5x - 8(2x - 5) = 7$$
$$5x - 16x + 40 = 7$$
$$-11x + 40 = 7$$
$$-11x + 40 - 40 = 7 - 40$$
$$-11x = -33$$
$$-\frac{1}{11}(-11x) = -\frac{1}{11}(-33)$$
$$x = 3$$

51. Solving the equation:
$$7(2y - 1) - 6y = -1$$
$$14y - 7 - 6y = -1$$
$$8y - 7 = -1$$
$$8y - 7 + 7 = -1 + 7$$
$$8y = 6$$
$$\frac{1}{8}(8y) = \frac{1}{8}(6)$$
$$y = \frac{3}{4}$$

53. Solving the equation:
$$0.2x + 0.5(12 - x) = 3.6$$
$$0.2x + 6 - 0.5x = 3.6$$
$$-0.3x + 6 = 3.6$$
$$-0.3x + 6 - 6 = 3.6 - 6$$
$$-0.3x = -2.4$$
$$x = \frac{-2.4}{-0.3} = 8$$

55. Solving the equation:
$$0.5x + 0.2(18 - x) = 5.4$$
$$0.5x + 3.6 - 0.2x = 5.4$$
$$0.3x + 3.6 = 5.4$$
$$0.3x + 3.6 - 3.6 = 5.4 - 3.6$$
$$0.3x = 1.8$$
$$x = \frac{1.8}{0.3} = 6$$

57. Solving the equation:
$$x + (x + 3)(-3) = x - 3$$
$$x - 3x - 9 = x - 3$$
$$-2x - 9 = x - 3$$
$$-2x - x - 9 = x - x - 3$$
$$-3x - 9 = -3$$
$$-3x - 9 + 9 = -3 + 9$$
$$-3x = 6$$
$$-\frac{1}{3}(-3x) = -\frac{1}{3}(6)$$
$$x = -2$$

59. Solving the equation:
$$5(x + 2) + 3(x - 1) = -9$$
$$5x + 10 + 3x - 3 = -9$$
$$8x + 7 = -9$$
$$8x + 7 - 7 = -9 - 7$$
$$8x = -16$$
$$\frac{1}{8}(8x) = \frac{1}{8}(-16)$$
$$x = -2$$

61. Solving the equation:
$$3(x - 3) + 2(2x) = 5$$
$$3x - 9 + 4x = 5$$
$$7x - 9 = 5$$
$$7x - 9 + 9 = 5 + 9$$
$$7x = 14$$
$$\frac{1}{7}(7x) = \frac{1}{7}(14)$$
$$x = 2$$

63. Solving the equation:
$$5(y + 2) = 4(y + 1)$$
$$5y + 10 = 4y + 4$$
$$-4y + 5y + 10 = -4y + 4y + 4$$
$$y + 10 = 4$$
$$y + 10 - 10 = 4 - 10$$
$$y = -6$$

65. Solving the equation:
$$3x+2(x-2)=6$$
$$3x+2x-4=6$$
$$5x-4=6$$
$$5x-4+4=6+4$$
$$5x=10$$
$$\frac{1}{5}(5x)=\frac{1}{5}(10)$$
$$x=2$$

67. Solving the equation:
$$50(x-5)=30(x+5)$$
$$50x-250=30x+150$$
$$50x-30x-250=30x-30x+150$$
$$20x-250=150$$
$$20x-250+250=150+250$$
$$20x=400$$
$$\frac{1}{20}(20x)=\frac{1}{20}(400)$$
$$x=20$$

69. Solving the equation:
$$0.08x+0.09(x+2{,}000)=860$$
$$0.08x+0.09x+180=860$$
$$0.17x+180=860$$
$$0.17x+180-180=860-180$$
$$0.17x=680$$
$$x=\frac{680}{0.17}=4{,}000$$

71. Solving the equation:
$$0.10x+0.12(x+500)=214$$
$$0.10x+0.12x+60=214$$
$$0.22x+60=214$$
$$0.22x+60-60=214-60$$
$$0.22x=154$$
$$x=\frac{154}{0.22}=700$$

73. Solving the equation:
$$5x+10(x+8)=245$$
$$5x+10x+80=245$$
$$15x+80=245$$
$$15x+80-80=245-80$$
$$15x=165$$
$$x=\frac{165}{15}=11$$

75. Solving the equation:
$$5x+10(x+3)+25(x+5)=435$$
$$5x+10x+30+25x+125=435$$
$$40x+155=435$$
$$40x+155-155=435-155$$
$$40x=280$$
$$x=\frac{280}{40}=7$$

77. Solving the equation:
$$3x-6=3(x+4)$$
$$3x-6=3x+12$$
$$3x-3x-6=3x-3x+12$$
$$-6=12$$
Since this statement is false, there is no solution. This is a contradiction.

79. Solving the equation:
$$2(4t-1)+3=5t+4+3t$$
$$8t-2+3=8t+4$$
$$8t+1=8t+4$$
$$8t-8t+1=8t-8t+4$$
$$1=4$$
Since this statement is false, there is no solution. This is a contradiction.

81. **a.** Solving the equation:
$$4x-5=0$$
$$4x-5+5=0+5$$
$$4x=5$$
$$\frac{1}{4}(4x)=\frac{1}{4}(5)$$
$$x=\frac{5}{4}=1.25$$

b. Solving the equation:
$$4x-5=25$$
$$4x-5+5=25+5$$
$$4x=30$$
$$\frac{1}{4}(4x)=\frac{1}{4}(30)$$
$$x=\frac{15}{2}=7.5$$

c. Adding: $(4x-5)+(2x+25)=6x+20$

d. Solving the equation:
$$4x-5=2x+25$$
$$4x-2x-5=2x-2x+25$$
$$2x-5=25$$
$$2x-5+5=25+5$$
$$2x=30$$
$$\frac{1}{2}(2x)=\frac{1}{2}(30)$$
$$x=15$$

e. Multiplying: $4(x-5)=4x-20$

f. Solving the equation:
$$4(x-5)=2x+25$$
$$4x-20=2x+25$$
$$4x-2x-20=2x-2x+25$$
$$2x-20=25$$
$$2x-20+20=25+20$$
$$2x=45$$
$$\frac{1}{2}(2x)=\frac{1}{2}(45)$$
$$x=\frac{45}{2}=22.5$$

83. Solving the equation:
$$40=2x+12$$
$$2x+12=40$$
$$2x=28$$
$$x=14$$

85. Solving the equation:
$$12+2y=6$$
$$2y=-6$$
$$y=-3$$

87. Solving the equation:
$$24x=6$$
$$x=\frac{6}{24}=\frac{1}{4}$$

89. Solving the equation:
$$70=x\cdot 210$$
$$x=\frac{70}{210}=\frac{1}{3}$$

91. Simplifying: $\frac{1}{2}(-3x+6)=\frac{1}{2}(-3x)+\frac{1}{2}(6)=-\frac{3}{2}x+3$

1.5 Formulas

1. Using the perimeter formula:
$$P = 2l + 2w$$
$$300 = 2l + 2(50)$$
$$300 = 2l + 100$$
$$200 = 2l$$
$$l = 100$$
The length is 100 feet.

3. Substituting $x = 3$:
$$2(3) + 3y = 6$$
$$6 + 3y = 6$$
$$6 + (-6) + 3y = 6 + (-6)$$
$$3y = 0$$
$$y = 0$$

5. Substituting $x = 0$:
$$2(0) + 3y = 6$$
$$0 + 3y = 6$$
$$3y = 6$$
$$y = 2$$

7. Substituting $y = 2$:
$$2x - 5(2) = 20$$
$$2x - 10 = 20$$
$$2x - 10 + 10 = 20 + 10$$
$$2x = 30$$
$$x = 15$$

9. Substituting $y = 0$:
$$2x - 5(0) = 20$$
$$2x - 0 = 20$$
$$2x = 20$$
$$x = 10$$

11. Substituting $x = -2$: $y = (-2+1)^2 - 3 = (-1)^2 - 3 = 1 - 3 = -2$

13. Substituting $x = 1$: $y = (1+1)^2 - 3 = (2)^2 - 3 = 4 - 3 = 1$

15. a. Substituting $x = 10$: $y = \frac{20}{10} = 2$
    b. Substituting $x = 5$: $y = \frac{20}{5} = 4$

17. a. Substituting $y = 15$ and $x = 3$:
$$15 = K(3)$$
$$K = \frac{15}{3} = 5$$
   b. Substituting $y = 72$ and $x = 4$:
$$72 = K(4)$$
$$K = \frac{72}{4} = 18$$

19. Solving for l:
$$lw = A$$
$$\frac{lw}{w} = \frac{A}{w}$$
$$l = \frac{A}{w}$$

21. Solving for h:
$$lwh = V$$
$$\frac{lwh}{lw} = \frac{V}{lw}$$
$$h = \frac{V}{lw}$$

23. Solving for a:
$$a + b + c = P$$
$$a + b + c - b - c = P - b - c$$
$$a = P - b - c$$

25. Solving for x:
$$x - 3y = -1$$
$$x - 3y + 3y = -1 + 3y$$
$$x = 3y - 1$$

27. Solving for y:
$$-3x + y = 6$$
$$-3x + 3x + y = 6 + 3x$$
$$y = 3x + 6$$

29. Solving for y:
$$2x + 3y = 6$$
$$-2x + 2x + 3y = -2x + 6$$
$$3y = -2x + 6$$
$$\frac{1}{3}(3y) = \frac{1}{3}(-2x + 6)$$
$$y = -\frac{2}{3}x + 2$$

31. Solving for y:
$$y - 3 = -2(x + 4)$$
$$y - 3 = -2x - 8$$
$$y = -2x - 5$$

33. Solving for y:
$$y - 3 = -\frac{2}{3}(x + 3)$$
$$y - 3 = -\frac{2}{3}x - 2$$
$$y = -\frac{2}{3}x + 1$$

35. Solving for w:
$$2l + 2w = P$$
$$2l - 2l + 2w = P - 2l$$
$$2w = P - 2l$$
$$\frac{2w}{2} = \frac{P - 2l}{2}$$
$$w = \frac{P - 2l}{2}$$

37. Solving for v:
$$vt + 16t^2 = h$$
$$vt + 16t^2 - 16t^2 = h - 16t^2$$
$$vt = h - 16t^2$$
$$\frac{vt}{t} = \frac{h - 16t^2}{t}$$
$$v = \frac{h - 16t^2}{t}$$

39. Solving for h:
$$\pi r^2 + 2\pi rh = A$$
$$\pi r^2 - \pi r^2 + 2\pi rh = A - \pi r^2$$
$$2\pi rh = A - \pi r^2$$
$$\frac{2\pi rh}{2\pi r} = \frac{A - \pi r^2}{2\pi r}$$
$$h = \frac{A - \pi r^2}{2\pi r}$$

41. a. Solving for y:
$$\frac{y - 1}{x} = \frac{3}{5}$$
$$5y - 5 = 3x$$
$$5y = 3x + 5$$
$$y = \frac{3}{5}x + 1$$

b. Solving for y:
$$\frac{y - 2}{x} = \frac{1}{2}$$
$$2y - 4 = x$$
$$2y = x + 4$$
$$y = \frac{1}{2}x + 2$$

c. Solving for y:
$$\frac{y - 3}{x} = 4$$
$$y - 3 = 4x$$
$$y = 4x + 3$$

43. Solving for y:
$$\frac{x}{7} - \frac{y}{3} = 1$$
$$-\frac{x}{7} + \frac{x}{7} - \frac{y}{3} = -\frac{x}{7} + 1$$
$$-\frac{y}{3} = -\frac{x}{7} + 1$$
$$-3\left(-\frac{y}{3}\right) = -3\left(-\frac{x}{7} + 1\right)$$
$$y = \frac{3}{7}x - 3$$

45. Solving for y:
$$-\frac{1}{4}x + \frac{1}{8}y = 1$$
$$-\frac{1}{4}x + \frac{1}{4}x + \frac{1}{8}y = 1 + \frac{1}{4}x$$
$$\frac{1}{8}y = \frac{1}{4}x + 1$$
$$8\left(\frac{1}{8}y\right) = 8\left(\frac{1}{4}x + 1\right)$$
$$y = 2x + 8$$

47. The complement of 30° is 90° − 30° = 60°, and the supplement is 180° − 30° = 150°.

49. The complement of 45° is 90° − 45° = 45°, and the supplement is 180° − 45° = 135°.

51. Translating into an equation and solving:
$$x = 0.25 \cdot 40$$
$$x = 10$$
The number 10 is 25% of 40.

53. Translating into an equation and solving:
$$x = 0.12 \cdot 2{,}000$$
$$x = 240$$
The number 240 is 12% of 2,000.

55. Translating into an equation and solving:
$$x \cdot 28 = 7$$
$$28x = 7$$
$$\frac{1}{28}(28x) = \frac{1}{28}(7)$$
$$x = 0.25 = 25\%$$
The number 7 is 25% of 28.

57. Translating into an equation and solving:
$$x \cdot 40 = 14$$
$$40x = 14$$
$$\frac{1}{40}(40x) = \frac{1}{40}(14)$$
$$x = 0.35 = 35\%$$
The number 14 is 35% of 40.

59. Translating into an equation and solving:
$$0.50 \cdot x = 32$$
$$\frac{0.50x}{0.50} = \frac{32}{0.50}$$
$$x = 64$$
The number 32 is 50% of 64.

61. Translating into an equation and solving:
$$0.12 \cdot x = 240$$
$$\frac{0.12x}{0.12} = \frac{240}{0.12}$$
$$x = 2{,}000$$
The number 240 is 12% of 2,000.

63. Substituting $F = 212$: $C = \frac{5}{9}(212 - 32) = \frac{5}{9}(180) = 100°C$. This value agrees with the information in Table 1.

65. Substituting $F = 68$: $C = \frac{5}{9}(68 - 32) = \frac{5}{9}(36) = 20°C$. This value agrees with the information in Table 1.

67. Solving for C:
$$\frac{9}{5}C + 32 = F$$
$$\frac{9}{5}C + 32 - 32 = F - 32$$
$$\frac{9}{5}C = F - 32$$
$$\frac{5}{9}\left(\frac{9}{5}C\right) = \frac{5}{9}(F - 32)$$
$$C = \frac{5}{9}(F - 32)$$

69. Budd's estimate would be: $F = 2(30) + 30 = 60 + 30 = 90°F$

The actual conversion would be: $F = \dfrac{9}{5}(30) + 32 = 54 + 32 = 86°F$

Budd's estimate is 4°F too high.

71. Solving for r:
$$2\pi r = C$$
$$2 \cdot \dfrac{22}{7} r = 44$$
$$\dfrac{44}{7} r = 44$$
$$\dfrac{7}{44}\left(\dfrac{44}{7} r\right) = \dfrac{7}{44}(44)$$
$$r = 7 \text{ meters}$$

73. Solving for r:
$$2\pi r = C$$
$$2 \cdot 3.14 r = 9.42$$
$$6.28 r = 9.42$$
$$\dfrac{6.28 r}{6.28} = \dfrac{9.42}{6.28}$$
$$r = \dfrac{3}{2} = 1.5 \text{ inches}$$

75. Solving for h:
$$\pi r^2 h = V$$
$$\dfrac{22}{7}\left(\dfrac{7}{22}\right)^2 h = 42$$
$$\dfrac{7}{22} h = 42$$
$$\dfrac{22}{7}\left(\dfrac{7}{22} h\right) = \dfrac{22}{7}(42)$$
$$h = 132 \text{ feet}$$

77. Solving for h:
$$\pi r^2 h = V$$
$$3.14(3)^2 h = 6.28$$
$$28.26 h = 6.28$$
$$\dfrac{28.26 h}{28.26} = \dfrac{6.28}{28.26}$$
$$h = \dfrac{2}{9} \text{ centimeters}$$

79. We need to find what percent of 150 is 90:
$$x \cdot 150 = 90$$
$$\dfrac{1}{150}(150x) = \dfrac{1}{150}(90)$$
$$x = 0.60 = 60\%$$
So 60% of the calories in one serving of vanilla ice cream are fat calories.

81. We need to find what percent of 98 is 26:
$$x \cdot 98 = 26$$
$$\dfrac{1}{98}(98x) = \dfrac{1}{98}(26)$$
$$x \approx 0.265 = 26.5\%$$
So 26.5% of one serving of frozen yogurt are carbohydrates.

83. The sum of 4 and 1 is 5.

85. The difference of 6 and 2 is 4.

87. The difference of a number and 15 is −12.

89. The sum of a number and 3 is four times the difference of that number and 3.

91. An equivalent expression is: $2(6+3) = 18$

93. An equivalent expression is: $2(5) + 3 = 13$

95. An equivalent expression is: $x + 5 = 13$

97. An equivalent expression is: $5(x+7) = 30$

1.6 Applications

1. Let x represent the number. The equation is:
$$x + 5 = 13$$
$$x = 8$$
The number is 8.

3. Let x represent the number. The equation is:
$$2x + 4 = 14$$
$$2x = 10$$
$$x = 5$$
The number is 5.

5. Let x represent the number. The equation is:
$$5(x + 7) = 30$$
$$5x + 35 = 30$$
$$5x = -5$$
$$x = -1$$
The number is -1.

7. Let x and $x + 2$ represent the two numbers. The equation is:
$$x + x + 2 = 8$$
$$2x + 2 = 8$$
$$2x = 6$$
$$x = 3$$
$$x + 2 = 5$$
The two numbers are 3 and 5.

9. Let x and $3x - 4$ represent the two numbers. The equation is:
$$(x + 3x - 4) + 5 = 25$$
$$4x + 1 = 25$$
$$4x = 24$$
$$x = 6$$
$$3x - 4 = 3(6) - 4 = 14$$
The two numbers are 6 and 14.

11. Completing the table:

	Four Years Ago	Now
Shelly	$x + 3 - 4 = x - 1$	$x + 3$
Michele	$x - 4$	x

    The equation is:
    $$x - 1 + x - 4 = 67$$
    $$2x - 5 = 67$$
    $$2x = 72$$
    $$x = 36$$
    $$x + 3 = 39$$
    Shelly is 39 and Michele is 36.

13. Completing the table:

	Three Years Ago	Now
Cody	$2x - 3$	$2x$
Evan	$x - 3$	x

    The equation is:
    $$2x - 3 + x - 3 = 27$$
    $$3x - 6 = 27$$
    $$3x = 33$$
    $$x = 11$$
    $$2x = 22$$
    Evan is 11 and Cody is 22.

15. Completing the table:

	Five Years Ago	Now
Fred	$x+4-5=x-1$	$x+4$
Barney	$x-5$	x

The equation is:
$$x-1+x-5=48$$
$$2x-6=48$$
$$2x=54$$
$$x=27$$
$$x+4=31$$
Barney is 27 and Fred is 31.

17. Completing the table:

	Now	Three Years from Now
Jack	$2x$	$2x+3$
Lacy	x	$x+3$

The equation is:
$$2x+3+x+3=54$$
$$3x+6=54$$
$$3x=48$$
$$x=16$$
$$2x=32$$
Lacy is 16 and Jack is 32.

19. Completing the table:

	Now	Two Years from Now
Pat	$x+20$	$x+20+2=x+22$
Patrick	x	$x+2$

The equation is:
$$x+22=2(x+2)$$
$$x+22=2x+4$$
$$22=x+4$$
$$x=18$$
$$x+20=38$$
Patrick is 18 and Pat is 38.

21. Using the formula $P=4s$, the equation is:
$$4s=36$$
$$s=9$$
The length of each side is 9 inches.

23. Using the formula $P=4s$, the equation is:
$$4s=60$$
$$s=15$$
The length of each side is 15 feet.

25. Let x, $3x$, and $x+7$ represent the sides of the triangle. The equation is:
$$x+3x+x+7=62$$
$$5x+7=62$$
$$5x=55$$
$$x=11$$
$$x+7=18$$
$$3x=33$$
The sides are 11 feet, 18 feet, and 33 feet.

27. Let x, $2x$, and $2x - 12$ represent the sides of the triangle. The equation is:
$$x + 2x + 2x - 12 = 53$$
$$5x - 12 = 53$$
$$5x = 65$$
$$x = 13$$
$$2x = 26$$
$$2x - 12 = 14$$
The sides are 13 feet, 14 feet, and 26 feet.

29. Let w represent the width and $w + 5$ represent the length. The equation is:
$$2w + 2(w + 5) = 34$$
$$2w + 2w + 10 = 34$$
$$4w + 10 = 34$$
$$4w = 24$$
$$w = 6$$
$$w + 5 = 11$$
The length is 11 inches and the width is 6 inches.

31. Let w represent the width and $2w + 7$ represent the length. The equation is:
$$2w + 2(2w + 7) = 68$$
$$2w + 4w + 14 = 68$$
$$6w + 14 = 68$$
$$6w = 54$$
$$w = 9$$
$$2w + 7 = 2(9) + 7 = 25$$
The length is 25 meters and the width is 9 meters.

33. Let w represent the width and $3w + 6$ represent the length. The equation is:
$$2w + 2(3w + 6) = 36$$
$$2w + 6w + 12 = 36$$
$$8w + 12 = 36$$
$$8w = 24$$
$$w = 3$$
$$3w + 6 = 3(3) + 6 = 15$$
The length is 15 feet and the width is 3 feet.

35. Completing the table:

	Dimes	Quarters
Number	x	$x + 5$
Value (cents)	$10(x)$	$25(x + 5)$

The equation is:
$$10(x) + 25(x + 5) = 440$$
$$10x + 25x + 125 = 440$$
$$35x + 125 = 440$$
$$35x = 315$$
$$x = 9$$
$$x + 5 = 14$$
Marissa has 9 dimes and 14 quarters.

37. Completing the table:

	Nickels	Quarters
Number	$x + 15$	x
Value (cents)	$5(x + 15)$	$25(x)$

The equation is:
$$5(x + 15) + 25(x) = 435$$
$$5x + 75 + 25x = 435$$
$$30x + 75 = 435$$
$$30x = 360$$
$$x = 12$$
$$x + 15 = 27$$
Tanner has 12 quarters and 27 nickels.

39. Completing the table:

	Nickels	Dimes
Number	x	$x+9$
Value (cents)	$5(x)$	$10(x+9)$

The equation is:
$$5(x)+10(x+9)=210$$
$$5x+10x+90=210$$
$$15x+90=210$$
$$15x=120$$
$$x=8$$
$$x+9=17$$
Sue has 8 nickels and 17 dimes.

41. Completing the table:

	Nickels	Dimes	Quarters
Number	x	$x+3$	$x+5$
Value (cents)	$5(x)$	$10(x+3)$	$25(x+5)$

The equation is:
$$5(x)+10(x+3)+25(x+5)=435$$
$$5x+10x+30+25x+125=435$$
$$40x+155=435$$
$$40x=280$$
$$x=7$$
$$x+3=10$$
$$x+5=12$$
Katie has 7 nickels, 10 dimes, and 12 quarters.

43. Completing the table:

	Nickels	Dimes	Quarters
Number	x	$x+6$	$2x$
Value (cents)	$5(x)$	$10(x+6)$	$25(2x)$

The equation is:
$$5(x)+10(x+6)+25(2x)=255$$
$$5x+10x+60+50x=255$$
$$65x+60=255$$
$$65x=195$$
$$x=3$$
$$x+6=9$$
$$2x=6$$
Cory has 3 nickels, 9 dimes, and 6 quarters.

45. Simplifying: $x+2x+2x=5x$

47. Simplifying: $x+0.075x=1.075x$

49. Simplifying: $0.09(x+2{,}000)=0.09x+180$

51. Solving the equation:
$$0.05x+0.06(x-1{,}500)=570$$
$$0.05x+0.06x-90=570$$
$$0.11x=660$$
$$x=6{,}000$$

53. Solving the equation:
$$x+2x+3x=180$$
$$6x=180$$
$$x=30$$

1.7 More Applications

1. Let x and $x + 1$ represent the two numbers. The equation is:
$$x + x + 1 = 11$$
$$2x + 1 = 11$$
$$2x = 10$$
$$x = 5$$
$$x + 1 = 6$$
The numbers are 5 and 6.

3. Let x and $x + 1$ represent the two numbers. The equation is:
$$x + x + 1 = -9$$
$$2x + 1 = -9$$
$$2x = -10$$
$$x = -5$$
$$x + 1 = -4$$
The numbers are −5 and −4.

5. Let x and $x + 2$ represent the two numbers. The equation is:
$$x + x + 2 = 28$$
$$2x + 2 = 28$$
$$2x = 26$$
$$x = 13$$
$$x + 2 = 15$$
The numbers are 13 and 15.

7. Let x and $x + 2$ represent the two numbers. The equation is:
$$x + x + 2 = 106$$
$$2x + 2 = 106$$
$$2x = 104$$
$$x = 52$$
$$x + 2 = 54$$
The numbers are 52 and 54.

9. Let x and $x + 2$ represent the two numbers. The equation is:
$$x + x + 2 = -30$$
$$2x + 2 = -30$$
$$2x = -322$$
$$x = -16$$
$$x + 2 = -14$$
The numbers are −16 and −14.

11. Let x, $x + 2$, and $x + 4$ represent the three numbers. The equation is:
$$x + x + 2 + x + 4 = 57$$
$$3x + 6 = 57$$
$$3x = 51$$
$$x = 17$$
$$x + 2 = 19$$
$$x + 4 = 21$$
The numbers are 17, 19, and 21.

13. Let x, $x+2$, and $x+4$ represent the three numbers. The equation is:
$$x + x + 2 + x + 4 = 132$$
$$3x + 6 = 132$$
$$3x = 126$$
$$x = 42$$
$$x + 2 = 44$$
$$x + 4 = 46$$
The numbers are 42, 44, and 46.

15. Completing the table:

	Dollars Invested at 8%	Dollars Invested at 9%
Number of	x	$x + 2,000$
Interest on	$0.08(x)$	$0.09(x + 2,000)$

The equation is:
$$0.08(x) + 0.09(x + 2,000) = 860$$
$$0.08x + 0.09x + 180 = 860$$
$$0.17x + 180 = 860$$
$$0.17x = 680$$
$$x = 4,000$$
$$x + 2,000 = 6,000$$
You have $4,000 invested at 8% and $6,000 invested at 9%.

17. Completing the table:

	Dollars Invested at 10%	Dollars Invested at 12%
Number of	x	$x + 500$
Interest on	$0.10(x)$	$0.12(x + 500)$

The equation is:
$$0.10(x) + 0.12(x + 500) = 214$$
$$0.10x + 0.12x + 60 = 214$$
$$0.22x + 60 = 214$$
$$0.22x = 154$$
$$x = 700$$
$$x + 500 = 1,200$$
Tyler has $700 invested at 10% and $1,200 invested at 12%.

19. Completing the table:

	Dollars Invested at 8%	Dollars Invested at 9%	Dollars Invested at 10%
Number of	x	$2x$	$3x$
Interest on	$0.08(x)$	$0.09(2x)$	$0.10(3x)$

The equation is:
$$0.08(x) + 0.09(2x) + 0.10(3x) = 280$$
$$0.08x + 0.18x + 0.30x = 280$$
$$0.56x = 280$$
$$x = 500$$
$$2x = 1{,}000$$
$$3x = 1{,}500$$
She has $500 invested at 8%, $1,000 invested at 9%, and $1,500 invested at 10%.

21. Let x represent the measure of the two equal angles, so $x + x = 2x$ represents the measure of the third angle. Since the sum of the three angles is 180°, the equation is:
$$x + x + 2x = 180°$$
$$4x = 180°$$
$$x = 45°$$
$$2x = 90°$$
The measures of the three angles are 45°, 45°, and 90°.

23. Let x represent the measure of the largest angle. Then $\frac{1}{5}x$ represents the measure of the smallest angle, and $2\left(\frac{1}{5}x\right) = \frac{2}{5}x$ represents the measure of the other angle. Since the sum of the three angles is 180°, the equation is:
$$x + \frac{1}{5}x + \frac{2}{5}x = 180°$$
$$\frac{5}{5}x + \frac{1}{5}x + \frac{2}{5}x = 180°$$
$$\frac{8}{5}x = 180°$$
$$x = 112.5°$$
$$\frac{1}{5}x = 22.5°$$
$$\frac{2}{5}x = 45°$$
The measures of the three angles are 22.5°, 45°, and 112.5°.

25. Let x represent the measure of the other acute angle, and 90° is the measure of the right angle. Since the sum of the three angles is 180°, the equation is:
$$x + 37° + 90° = 180°$$
$$x + 127° = 180°$$
$$x = 53°$$
The other two angles are 53° and 90°.

27. Let x represent the measure of the smallest angle, so $x + 20$ represents the measure of the second angle and $2x$ represents the measure of the third angle. Since the sum of the three angles is $180°$, the equation is:
$$x + x + 20 + 2x = 180°$$
$$4x + 20 = 180°$$
$$4x = 160°$$
$$x = 40°$$
$$x + 20 = 60°$$
$$2x = 80°$$
The measures of the three angles are $40°, 60°,$ and $80°$.

29. Completing the table:

	Adult	Child
Number	x	$x+6$
Income	$6(x)$	$4(x+6)$

The equation is:
$$6(x) + 4(x+6) = 184$$
$$6x + 4x + 24 = 184$$
$$10x + 24 = 184$$
$$10x = 160$$
$$x = 16$$
$$x + 6 = 22$$
Miguel sold 16 adult and 22 children's tickets.

31. Let x represent the total minutes for the call. Then $0.41 is charged for the first minute, and $0.32 is charged for the additional $x - 1$ minutes. The equation is:
$$0.41(1) + 0.32(x-1) = 5.21$$
$$0.41 + 0.32x - 0.32 = 5.21$$
$$0.32x + 0.09 = 5.21$$
$$0.32x = 5.12$$
$$x = 16$$
The call was 16 minutes long.

33. Let x represent the hours JoAnn worked that week. Then $12/hour is paid for the first 35 hours and $18/hour is paid for the additional $x - 35$ hours. The equation is:
$$12(35) + 18(x - 35) = 492$$
$$420 + 18x - 630 = 492$$
$$18x - 210 = 492$$
$$18x = 702$$
$$x = 39$$
JoAnn worked 39 hours that week.

35. Let x and $x + 2$ represent the two office numbers. The equation is:
$$x + x + 2 = 14,660$$
$$2x + 2 = 14,660$$
$$2x = 14,658$$
$$x = 7329$$
$$x + 2 = 7331$$
They are in offices 7329 and 7331.

37. Let x represent Kendra's age and $x + 2$ represent Marissa's age. The equation is:
$$x + 2 + 2x = 26$$
$$3x + 2 = 26$$
$$3x = 24$$
$$x = 8$$
$$x + 2 = 10$$
Kendra is 8 years old and Marissa is 10 years old.

39. For Jeff, the total time traveled is $\frac{425 \text{ miles}}{55 \text{ miles/hour}} \approx 7.72 \text{ hours} \approx 463 \text{ minutes}$. Since he left at 11:00 AM, he will arrive at 6:43 PM. For Carla, the total time traveled is $\frac{425 \text{ miles}}{65 \text{ miles/hour}} \approx 6.54 \text{ hours} \approx 392 \text{ minutes}$. Since she left at 1:00 PM, she will arrive at 7:32 PM. Thus Jeff will arrive in Lake Tahoe first.

41. Since $\frac{1}{5}$ mile $= 0.2$ mile, the taxi charge is $1.25 for the first $\frac{1}{5}$ mile and $0.25 per fifth mile for the remaining $7.5 - 0.2 = 7.3$ miles. Since 7.3 miles $= \frac{7.3}{0.2} = 36.5$ fifths, the total charge is: $\$1.25 + \$0.25(36.5) \approx \$10.38$

43. Let w represent the width and $w + 2$ represent the length. The equation is:
$$2w + 2(w + 2) = 44$$
$$2w + 2w + 4 = 44$$
$$4w + 4 = 44$$
$$4w = 40$$
$$w = 10$$
$$w + 2 = 12$$
The length is 12 meters and the width is 10 meters.

45. Let $x, x + 1$, and $x + 2$ represent the measures of the three angles. Since the sum of the three angles is 180°, the equation is:
$$x + x + 1 + x + 2 = 180°$$
$$3x + 3 = 180°$$
$$3x = 177°$$
$$x = 59°$$
$$x + 1 = 60°$$
$$x + 2 = 61°$$
The measures of the three angles are 59°, 60°, and 61°.

47. If all 36 people are Elk's Lodge members (which would be the least amount), the cost of the lessons would be $\$3(36) = \$108$. Since half of the money is paid to Ike and Nancy, the least amount they could make is $\frac{1}{2}(\$108) = \$54$.

49. Yes. The total receipts were $160, which is possible if there were 10 Elk's members and 26 nonmembers. Computing the total receipts: $10(\$3) + 26(\$5) = \$30 + \$130 = \$160$

51. a. Solving the equation:
$$x - 3 = 6$$
$$x = 9$$
b. Solving the equation:
$$x + 3 = 6$$
$$x = 3$$
c. Solving the equation:
$$-x - 3 = 6$$
$$-x = 9$$
$$x = -9$$
d. Solving the equation:
$$-x + 3 = 6$$
$$-x = 3$$
$$x = -3$$

53.

a. Solving the equation:
$$\frac{x}{4} = -2$$
$$x = -2(4) = -8$$

b. Solving the equation:
$$-\frac{x}{4} = -2$$
$$x = -2(-4) = 8$$

c. Solving the equation:
$$\frac{x}{4} = 2$$
$$x = 2(4) = 8$$

d. Solving the equation:
$$-\frac{x}{4} = 2$$
$$x = 2(-4) = -8$$

55. Solving the equation:
$$2.5x - 3.48 = 4.9x + 2.07$$
$$-2.4x - 3.48 = 2.07$$
$$-2.4x = 5.55$$
$$x = -2.3125$$

57. Solving the equation:
$$3(x-4) = -2$$
$$3x - 12 = -2$$
$$3x = 10$$
$$x = \frac{10}{3}$$

1.8 Linear Inequalities

1. Solving the inequality:
$$x - 5 < 7$$
$$x - 5 + 5 < 7 + 5$$
$$x < 12$$
The solution is $\{x \mid x < 12\}$. Graphing the inequality:

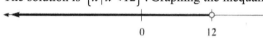

3. Solving the inequality:
$$a - 4 \leq 8$$
$$a - 4 + 4 \leq 8 + 4$$
$$a \leq 12$$
The solution is $\{a \mid a \leq 12\}$. Graphing the inequality:

5. Solving the inequality:
$$x - 4.3 > 8.7$$
$$x - 4.3 + 4.3 > 8.7 + 4.3$$
$$x > 13$$
The solution is $\{x \mid x > 13\}$. Graphing the inequality:

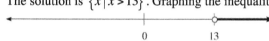

7. Solving the inequality:
$$y + 6 \geq 10$$
$$y + 6 + (-6) \geq 10 + (-6)$$
$$y \geq 4$$
The solution is $\{y \mid y \geq 4\}$. Graphing the inequality:

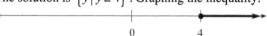

9. Solving the inequality:
$$2 < x - 7$$
$$2 + 7 < x - 7 + 7$$
$$9 < x$$
$$x > 9$$
The solution is $\{x \mid x > 9\}$. Graphing the inequality:

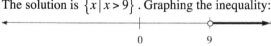

11. Solving the inequality:
$$3x < 6$$
$$\frac{1}{3}(3x) < \frac{1}{3}(6)$$
$$x < 2$$
The solution is $\{x \mid x < 2\}$. Graphing the inequality:

13. Solving the inequality:
$$5a \le 25$$
$$\frac{1}{5}(5a) \le \frac{1}{5}(25)$$
$$a \le 5$$
The solution is $\{a \mid a \le 5\}$. Graphing the inequality:

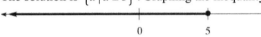

15. Solving the inequality:
$$\frac{x}{3} > 5$$
$$3\left(\frac{x}{3}\right) > 3(5)$$
$$x > 15$$
The solution is $\{x \mid x > 15\}$. Graphing the inequality:

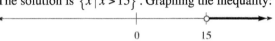

17. Solving the inequality:
$$-2x > 6$$
$$-\frac{1}{2}(-2x) < -\frac{1}{2}(6)$$
$$x < -3$$
The solution is $\{x \mid x < -3\}$. Graphing the inequality:

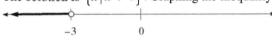

19. Solving the inequality:
$$-3x \ge -18$$
$$-\frac{1}{3}(-3x) \le -\frac{1}{3}(-18)$$
$$x \le 6$$
The solution is $\{x \mid x \le 6\}$. Graphing the inequality:

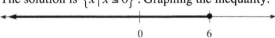

21. Solving the inequality:
$$-\frac{x}{5} \le 10$$
$$-5\left(-\frac{x}{5}\right) \ge -5(10)$$
$$x \ge -50$$
The solution is $\{x \mid x \ge -50\}$. Graphing the inequality:

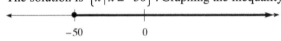

23. Solving the inequality:
$$-\frac{2}{3}y > 4$$
$$-\frac{3}{2}\left(-\frac{2}{3}y\right) < -\frac{3}{2}(4)$$
$$y < -6$$
The solution is $\{y \mid y < -6\}$. Graphing the inequality:

25. Solving the inequality:
$$2x - 3 < 9$$
$$2x - 3 + 3 < 9 + 3$$
$$2x < 12$$
$$\frac{1}{2}(2x) < \frac{1}{2}(12)$$
$$x < 6$$
The solution is $\{x \mid x < 6\}$. Graphing the inequality:

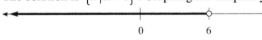

27. Solving the inequality:
$$-\frac{1}{5}y - \frac{1}{3} \le \frac{2}{3}$$
$$-\frac{1}{5}y - \frac{1}{3} + \frac{1}{3} \le \frac{2}{3} + \frac{1}{3}$$
$$-\frac{1}{5}y \le 1$$
$$-5\left(-\frac{1}{5}y\right) \ge -5(1)$$
$$y \ge -5$$
The solution is $\{y \mid y \ge -5\}$. Graphing the inequality:

29. Solving the inequality:

$$-7.2x + 1.8 > -19.8$$
$$-7.2x + 1.8 - 1.8 > -19.8 - 1.8$$
$$-7.2x > -21.6$$
$$\frac{-7.2x}{-7.2} < \frac{-21.6}{-7.2}$$
$$x < 3$$

The solution is $\{x \mid x < 3\}$. Graphing the inequality:

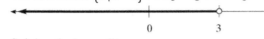

31. Solving the inequality:

$$\frac{2}{3}x - 5 \leq 7$$
$$\frac{2}{3}x - 5 + 5 \leq 7 + 5$$
$$\frac{2}{3}x \leq 12$$
$$\frac{3}{2}\left(\frac{2}{3}x\right) \leq \frac{3}{2}(12)$$
$$x \leq 18$$

The solution is $\{x \mid x \leq 18\}$. Graphing the inequality:

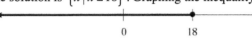

33. Solving the inequality:

$$-\frac{2}{5}a - 3 > 5$$
$$-\frac{2}{5}a - 3 + 3 > 5 + 3$$
$$-\frac{2}{5}a > 8$$
$$-\frac{5}{2}\left(-\frac{2}{5}a\right) < -\frac{5}{2}(8)$$
$$a < -20$$

The solution is $\{a \mid a < -20\}$. Graphing the inequality:

35. Solving the inequality:

$$5 - \frac{3}{5}y > -10$$
$$-5 + 5 - \frac{3}{5}y > -5 + (-10)$$
$$-\frac{3}{5}y > -15$$
$$-\frac{5}{3}\left(-\frac{3}{5}y\right) < -\frac{5}{3}(-15)$$
$$y < 25$$

The solution is $\{y \mid y < 25\}$. Graphing the inequality:

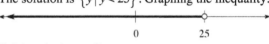

37. Solving the inequality:

$$0.3(a + 1) \leq 1.2$$
$$0.3a + 0.3 \leq 1.2$$
$$0.3a + 0.3 + (-0.3) \leq 1.2 + (-0.3)$$
$$0.3a \leq 0.9$$
$$\frac{0.3a}{0.3} \leq \frac{0.9}{0.3}$$
$$a \leq 3$$

The solution is $\{a \mid a \leq 3\}$. Graphing the inequality:

39. Solving the inequality:

$$2(5 - 2x) \leq -20$$
$$10 - 4x \leq -20$$
$$-10 + 10 - 4x \leq -10 + (-20)$$
$$-4x \leq -30$$
$$-\frac{1}{4}(-4x) \geq -\frac{1}{4}(-30)$$
$$x \geq \frac{15}{2}$$

The solution is $\left\{x \mid x \geq \frac{15}{2}\right\}$. Graphing the inequality:

41. Solving the inequality:
$$3x-5 > 8x$$
$$-3x+3x-5 > -3x+8x$$
$$-5 > 5x$$
$$\frac{1}{5}(-5) > \frac{1}{5}(5x)$$
$$-1 > x$$
$$x < -1$$

The solution is $\{x \mid x < -1\}$. Graphing the inequality:

43. First multiply by 6 to clear the inequality of fractions:
$$\frac{1}{3}y - \frac{1}{2} \leq \frac{5}{6}y + \frac{1}{2}$$
$$6\left(\frac{1}{3}y - \frac{1}{2}\right) \leq 6\left(\frac{5}{6}y + \frac{1}{2}\right)$$
$$2y - 3 \leq 5y + 3$$
$$-5y + 2y - 3 \leq -5y + 5y + 3$$
$$-3y - 3 \leq 3$$
$$-3y - 3 + 3 \leq 3 + 3$$
$$-3y \leq 6$$
$$-\frac{1}{3}(-3y) \geq -\frac{1}{3}(6)$$
$$y \geq -2$$

The solution is $\{y \mid y > -2\}$. Graphing the inequality:

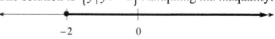

45. Solving the inequality:
$$-2.8x + 8.4 < -14x - 2.8$$
$$-2.8x + 14x + 8.4 < -14x + 14x - 2.8$$
$$11.2x + 8.4 < -2.8$$
$$11.2x + 8.4 - 8.4 < -2.8 - 8.4$$
$$11.2x < -11.2$$
$$\frac{11.2x}{11.2} < \frac{-11.2}{11.2}$$
$$x < -1$$

The solution is $\{x \mid x < -1\}$. Graphing the inequality:

47. Solving the inequality:
$$3(m-2) - 4 \geq 7m + 14$$
$$3m - 6 - 4 \geq 7m + 14$$
$$3m - 10 \geq 7m + 14$$
$$-7m + 3m - 10 \geq -7m + 7m + 14$$
$$-4m - 10 \geq 14$$
$$-4m - 10 + 10 \geq 14 + 10$$
$$-4m \geq 24$$
$$-\frac{1}{4}(-4m) \leq -\frac{1}{4}(24)$$
$$m \leq -6$$

The solution is $\{m \mid m \leq -6\}$. Graphing the inequality:

49. Solving the inequality:
$$3 - 4(x-2) \leq -5x + 6$$
$$3 - 4x + 8 \leq -5x + 6$$
$$-4x + 11 \leq -5x + 6$$
$$-4x + 5x + 11 \leq -5x + 5x + 6$$
$$x + 11 \leq 6$$
$$x + 11 + (-11) \leq 6 + (-11)$$
$$x \leq -5$$

The solution is $\{x \mid x \leq -5\}$. Graphing the inequality:

51. Solving for y:
$$3x + 2y < 6$$
$$2y < -3x + 6$$
$$y < -\frac{3}{2}x + 3$$

53. Solving for y:
$$2x - 5y > 10$$
$$-5y > -2x + 10$$
$$y < \frac{2}{5}x - 2$$

55. Solving for y:
$$-3x + 7y \leq 21$$
$$7y \leq 3x + 21$$
$$y \leq \frac{3}{7}x + 3$$

57. Solving for y:
$$2x - 4y \geq -4$$
$$-4y \geq -2x - 4$$
$$y \leq \frac{1}{2}x + 1$$

59.
  a. Evaluating when $x = 0$:
$$-5x + 3 = -5(0) + 3 = 0 + 3 = 3$$

  b. Solving the equation:
$$-5x + 3 = -7$$
$$-5x + 3 - 3 = -7 - 3$$
$$-5x = -10$$
$$-\frac{1}{5}(-5x) = -\frac{1}{5}(-10)$$
$$x = 2$$

  c. Substituting $x = 0$:
$$-5(0) + 3 < -7$$
$$3 < -7 \text{ (false)}$$
No, $x = 0$ is not a solution to the inequality.

  d. Solving the inequality:
$$-5x + 3 < -7$$
$$-5x + 3 - 3 < -7 - 3$$
$$-5x < -10$$
$$-\frac{1}{5}(-5x) > -\frac{1}{5}(-10)$$
$$x > 2$$
The solution is $\{x \mid x > 2\}$.

61. The inequality is $x < 3$.

63. The inequality is $x \leq 3$.

65. Let x and $x + 1$ represent the integers. Solving the inequality:
$$x + x + 1 \geq 583$$
$$2x + 1 \geq 583$$
$$2x \geq 582$$
$$x \geq 291$$
The two numbers are at least 291.

67. Let x represent the number. Solving the inequality:
$$2x + 6 < 10$$
$$2x < 4$$
$$x < 2$$
The solution is $\{x \mid x < 2\}$.

69. Let x represent the number. Solving the inequality:
$$4x > x - 8$$
$$3x > -8$$
$$x > -\frac{8}{3}$$
The solution is $\left\{x \mid x > -\frac{8}{3}\right\}$.

71. Let w represent the width, so $3w$ represents the length. Using the formula for perimeter:
$$2(w) + 2(3w) \geq 48$$
$$2w + 6w \geq 48$$
$$8w \geq 48$$
$$w \geq 6$$
The width is at least 6 meters.

73. Let x, $x + 2$, and $x + 4$ represent the sides of the triangle. The inequality is:
$$x + (x+2) + (x+4) > 24$$
$$3x + 6 > 24$$
$$3x > 18$$
$$x > 6$$
The shortest side is an even number greater than 6 inches (greater than or equal to 8 inches).

75. Solving the inequality:
$$2x - 1 \geq 3$$
$$2x - 1 + 1 \geq 3 + 1$$
$$2x \geq 4$$
$$x \geq 2$$
The solution is $\{x \mid x \geq 2\}$.

77. Solving the inequality:
$$-2x > -8$$
$$x < 4$$
The solution is $\{x \mid x < 4\}$.

79. Solving the inequality:
$$4x + 1 < -3$$
$$4x + 1 - 1 < -3 - 1$$
$$4x < -4$$
$$x < -1$$
The solution is $\{x \mid x < -1\}$.

1.9 Compound Inequalities

1. Graphing the solution set:

3. Graphing the solution set:

5. Graphing the solution set:

7. Graphing the solution set:

9. Graphing the solution set:

11. Graphing the solution set:

13. Graphing the solution set:

15. Graphing the solution set:

17. Solving the compound inequality:
$$3x - 1 < 5 \quad \text{or} \quad 5x - 5 > 10$$
$$3x < 6 \qquad\qquad 5x > 15$$
$$x < 2 \qquad\qquad x > 3$$
Graphing the solution set:
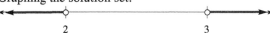

19. Solving the compound inequality:
$$x - 2 > -5 \quad \text{and} \quad x + 7 < 13$$
$$x > -3 \qquad\qquad x < 6$$
Graphing the solution set $-3 < x < 6$:
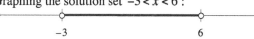

21. Solving the compound inequality:
$$11x < 22 \quad \text{or} \quad 12x > 36$$
$$x < 2 \qquad\qquad x > 3$$
Graphing the solution set:
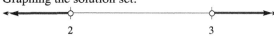

23. Solving the compound inequality:
$$3x - 5 < 10 \quad \text{and} \quad 2x + 1 > -5$$
$$3x < 15 \qquad\qquad 2x > -6$$
$$x < 5 \qquad\qquad x > -3$$
Graphing the solution set $-3 < x < 5$:

25. Solving the compound inequality:
$$2x - 3 < 8 \quad \text{and} \quad 3x + 1 > -10$$
$$2x < 11 \qquad\qquad 3x > -11$$
$$x < \frac{11}{2} \qquad\qquad x > -\frac{11}{3}$$
Graphing the solution set $-\frac{11}{3} < x < \frac{11}{2}$:

27. Solving the compound inequality:
$$2x - 1 < 3 \quad \text{and} \quad 3x - 2 > 1$$
$$2x < 4 \qquad\qquad 3x > 3$$
$$x < 2 \qquad\qquad x > 1$$
Graphing the solution set $1 < x < 2$:
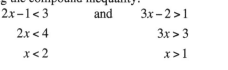

29. Solving the compound inequality:
$$-1 \le x - 5 \le 2$$
$$4 \le x \le 7$$
Graphing the solution set:
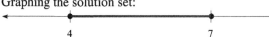

31. Solving the compound inequality:
$$-4 \le 2x \le 6$$
$$-2 \le x \le 3$$
Graphing the solution set:

33. Solving the compound inequality:
$$-3 < 2x + 1 < 5$$
$$-4 < 2x < 4$$
$$-2 < x < 2$$
Graphing the solution set:

35. Solving the compound inequality:
$$0 \le 3x + 2 \le 7$$
$$-2 \le 3x \le 5$$
$$-\frac{2}{3} \le x \le \frac{5}{3}$$
Graphing the solution set:
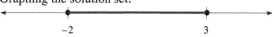

37. Solving the compound inequality:
$$-7 < 2x + 3 < 11$$
$$-10 < 2x < 8$$
$$-5 < x < 4$$
Graphing the solution set:
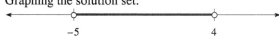

39. Solving the compound inequality:
$$-1 \le 4x + 5 \le 9$$
$$-6 \le 4x \le 4$$
$$-\frac{3}{2} \le x \le 1$$
Graphing the solution set:
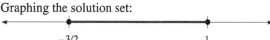

41. The inequality is $-2 < x < 3$.

43. The inequality is $x \le -2$ or $x \ge 3$.

45. **a.** The three inequalities are $2x + x > 10$, $2x + 10 > x$, and $x + 10 > 2x$.

   b. For $x + 10 > 2x$ we have $x < 10$. For $3x > 10$ we have $x > \frac{10}{3}$. The compound inequality is $\frac{10}{3} < x < 10$.

47. Graphing the inequality:

50   266

49. Let x represent the number. Solving the inequality:
$$5 < 2x - 3 < 7$$
$$8 < 2x < 10$$
$$4 < x < 5$$
The number is between 4 and 5.

51. Let w represent the width and $w + 4$ represent the length.
   a. Written as an inequality, the perimeter is: $20 < P < 30$
   b. Using the perimeter formula:
   $$20 < 2w + 2(w+4) < 30$$
   $$20 < 2w + 2w + 8 < 30$$
   $$20 < 4w + 8 < 30$$
   $$12 < 4w < 22$$
   $$3 < w < \frac{11}{2}$$
   c. Since $l = w + 4$, the inequality is: $7 < l < \frac{19}{2}$

53. Finding the number: $0.25(32) = 8$

55. Finding the number: $0.20(120) = 24$

57. Finding the percent:
$$p \cdot 36 = 9$$
$$p = \frac{9}{36} = 0.25 = 25\%$$

59. Finding the percent:
$$p \cdot 50 = 5$$
$$p = \frac{5}{50} = 0.10 = 10\%$$

61. Finding the number:
$$0.20n = 16$$
$$n = \frac{16}{0.20} = 80$$

63. Finding the number:
$$0.02n = 8$$
$$n = \frac{8}{0.02} = 400$$

65. Simplifying the expression: $-|-5| = -(5) = -5$

67. Simplifying the expression: $-3 - 4(-2) = -3 + 8 = 5$

69. Simplifying the expression: $5|3-8| - 6|2-5| = 5|-5| - 6|-3| = 5(5) - 6(3) = 25 - 18 = 7$

71. Simplifying the expression: $5 - 2[-3(5-7) - 8] = 5 - 2[-3(-2) - 8] = 5 - 2(6-8) = 5 - 2(-2) = 5 + 4 = 9$

73. The expression is: $-3 - (-9) = -3 + 9 = 6$

75. Applying the distributive property: $\frac{1}{2}(4x - 6) = \frac{1}{2} \cdot 4x - \frac{1}{2} \cdot 6 = 2x - 3$

77. The integers are: $-3, 0, 2$

Chapter 1 Test

1. Simplifying: $5y - 3 - 6y + 4 = 5y - 6y - 3 + 4 = -y + 1$
2. Simplifying: $3x - 4 + x + 3 = 3x + x - 4 + 3 = 4x - 1$
3. Simplifying: $4 - 2(y - 3) - 6 = 4 - 2y + 6 - 6 = -2y + 4$
4. Simplifying: $3(3x - 4) - 2(4x + 5) = 9x - 12 - 8x - 10 = 9x - 8x - 12 - 10 = x - 22$
5. Evaluating when $x = -3$: $3x + 12 + 2x = 5x + 12 = 5(-3) + 12 = -15 + 12 = -3$
6. Evaluating when $x = -2$ and $y = -4$: $x^2 - 3xy + y^2 = (-2)^2 - 3(-2)(-4) + (-4)^2 = 4 - 24 + 16 = -4$
7. a. Completing the table:

n	1	2	3	4
$(n+2)^2$	9	16	25	36

    b. Completing the table:

n	1	2	3	4
$n^2 + 2$	3	6	11	18

8. Solving the equation:
$$3x - 2 = 7$$
$$3x - 2 + 2 = 7 + 2$$
$$3x = 9$$
$$x = 3$$

9. Solving the equation:
$$4y + 15 = y$$
$$4y - 4y + 15 = y - 4y$$
$$15 = -3y$$
$$y = -5$$

10. First clear the equation of fractions by multiplying by 12:
$$\frac{1}{4}x - \frac{1}{12} = \frac{1}{3}x + \frac{1}{6}$$
$$12\left(\frac{1}{4}x - \frac{1}{12}\right) = 12\left(\frac{1}{3}x + \frac{1}{6}\right)$$
$$3x - 1 = 4x + 2$$
$$3x - 4x - 1 = 4x - 4x + 2$$
$$-x - 1 = 2$$
$$-x - 1 + 1 = 2 + 1$$
$$-x = 3$$
$$x = -3$$

11. Solving the equation:
$$-3(3 - 2x) - 7 = 8$$
$$-9 + 6x - 7 = 8$$
$$6x - 16 = 8$$
$$6x - 16 + 16 = 8 + 16$$
$$6x = 24$$
$$x = 4$$

12. Solving the equation:
$$3x - 9 = -6$$
$$3x - 9 + 9 = -6 + 9$$
$$3x = 3$$
$$x = 1$$

13. Solving the equation:
$$0.05 + 0.07(100 - x) = 3.2$$
$$0.05 + 7 - 0.07x = 3.2$$
$$-0.07x + 7.05 = 3.2$$
$$-0.07x + 7.05 - 7.05 = 3.2 - 7.05$$
$$-0.07x = -3.85$$
$$x = \frac{-3.85}{-0.07} = 55$$

14. Solving the equation:
$$4(t - 3) + 2(t + 4) = 2t - 16$$
$$4t - 12 + 2t + 8 = 2t - 16$$
$$6t - 4 = 2t - 16$$
$$6t - 2t - 4 = 2t - 2t - 16$$
$$4t - 4 = -16$$
$$4t - 4 + 4 = -16 + 4$$
$$4t = -12$$
$$t = -3$$

15. Solving the equation:
$$4x - 2(3x - 1) = 2x - 8$$
$$4x - 6x + 2 = 2x - 8$$
$$-2x + 2 = 2x - 8$$
$$-2x - 2x + 2 = 2x - 2x - 8$$
$$-4x + 2 = -8$$
$$-4x + 2 - 2 = -8 - 2$$
$$-4x = -10$$
$$x = \frac{10}{4} = \frac{5}{2}$$

16. Writing the equation: $x = 0.40 \cdot 56$

17. Writing the equation: $0.24 \cdot x = 720$

18. Substituting $x = 4$:
$$3(4) - 4y = 16$$
$$12 - 4y = 16$$
$$-4y = 4$$
$$y = -1$$

19. Substituting $y = 2$:
$$3x - 4(2) = 16$$
$$3x - 8 = 16$$
$$3x = 24$$
$$x = 8$$

20. Solving for y:
$$2x + 6y = 12$$
$$6y = -2x + 12$$
$$y = -\frac{1}{3}x + 2$$

21. Solving for v:
$$v^2 + 2ad = x^2$$
$$2ad = x^2 - v^2$$
$$a = \frac{x^2 - v^2}{2d}$$

22. Completing the table:

	Five Years Ago	Now
Paul	$2x - 5$	$2x$
Becca	$x - 5$	x

The equation is:
$$2x - 5 + x - 5 = 44$$
$$3x - 10 = 44$$
$$3x = 54$$
$$x = 18$$
$$2x = 36$$
Paul is 36 years old and Becca is 18 years old.

23. Let w represent the width and $3w - 5$ represent the length. Using the perimeter formula:
$$2(w) + 2(3w - 5) = 150$$
$$2w + 6w - 10 = 150$$
$$8w - 10 = 150$$
$$8w = 160$$
$$w = 20$$
$$3w - 5 = 55$$
The width is 20 cm and the length is 55 cm.

24. Completing the table:

	Dimes	Nickels
Number	$x + 8$	x
Value (cents)	$10(x+8)$	$5(x)$

The equation is:
$$10(x+8) + 5(x) = 170$$
$$10x + 80 + 5x = 170$$
$$15x + 80 = 170$$
$$15x = 90$$
$$x = 6$$
$$x + 8 = 14$$
He has 6 nickels and 14 dimes in his collection.

25. Completing the table:

	Dollars Invested at 6%	Dollars Invested at 12%
Number of	x	$x + 500$
Interest on	$0.06(x)$	$0.12(x+500)$

The equation is:
$$0.06(x) + 0.12(x + 500) = 186$$
$$0.06x + 0.12x + 60 = 186$$
$$0.18x + 60 = 186$$
$$0.18x = 126$$
$$x = 700$$
$$x + 500 = 1,200$$
She invested $700 at 6% and $1,200 at 12%.

26. Solving the inequality:
$$\frac{1}{2}x - 2 > 3$$
$$\frac{1}{2}x > 5$$
$$x > 10$$
The solution is $\{x \mid x > 10\}$. Graphing the inequality:

27. Solving the inequality:
$$-6y \leq 24$$
$$-\frac{1}{6}(-6y) \geq -\frac{1}{6}(24)$$
$$y \geq -4$$
The solution is $\{y \mid y \geq -4\}$. Graphing the inequality:

28. Solving the inequality:
$$0.3 - 0.2x < 1.1$$
$$-0.2x < 0.8$$
$$\frac{-0.2x}{-0.2} > \frac{0.8}{-0.2}$$
$$x > -4$$

The solution is $\{x \mid x > -4\}$. Graphing the inequality:

29. Solving the inequality:
$$3 - 2(n - 1) \geq 9$$
$$3 - 2n + 2 \geq 9$$
$$-2n + 5 \geq 9$$
$$-2n \geq 4$$
$$-\frac{1}{2}(-2n) \leq -\frac{1}{2}(4)$$
$$n \leq -2$$
The solution is $\{n \mid n \leq -2\}$. Graphing the inequality:

30. Solving the compound inequality:

$$5x - 3 < 2x \quad \text{or} \quad 2x > 6$$
$$3x < 3 \quad \quad\quad x > 3$$
$$x < 1 \quad \quad\quad x > 3$$

Graphing the solution set:

31. Solving the compound inequality:

$$-3 \le 2x - 7 \le 9$$
$$4 \le 2x \le 16$$
$$2 \le x \le 8$$

Graphing the solution set $2 \le x \le 8$:

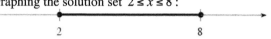

Chapter 2
Linear Equations and Inequalities in Two Variables

2.1 Paired Data and Graphing Ordered Pairs

1. The point lies in quadrant I:

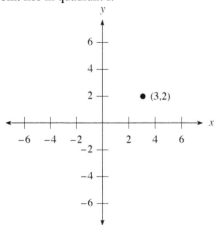

3. The point lies in quadrant II:

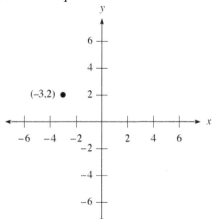

5. The point lies in quadrant I:

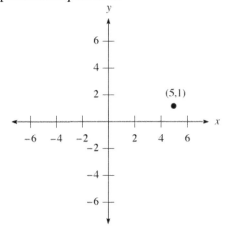

7. The point lies in quadrant I:

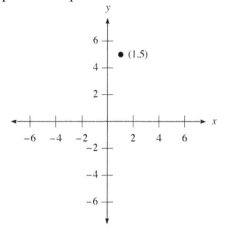

9. The point lies in quadrant II:

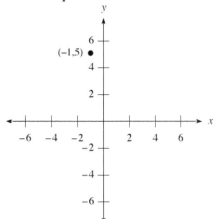

11. The point lies in quadrant I:

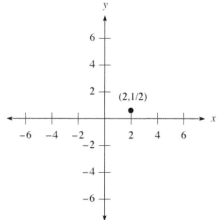

13. The point lies in quadrant III:

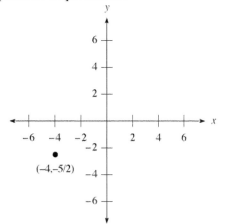

15. The point is not in a quadrant:

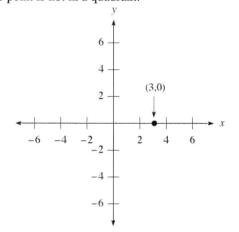

17. The point is not in a quadrant:

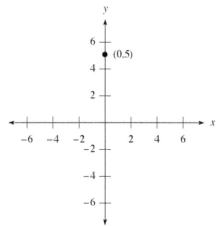

19. The coordinates are $(-4, 4)$.

21. The coordinates are $(-4, 2)$.

23. The coordinates are $(-3, 0)$.

25. The coordinates are $(2, -2)$.

27. The coordinates are $(-5, -5)$.

Essentials of Elementary and Intermediate Algebra: A Combined Course
Student Solutions Manual, Problem Set 2.1

29. Graphing the line:

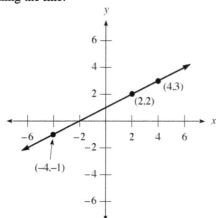

Yes, the point $(2,2)$ lies on the line.

31. Graphing the line:

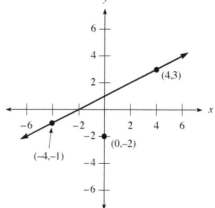

No, the point $(0,-2)$ does not lie on the line.

33. Graphing the line:

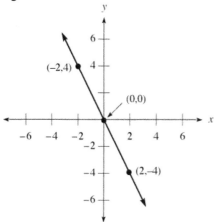

Yes, the point $(0,0)$ lies on the line.

35. Graphing the line:

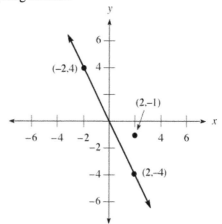

No, the point $(2,-1)$ does not lie on the line.

37. Graphing the line:

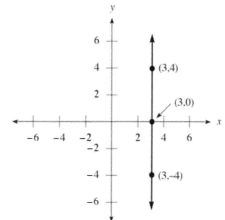

Yes, the point $(3,0)$ lies on the line.

39. No, the x-coordinate of every point on this line is 3.

41. Graphing the line:

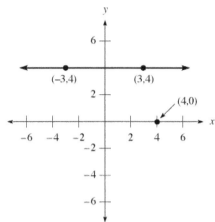

No, the point $(4,0)$ does not lie on the line.

43. No, the y-coordinate of every point on this line is 4.

45.
 a. Three ordered pairs on the graph are (5, 40), (10, 80), and (20, 160).
 b. She will earn $320 for working 40 hours.
 c. If her check is $240, she worked 30 hours that week.
 d. No. She should be paid $280 for working 35 hours, not $260.

47. Sketching a line graph:

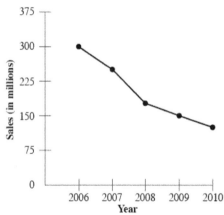

49. Five ordered pairs on the graph are (1985, 20.2), (1990, 34.4), (1995, 44.8), (2000, 65.4), and (2005, 104).

51. Point A is (6 – 5,2) = (1,2), and point B is (6,2 + 5) = (6,7).

53. Point A is (7 – 5, 2) = (2, 2), point B is (2, 2 + 3) = (2, 5), and point C is (2 + 5, 5) = (7, 5).

55.
 a. Substituting $y = 4$:
 $$2x + 3(4) = 6$$
 $$2x + 12 = 6$$
 $$2x = -6$$
 $$x = -3$$

 b. Substituting $y = -2$:
 $$2x + 3(-2) = 6$$
 $$2x - 6 = 6$$
 $$2x = 12$$
 $$x = 6$$

 c. Substituting $x = 3$:
 $$2(3) + 3y = 6$$
 $$6 + 3y = 6$$
 $$3y = 0$$
 $$y = 0$$

 d. Substituting $x = 9$:
 $$2(9) + 3y = 6$$
 $$18 + 3y = 6$$
 $$3y = -12$$
 $$y = -4$$

57. a. Substituting $y = 7$:
$$2x - 1 = 7$$
$$2x = 8$$
$$x = 4$$

b. Substituting $y = 3$:
$$2x - 1 = 3$$
$$2x = 4$$
$$x = 2$$

c. Substituting $x = 0$: $y = 2(0) - 1 = 0 - 1 = -1$

d. Substituting $x = 5$: $y = 2(5) - 1 = 10 - 1 = 9$

2.2 Solutions to Linear Equations in Two Variables

1. Substituting $x = 0$, $y = 0$, and $y = -6$:
$$2(0) + y = 6 \qquad 2x + 0 = 6 \qquad 2x + (-6) = 6$$
$$0 + y = 6 \qquad 2x = 6 \qquad 2x = 12$$
$$y = 6 \qquad x = 3 \qquad x = 6$$
The ordered pairs are $(0,6)$, $(3,0)$, and $(6,-6)$.

3. Substituting $x = 0$, $y = 0$, and $x = -4$:
$$3(0) + 4y = 12 \qquad 3x + 4(0) = 12 \qquad 3(-4) + 4y = 12$$
$$0 + 4y = 12 \qquad 3x + 0 = 12 \qquad -12 + 4y = 12$$
$$4y = 12 \qquad 3x = 12 \qquad 4y = 24$$
$$y = 3 \qquad x = 4 \qquad y = 6$$
The ordered pairs are $(0,3)$, $(4,0)$, and $(-4,6)$.

5. Substituting $x = 1$, $y = 0$, and $x = 5$:
$$y = 4(1) - 3 \qquad 0 = 4x - 3 \qquad y = 4(5) - 3$$
$$y = 4 - 3 \qquad 3 = 4x \qquad y = 20 - 3$$
$$y = 1 \qquad x = \frac{3}{4} \qquad y = 17$$
The ordered pairs are $(1,1)$, $\left(\frac{3}{4}, 0\right)$, and $(5,17)$.

7. Substituting $x = 2$, $y = 6$, and $x = 0$:
$$y = 7(2) - 1 \qquad 6 = 7x - 1 \qquad y = 7(0) - 1$$
$$y = 14 - 1 \qquad 7 = 7x \qquad y = 0 - 1$$
$$y = 13 \qquad x = 1 \qquad y = -1$$
The ordered pairs are $(2,13)$, $(1,6)$, and $(0,-1)$.

9. Substituting $y = 4$, $y = -3$, and $y = 0$ results (in each case) in $x = -5$. The ordered pairs are $(-5,4)$, $(-5,-3)$, and $(-5,0)$.

11. Completing the table:

x	y
1	3
-3	-9
4	12
6	18

13. Completing the table:

x	y
0	0
$-\frac{1}{2}$	-2
-3	-12
3	12

15. Completing the table:

x	y
2	3
3	2
5	0
9	-4

17. Completing the table:

x	y
2	0
3	2
1	-2
-3	-10

19. Completing the table:

x	y
0	-1
-1	-7
-3	-19
$\frac{3}{2}$	8

21. Substituting each ordered pair into the equation:
$(2,3): 2(2)-5(3)=4-15=-11 \neq 10$
$(0,-2): 2(0)-5(-2)=0+10=10$
$\left(\frac{5}{2},1\right): 2\left(\frac{5}{2}\right)-5(1)=5-5=0 \neq 10$

Only the ordered pair $(0,-2)$ is a solution.

23. Substituting each ordered pair into the equation:
$(1,5): 7(1)-2=7-2=5$
$(0,-2): 7(0)-2=0-2=-2$
$(-2,-16): 7(-2)-2=-14-2=-16$

All the ordered pairs $(1,5)$, $(0,-2)$ and $(-2,-16)$ are solutions.

25. Substituting each ordered pair into the equation:
$(1,6): 6(1)=6$
$(-2,12): 6(-2)=-12 \neq 12$
$(0,0): 6(0)=0$

The ordered pairs $(1,6)$ and $(0,0)$ are solutions.

27. Substituting each ordered pair into the equation:
$(1,1): 1+1=2 \neq 0$
$(2,-2): 2+(-2)=0$
$(3,3): 3+3=6 \neq 0$

Only the ordered pair $(2,-2)$ is a solution.

29. Since $x = 3$, the ordered pair $(5,3)$ cannot be a solution. The ordered pairs $(3,0)$ and $(3,-3)$ are solutions.

31. Substituting $w = 3$:
$2l+2(3)=30$
$2l+6=30$
$2l=24$
$l=12$

The length is 12 inches.

33. a. This is correct, since; $y = 12(5) = \$60$
    b. This is not correct, since: $y = 12(9) = \$108$. Her check should be for $108.
    c. This is not correct, since: $y = 12(7) = \$84$. Her check should be for $84.
    d. This is correct, since: $y = 12(14) = \$168$

35. a. Substituting $t = 5$: $V = -45,000(5) + 600,000 = -225,000 + 600,000 = \$375,000$
    b. Solving when $V = 330,000$:
    $$-45,000t + 600,000 = 330,000$$
    $$-45,000t = -270,000$$
    $$t = 6$$
    The crane will be worth $330,000 at the end of 6 years.
    c. Substituting $t = 9$: $V = -45,000(9) + 600,000 = -405,000 + 600,000 = \$195,000$
    No, the crane will be worth $195,000 after 9 years.
    d. The crane cost $600,000 (the value when $t = 0$).

37. Substituting $x = 4$:
$$3(4) + 2y = 6$$
$$12 + 2y = 6$$
$$2y = -6$$
$$y = -3$$

39. Substituting $x = 0$: $y = -\frac{1}{3}(0) + 2 = 0 + 2 = 2$

41. Substituting $x = 2$: $y = \frac{3}{2}(2) - 3 = 3 - 3 = 0$

43. Solving for y:
$$5x + y = 4$$
$$y = -5x + 4$$

45. Solving for y:
$$3x - 2y = 6$$
$$-2y = -3x + 6$$
$$y = \frac{3}{2}x - 3$$

2.3 Graphing Linear Equations in Two Variables

1. The ordered pairs are $(0,4)$, $(2,2)$, and $(4,0)$:

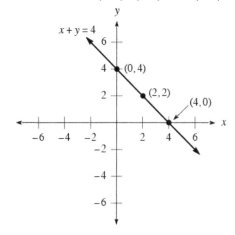

3. The ordered pairs are $(0,3)$, $(2,1)$, and $(4,-1)$:

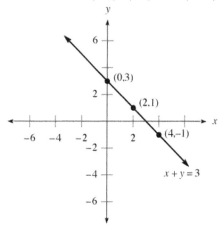

5. The ordered pairs are $(0,0)$, $(-2,-4)$, and $(2,4)$:

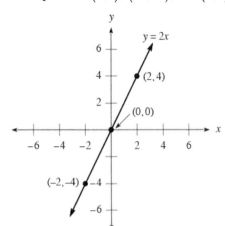

7. The ordered pairs are $(-3,-1)$, $(0,0)$, and $(3,1)$:

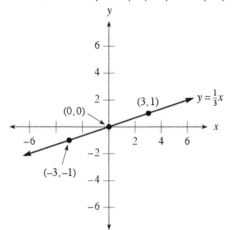

9. The ordered pairs are $(0,1)$, $(-1,-1)$, and $(1,3)$:

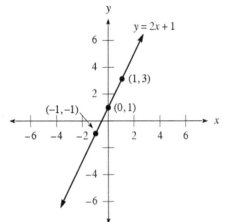

11. The ordered pairs are $(0,4)$, $(-1,4)$, and $(2,4)$:

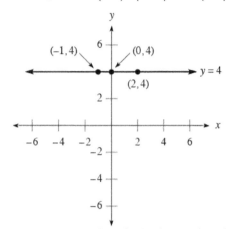

13. The ordered pairs are $(-2,2)$, $(0,3)$, and $(2,4)$:

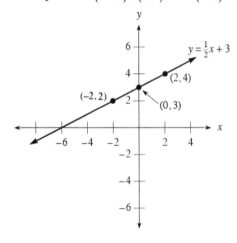

15. The ordered pairs are $(-3,3)$, $(0,1)$, and $(3,-1)$:

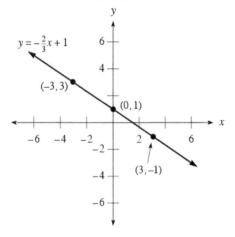

17. Solving for y:
$$2x + y = 3$$
$$y = -2x + 3$$

The ordered pairs are $(-1, 5)$, $(0, 3)$, and $(1, 1)$:

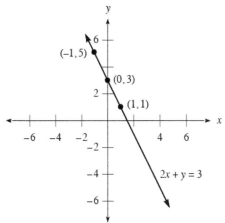

19. Solving for y:
$$3x + 2y = 6$$
$$2y = -3x + 6$$
$$y = -\frac{3}{2}x + 3$$

The ordered pairs are $(0, 3)$, $(2, 0)$, and $(4, -3)$:

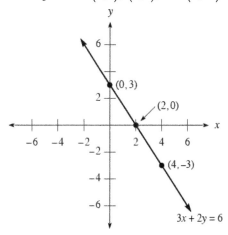

21. Solving for y:
$$-x + 2y = 6$$
$$2y = x + 6$$
$$y = \frac{1}{2}x + 3$$

The ordered pairs are $(-2, 2)$, $(0, 3)$, and $(2, 4)$:

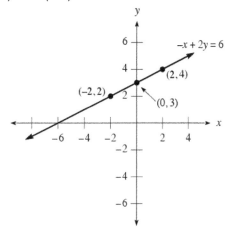

23. Three solutions are $(-4,2)$, $(0,0)$, and $(4,-2)$:

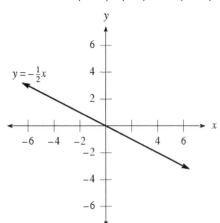

25. Three solutions are $(-1,-4)$, $(0,-1)$, and $(1,2)$:

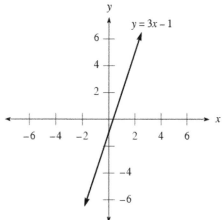

27. Solving for y:
$$-2x+y=1$$
$$y=2x+1$$

Three solutions are $(-2,-3)$, $(0,1)$, and $(2,5)$:

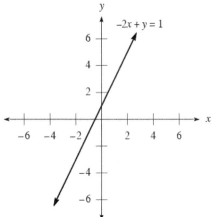

29. Solving for y:
$$3x+4y=8$$
$$4y=-3x+8$$
$$y=-\frac{3}{4}x+2$$

Three solutions are $(-4,5)$, $(0,2)$, and $(4,-1)$:

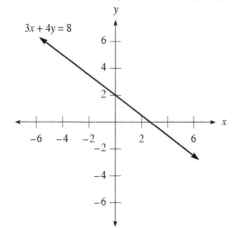

31. Three solutions are $(-2,-4)$, $(-2,0)$, and $(-2,4)$:

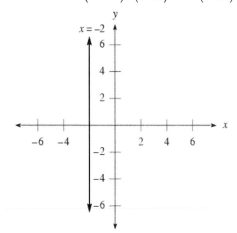

33. Three solutions are $(-4,2)$, $(0,2)$, and $(4,2)$:

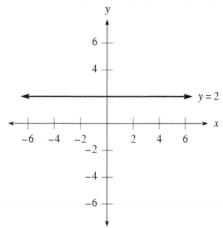

35. Graphing the equation:

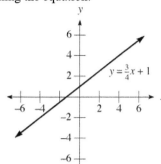

37. Graphing the equation:

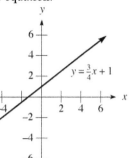

39. Graphing the equation:

41. Completing the table:

Equation	H, V, and/or O
$x = 3$	V
$y = 3$	H
$y = 3x$	O
$y = 0$	O,H

43. Completing the table:

Equation	H, V, and/or O
$x = -\dfrac{3}{5}$	V
$y = -\dfrac{3}{5}$	H
$y = -\dfrac{3}{5}x$	O
$x = 0$	O,V

45. Completing the table:

x	y
-4	-3
-2	-2
0	-1
2	0
6	2

47.
a. Solving the equation:
$$2x + 5 = 10$$
$$2x = 5$$
$$x = \frac{5}{2}$$

b. Substituting $y = 0$:
$$2x + 5(0) = 10$$
$$2x = 10$$
$$x = 5$$

c. Substituting $x = 0$:
$$2(0) + 5y = 10$$
$$5y = 10$$
$$y = 2$$

d. Graphing the line:

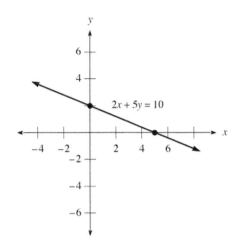

e. Solving for y:
$$2x + 5y = 10$$
$$5y = -2x + 10$$
$$y = -\frac{2}{5}x + 2$$

49. **a.** Substituting $y = 0$:
$$3x + 2(0) = 6$$
$$3x + 0 = 6$$
$$3x = 6$$
$$x = 2$$

b. Substituting $x = 0$:
$$3(0) + 2y = 6$$
$$0 + 2y = 6$$
$$2y = 6$$
$$y = 3$$

51. **a.** Substituting $y = 0$:
$$-x + 2(0) = 4$$
$$-x + 0 = 4$$
$$-x = 4$$
$$x = -4$$

b. Substituting $x = 0$:
$$-(0) + 2y = 4$$
$$0 + 2y = 4$$
$$2y = 4$$
$$y = 2$$

53. **a.** Substituting $y = 0$:
$$0 = -\frac{1}{3}x + 2$$
$$-\frac{1}{3}x = -2$$
$$x = 6$$

b. Substituting $x = 0$:
$$y = -\frac{1}{3}(0) + 2$$
$$y = 2$$

2.4 More on Graphing: Intercepts

1. To find the x-intercept, let $y = 0$:
    $$2x + 0 = 4$$
    $$2x = 4$$
    $$x = 2$$
    To find the y-intercept, let $x = 0$:
    $$2(0) + y = 4$$
    $$0 + y = 4$$
    $$y = 4$$
    Graphing the line:

    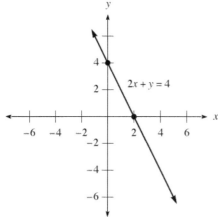

3. To find the x-intercept, let $y = 0$:
    $$-x + 0 = 3$$
    $$-x = 3$$
    $$x = -3$$
    To find the y-intercept, let $x = 0$:
    $$-0 + y = 3$$
    $$y = 3$$
    Graphing the line:

    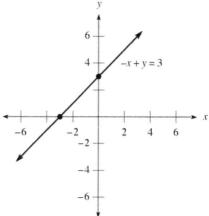

5. To find the x-intercept, let $y = 0$:
    $$-x + 2(0) = 2$$
    $$-x = 2$$
    $$x = -2$$
    To find the y-intercept, let $x = 0$:
    $$-0 + 2y = 2$$
    $$2y = 2$$
    $$y = 1$$
    Graphing the line:

    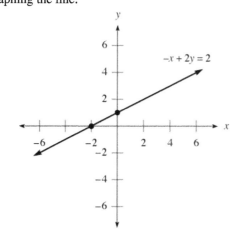

7. To find the x-intercept, let $y = 0$:
    $$5x + 2(0) = 10$$
    $$5x = 10$$
    $$x = 2$$
    To find the y-intercept, let $x = 0$:
    $$5(0) + 2y = 10$$
    $$2y = 10$$
    $$y = 5$$
    Graphing the line:

    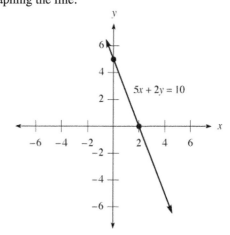

9. To find the x-intercept, let $y = 0$:
$$4x - 2(0) = 8$$
$$4x = 8$$
$$x = 2$$
To find the y-intercept, let $x = 0$:
$$4(0) - 2y = 8$$
$$-2y = 8$$
$$y = -4$$
Graphing the line:

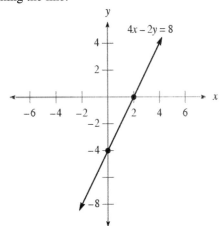

11. To find the x-intercept, let $y = 0$:
$$-4x + 5(0) = 20$$
$$-4x = 20$$
$$x = -5$$
To find the y-intercept, let $x = 0$:
$$-4(0) + 5y = 20$$
$$5y = 20$$
$$y = 4$$
Graphing the line:

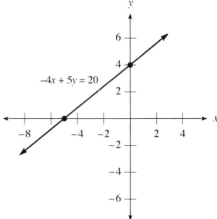

13. To find the x-intercept, let $y = 0$:
$$2x - 6 = 0$$
$$2x = 6$$
$$x = 3$$
To find the y-intercept, let $x = 0$:
$$y = 2(0) - 6 = -6$$
Graphing the line:

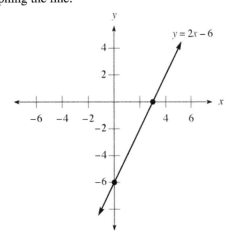

15. To find the x-intercept, let $y = 0$:
$$2x + 2 = 0$$
$$2x = -2$$
$$x = -1$$
To find the y-intercept, let $x = 0$:
$$y = 2(0) + 2 = 2$$
Graphing the line:

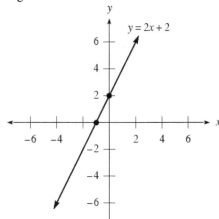

17. To find the x-intercept, let $y = 0$:

    $$2x - 1 = 0$$
    $$2x = 1$$
    $$x = \frac{1}{2}$$

    To find the y-intercept, let $x = 0$:

    $$y = 2(0) - 1 = -1$$

    Graphing the line:

    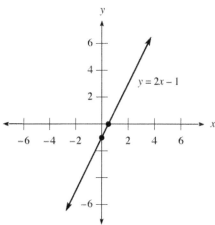

19. To find the x-intercept, let $y = 0$:

    $$\frac{1}{2}x + 3 = 0$$
    $$\frac{1}{2}x = -3$$
    $$x = -6$$

    To find the y-intercept, let $x = 0$:

    $$y = \frac{1}{2}(0) + 3 = 3$$

    Graphing the line:

    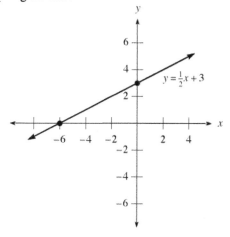

21. To find the x-intercept, let $y = 0$:

    $$-\frac{1}{3}x - 2 = 0$$
    $$-\frac{1}{3}x = 2$$
    $$x = -6$$

    To find the y-intercept, let $x = 0$: $y = -\frac{1}{3}(0) - 2 = -2$

    Graphing the line:

    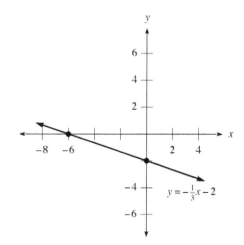

Essentials of Elementary and Intermediate Algebra: A Combined Course
Student Solutions Manual, Problem Set 2.4

23. Another point on the line is $(2,-4)$:

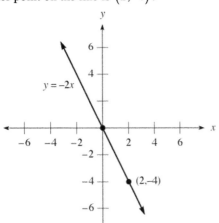

25. Another point on the line is $(3,-1)$:

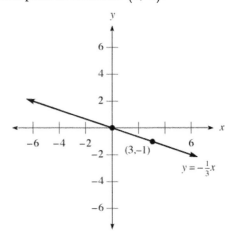

27. Another point on the line is $(3,2)$:

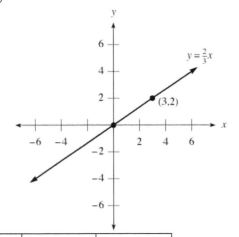

29. Completing the table:

Equation	x – intercept	y – intercept
$3x+4y=12$	4	3
$3x+4y=4$	$\frac{4}{3}$	1
$3x+4y=3$	1	$\frac{3}{4}$
$3x+4y=2$	$\frac{2}{3}$	$\frac{1}{2}$

31. Completing the table:

Equation	x – intercept	y – intercept
$x-3y=2$	2	$-\frac{2}{3}$
$y=\frac{1}{3}x-\frac{2}{3}$	2	$-\frac{2}{3}$
$x-3y=0$	0	0
$y=\frac{1}{3}x$	0	0

33. **a.** Solving the equation:
$$2x - 3 = -3$$
$$2x = 0$$
$$x = 0$$

b. Substituting $y = 0$:
$$2x - 3(0) = -3$$
$$2x - 0 = -3$$
$$2x = -3$$
$$x = -\frac{3}{2}$$

c. Substituting $x = 0$:
$$2(0) - 3y = -3$$
$$0 - 3y = -3$$
$$-3y = -3$$
$$y = 1$$

d. Graphing the equation:

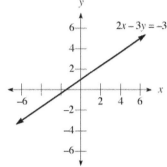

e. Solving for y:
$$2x - 3y = -3$$
$$-3y = -2x - 3$$
$$y = \frac{2}{3}x + 1$$

35. The x-intercept is 3 and the y-intercept is 5.

37. The x-intercept is -1 and the y-intercept is -3.

39. The y-intercept is -4:

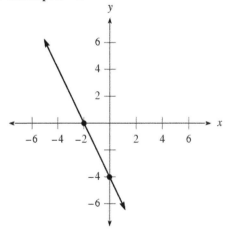

41. The x-intercept is -3:

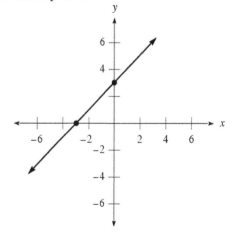

43. The *x*- and *y*-intercepts are both 3:

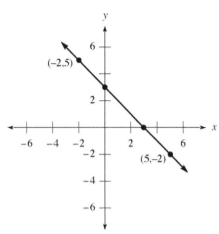

45. Completing the table:

x	y
−2	1
0	−1
−1	0
1	−2

47. The *x*-intercept is 3:

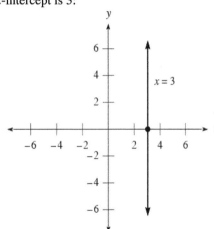

49. The *y*-intercept is 4:

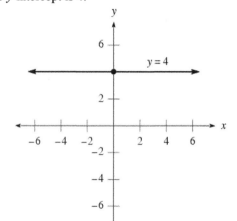

51. Sketching the graph:

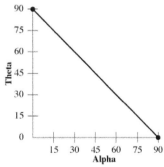

53. a. Evaluating: $\dfrac{5-2}{3-1} = \dfrac{3}{2}$

b. Evaluating: $\dfrac{2-5}{1-3} = \dfrac{-3}{-2} = \dfrac{3}{2}$

55. **a.** Evaluating when $x = 3$ and $y = 5$: $\dfrac{y-2}{x-1} = \dfrac{5-2}{3-1} = \dfrac{3}{2}$

b. Evaluating when $x = 3$ and $y = 5$: $\dfrac{2-y}{1-x} = \dfrac{2-5}{1-3} = \dfrac{-3}{-2} = \dfrac{3}{2}$

2.5 The Slope of a Line

1. The slope is given by: $m = \dfrac{4-1}{4-2} = \dfrac{3}{2}$

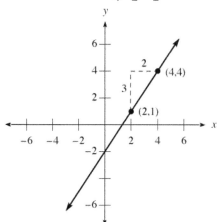

3. The slope is given by: $m = \dfrac{2-4}{5-1} = \dfrac{-2}{4} = -\dfrac{1}{2}$

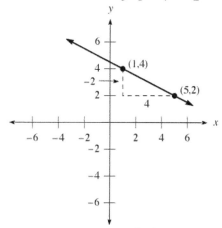

5. The slope is given by: $m = \dfrac{2-(-3)}{4-1} = \dfrac{5}{3}$

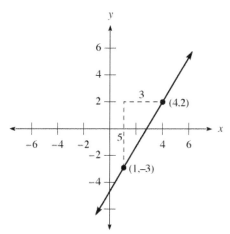

7. The slope is given by: $m = \dfrac{3-(-2)}{1-(-3)} = \dfrac{5}{4}$

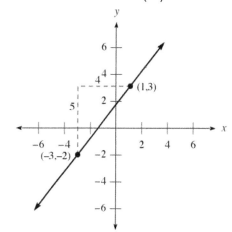

9. The slope is given by: $m = \dfrac{-2-2}{3-(-3)} = \dfrac{-4}{6} = -\dfrac{2}{3}$

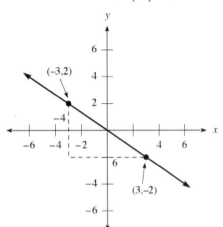

11. The slope is given by: $m = \dfrac{-2-(-5)}{3-2} = \dfrac{3}{1} = 3$

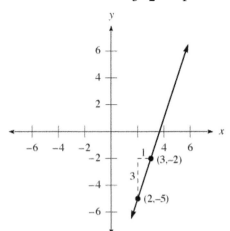

13. Graphing the line:

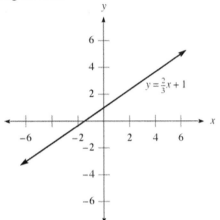

15. Graphing the line:

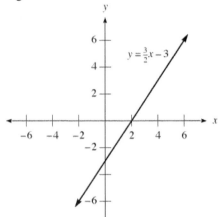

17. Graphing the line:

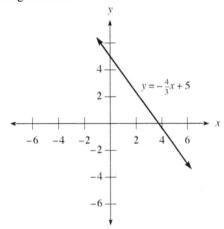

19. Graphing the line:

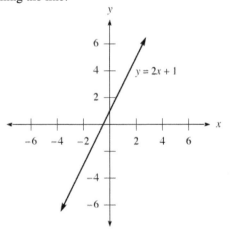

21. Graphing the line:

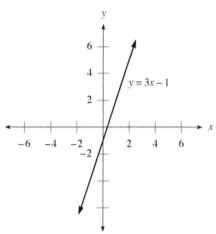

23. The y-intercept is 2, and the slope is given by: $m = \dfrac{5-(-1)}{1-(-1)} = \dfrac{6}{2} = 3$

25. The y-intercept is –2, and the slope is given by: $m = \dfrac{2-0}{2-1} = \dfrac{2}{1} = 2$

27. The slope is given by: $m = \dfrac{0-(-2)}{3-0} = \dfrac{2}{3}$

29. The slope is given by: $m = \dfrac{0-2}{4-0} = \dfrac{-2}{4} = -\dfrac{1}{2}$

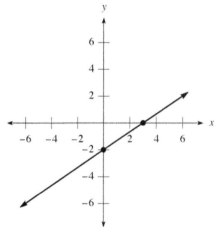
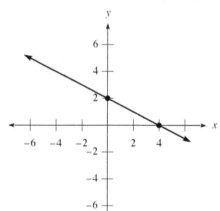

31. The slope is 2 and the y-intercept is –3:

33. The slope is $\dfrac{1}{2}$ and the y-intercept is 1:

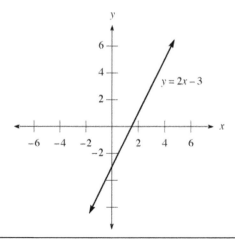
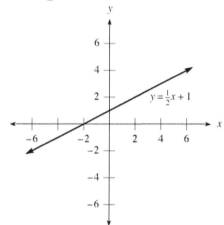

35. Using the slope formula:
$$\frac{y-2}{6-4} = 2$$
$$\frac{y-2}{2} = 2$$
$$y - 2 = 4$$
$$y = 6$$

37. The slopes are given in the table:

Equation	Slope
$x = 3$	undefined
$y = 3$	0
$y = 3x$	3

39. The slopes are given in the table:

Equation	Slope
$y = -\frac{2}{3}$	0
$x = -\frac{2}{3}$	undefined
$y = -\frac{2}{3}x$	$-\frac{2}{3}$

41. Finding the slopes:

$A: \dfrac{121-88}{1970-1960} = \dfrac{33}{10} = 3.3$

$B: \dfrac{152-121}{1980-1970} = \dfrac{31}{10} = 3.1$

$C: \dfrac{205-152}{1990-1980} = \dfrac{53}{10} = 5.3$

$D: \dfrac{224-205}{2000-1990} = \dfrac{19}{10} = 1.9$

43. Finding the slopes:

$A: \dfrac{250-300}{2007-2006} = -50$

$B: \dfrac{175-250}{2008-2007} = -75$

$C: \dfrac{125-150}{2010-2009} = -25$

45. Solving for y:
$$-2x + y = 4$$
$$y = 2x + 4$$

47. Solving for y:
$$2x + y = 3$$
$$y = -2x + 3$$

49. Solving for y:
$$4x - 5y = 20$$
$$-5y = -4x + 20$$
$$y = \frac{4}{5}x - 4$$

51. Solving for y:
$$-y - 3 = -2(x+4)$$
$$-y - 3 = -2x - 8$$
$$-y = -2x - 5$$
$$y = 2x + 5$$

53. Solving for y:
$$-y - 3 = -\frac{2}{3}(x+3)$$
$$-y - 3 = -\frac{2}{3}x - 2$$
$$-y = -\frac{2}{3}x + 1$$
$$y = \frac{2}{3}x - 1$$

55. Solving for y:
$$-\frac{y-1}{x} = \frac{3}{2}$$
$$-2y + 2 = 3x$$
$$-2y = 3x - 2$$
$$y = -\frac{3}{2}x + 1$$

2.6 Finding the Equation of a Line

1. The slope-intercept form is $y = \frac{2}{3}x + 1$.

3. The slope-intercept form is $y = \frac{3}{2}x - 1$.

5. The slope-intercept form is $y = -\frac{2}{3}x + 3$.

7. The slope-intercept form is $y = 2x - 4$.

9. Solving for y:
$$-2x + y = 4$$
$$y = 2x + 4$$
The slope is 2 and the y-intercept is 4:

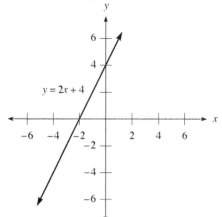

11. Solving for y:
$$3x + y = 3$$
$$y = -3x + 3$$
The slope is –3 and the y-intercept is 3:

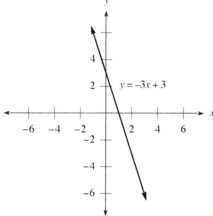

13. Solving for y:
$$3x + 2y = 6$$
$$2y = -3x + 6$$
$$y = -\frac{3}{2}x + 3$$
The slope is $-\frac{3}{2}$ and the y-intercept is 3:

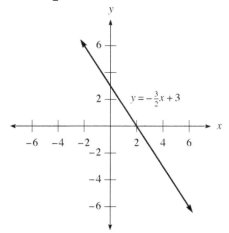

15. Solving for y:
$$4x - 5y = 20$$
$$-5y = -4x + 20$$
$$y = \frac{4}{5}x - 4$$
The slope is $\frac{4}{5}$ and the y-intercept is –4:

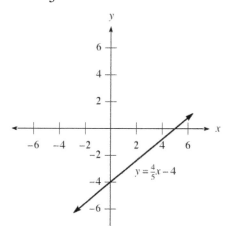

17. Solving for y:
$$-2x - 5y = 10$$
$$-5y = 2x + 10$$
$$y = -\frac{2}{5}x - 2$$

The slope is $-\frac{2}{5}$ and the y-intercept is –2:

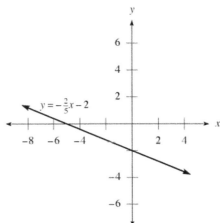

19. Using the point-slope formula:
$$y - (-5) = 2(x - (-2))$$
$$y + 5 = 2(x + 2)$$
$$y + 5 = 2x + 4$$
$$y = 2x - 1$$

21. Using the point-slope formula:
$$y - 1 = -\frac{1}{2}(x - (-4))$$
$$y - 1 = -\frac{1}{2}(x + 4)$$
$$y - 1 = -\frac{1}{2}x - 2$$
$$y = -\frac{1}{2}x - 1$$

23. Using the point-slope formula:
$$y - (-3) = \frac{3}{2}(x - 2)$$
$$y + 3 = \frac{3}{2}x - 3$$
$$y = \frac{3}{2}x - 6$$

25. Using the point-slope formula:
$$y - 4 = -3(x - (-1))$$
$$y - 4 = -3(x + 1)$$
$$y - 4 = -3x - 3$$
$$y = -3x + 1$$

27. Finding the slope: $m = \frac{-1 - (-4)}{1 - (-2)} = \frac{-1 + 4}{1 + 2} = \frac{3}{3} = 1$

Using the point-slope formula:
$$y - (-4) = 1(x - (-2))$$
$$y + 4 = x + 2$$
$$y = x - 2$$

29. Finding the slope: $m = \frac{1 - (-5)}{2 - (-1)} = \frac{1 + 5}{2 + 1} = \frac{6}{3} = 2$

Using the point-slope formula:
$$y - 1 = 2(x - 2)$$
$$y - 1 = 2x - 4$$
$$y = 2x - 3$$

31. Finding the slope: $m = \dfrac{6-(-2)}{3-(-3)} = \dfrac{6+2}{3+3} = \dfrac{8}{6} = \dfrac{4}{3}$

Using the point-slope formula:
$$y - 6 = \dfrac{4}{3}(x-3)$$
$$y - 6 = \dfrac{4}{3}x - 4$$
$$y = \dfrac{4}{3}x + 2$$

33. Finding the slope: $m = \dfrac{-5-(-1)}{3-(-3)} = \dfrac{-5+1}{3+3} = \dfrac{-4}{6} = -\dfrac{2}{3}$

Using the point-slope formula:
$$y - (-5) = -\dfrac{2}{3}(x-3)$$
$$y + 5 = -\dfrac{2}{3}x + 2$$
$$y = -\dfrac{2}{3}x - 3$$

35. The y-intercept is 3, and the slope is: $m = \dfrac{3-0}{0-(-1)} = \dfrac{3}{1} = 3$. The slope-intercept form is $y = 3x + 3$.

37. The y-intercept is -1, and the slope is: $m = \dfrac{0-(-1)}{4-0} = \dfrac{0+1}{4} = \dfrac{1}{4}$. The slope-intercept form is $y = \dfrac{1}{4}x - 1$.

39. **a.** Solving the equation:
$$-2x + 1 = 6$$
$$-2x = 5$$
$$x = -\dfrac{5}{2}$$

b. Writing in slope-intercept form:
$$-2x + y = 6$$
$$y = 2x + 6$$

c. Substituting $x = 0$:
$$-2(0) + y = 6$$
$$0 + y = 6$$
$$y = 6$$

d. Finding the slope:
$$-2x + y = 6$$
$$y = 2x + 6$$
The slope is 2.

e. Graphing the line:

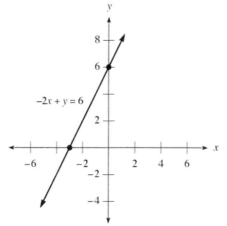

41. The slope is given by: $m = \dfrac{0-2}{3-0} = -\dfrac{2}{3}$. Since $b = 2$, the equation is $y = -\dfrac{2}{3}x + 2$.

43. The slope is given by: $m = \dfrac{0-(-5)}{-2-0} = -\dfrac{5}{2}$. Since $b = -5$, the equation is $y = -\dfrac{5}{2}x - 5$.

45. Since this is a vertical line, its equation is $x = 3$.

47. **a.** After 5 years, the copier is worth $6,000.
   b. The copier is worth $12,000 after 3 years.
   c. The slope of the line is $-3,000$.
   d. The copier is decreasing in value by $3,000 per year.
   e. The equation is $V = -3,000t + 21,000$.

49. Graphing the line:

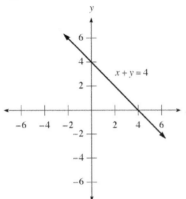

51. Graphing the line:

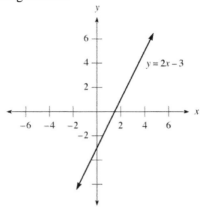

53. Graphing the line:

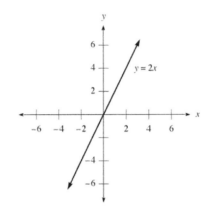

2.7 Linear Inequalities in Two Variables

1. Checking the point $(0,0)$:

$2(0)-3(0) = 0-0 < 6$ (true)

Graphing the linear inequality:

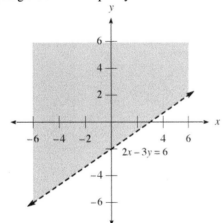

3. Checking the point $(0,0)$:

$0-2(0) = 0-0 \le 4$ (true)

Graphing the linear inequality:

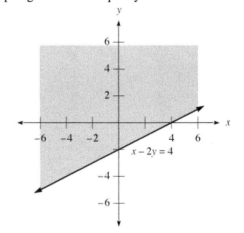

5. Checking the point $(0,0)$:
$0 - 0 \leq 2$ (true)
Graphing the linear inequality:
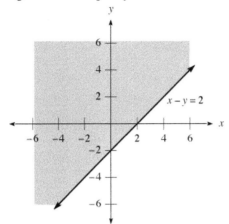

7. Checking the point $(0,0)$:
$3(0) - 4(0) = 0 - 0 \geq 12$ (false)
Graphing the linear inequality:
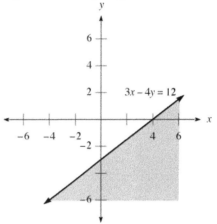

9. Checking the point $(0,0)$:
$5(0) - 0 = 0 - 0 \leq 5$ (true)
Graphing the linear inequality:
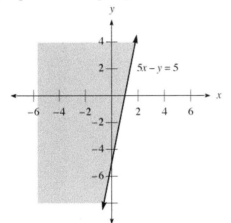

11. Checking the point $(0,0)$:
$2(0) + 6(0) = 0 + 0 \leq 12$ (true)
Graphing the linear inequality:
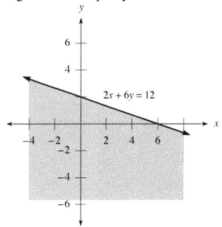

13. Graphing the linear inequality:
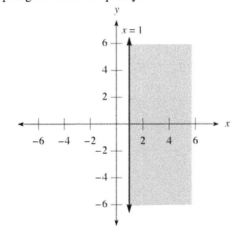

15. Graphing the linear inequality:
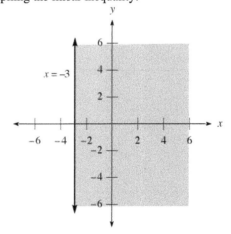

17. Graphing the linear inequality:

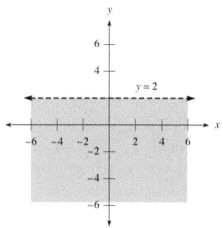

19. Checking the point $(0,0)$:

$2(0) + 0 = 0 + 0 > 3$ (false)

Graphing the linear inequality:

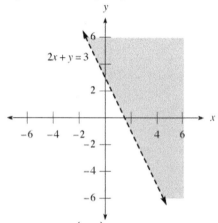

21. Checking the point $(0,0)$:

$0 \le 3(0) - 1$
$0 \le -1$ (false)

Graphing the linear inequality:

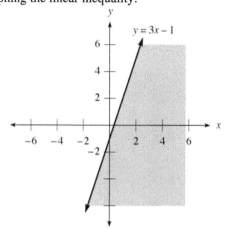

23. Checking the point $(0,0)$:

$0 \le -\frac{1}{2}(0) + 2$
$0 \le 2$ (true)

Graphing the linear inequality:

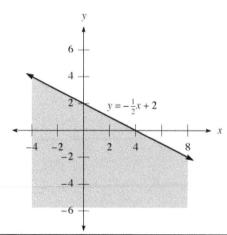

25. a. Solving the inequality:
$$4 + 3y < 12$$
$$3y < 8$$
$$y < \frac{8}{3}$$

b. Solving the inequality:
$$4 - 3y < 12$$
$$-3y < 8$$
$$y > -\frac{8}{3}$$

c. Solving for y:
$$4x + 3y = 12$$
$$3y = -4x + 12$$
$$y = -\frac{4}{3}x + 4$$

d. Graphing the inequality:

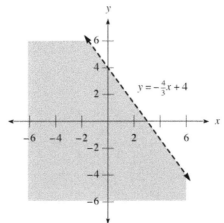

27. a. The slope is $\frac{2}{5}$ and the y-intercept is 2, so the equation is $y = \frac{2}{5}x + 2$.

b. Since the shading is below the line, the inequality is $y < \frac{2}{5}x + 2$.

c. Since the shading is above the line, the inequality is $y > \frac{2}{5}x + 2$.

29. Simplifying the expression: $7 - 3(2x - 4) - 8 = 7 - 6x + 12 - 8 = -6x + 11$

31. Solving the equation:
$$-\frac{3}{2}x = 12$$
$$-\frac{2}{3}\left(-\frac{3}{2}x\right) = -\frac{2}{3}(12)$$
$$x = -8$$

33. Solving the equation:
$$8 - 2(x + 7) = 2$$
$$8 - 2x - 14 = 2$$
$$-2x - 6 = 2$$
$$-2x = 8$$
$$x = -4$$

35. Solving for w:
$$2l + 2w = P$$
$$2w = P - 2l$$
$$w = \frac{P - 2l}{2}$$

37. Solving the inequality:
$$3 - 2x > 5$$
$$3 - 3 - 2x > 5 - 3$$
$$-2x > 2$$
$$-\frac{1}{2}(-2x) < -\frac{1}{2}(2)$$
$$x < -1$$
The solution set is $\{x \mid x < -1\}$. Graphing the inequality:

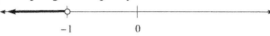

39. Solving for y:
$$3x - 2y \le 12$$
$$3x - 3x - 2y \le -3x + 12$$
$$-2y \le -3x + 12$$
$$-\frac{1}{2}(-2y) \ge -\frac{1}{2}(-3x + 12)$$
$$y \ge \frac{3}{2}x - 6$$

41. Let w represent the width and 3w + 5 represent the length. Using the perimeter formula:
$$2(w) + 2(3w + 5) = 26$$
$$2w + 6w + 10 = 26$$
$$8w + 10 = 26$$
$$8w = 16$$
$$w = 2$$
$$3w + 5 = 3(2) + 5 = 11$$
The width is 2 inches and the length is 11 inches.

Chapter 2 Test

1. Graphing the ordered pair:

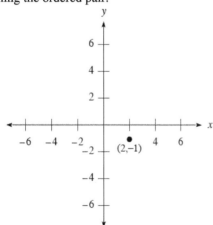

2. Graphing the ordered pair:

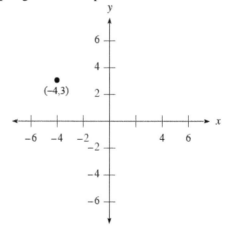

3. Graphing the ordered pair:

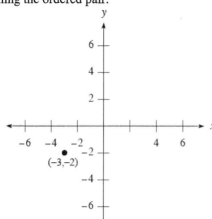

4. Graphing the ordered pair:

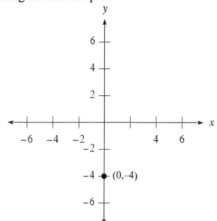

5. Substituting $x = 0, y = 0, x = 4$, and $y = -6$:

$3(0) - 2y = 6$ $3x - 2(0) = 6$ $3(4) - 2y = 6$ $3x - 2(-6) = 6$
$0 - 2y = 6$ $3x - 0 = 6$ $12 - 2y = 6$ $3x + 12 = 6$
$-2y = 6$ $3x = 6$ $-2y = -6$ $3x = -6$
$y = -3$ $x = 2$ $y = 3$ $x = -2$

The ordered pairs are $(0,-3), (2,0), (4,3)$, and $(-2,-6)$.

6. Substituting each ordered pair into the equation:

$(0,7)$: $-3(0) + 7 = 0 + 7 = 7$
$(2,-1)$: $-3(2) + 7 = -6 + 7 = 1 \neq -1$
$(4,-5)$: $-3(4) + 7 = -12 + 7 = -5$
$(-5,-3)$: $-3(-5) + 7 = 15 + 7 = 22 \neq -3$

The ordered pairs $(0,7)$ and $(4,-5)$ are solutions.

7. Graphing the line:

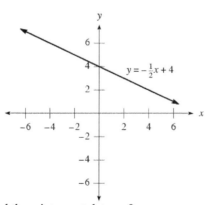

8. Graphing the line:

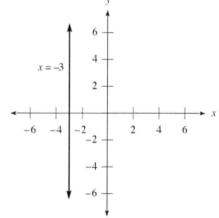

9. To find the x-intercept, let $y = 0$:

$8x - 4(0) = 16$
$8x = 16$
$x = 2$

To find the y-intercept, let $x = 0$:

$8(0) - 4y = 16$
$-4y = 16$
$y = -4$

The x-intercept is $(2,0)$ and the y-intercept is $(0,-4)$.

Essentials of Elementary and Intermediate Algebra: A Combined Course

10. To find the x-intercept, let $y = 0$:
$$0 = \frac{3}{2}x + 6$$
$$\frac{3}{2}x = -6$$
$$x = -4$$
The x-intercept is $(-4, 0)$ and the y-intercept is $(0, 6)$.

To find the y-intercept, let $x = 0$:
$$y = \frac{3}{2}(0) + 6$$
$$y = 6$$

11. There is no x-intercept, and the y-intercept is $(0, 3)$.

12. The slope is given by: $m = \dfrac{6-2}{-5-3} = \dfrac{4}{-8} = -\dfrac{1}{2}$

13. The slope is given by: $m = \dfrac{1-9}{7-0} = \dfrac{-8}{7} = -\dfrac{8}{7}$

14. The slope is given by: $m = \dfrac{4-1}{0-1} = \dfrac{3}{-1} = -3$

15. Since the line is vertical, the slope is undefined.

16. Since the line is horizontal, the slope is 0.

17. Using the point-slope formula:
$$y - 1 = -\frac{1}{2}(x - 4)$$
$$y - 1 = -\frac{1}{2}x + 2$$
$$y = -\frac{1}{2}x + 3$$

18. Since $m = 3$ and $b = -5$, the equation of the line is $y = 3x - 5$.

19. The slope is given by: $m = \dfrac{2-(-4)}{-6-3} = \dfrac{6}{-9} = -\dfrac{2}{3}$

Using the point-slope formula:
$$y - (-4) = -\frac{2}{3}(x - 3)$$
$$y + 4 = -\frac{2}{3}x + 2$$
$$y = -\frac{2}{3}x - 2$$

20. The slope is given by: $m = \dfrac{6-0}{-2-3} = -\dfrac{6}{5}$

Using the point-slope formula:
$$y - 0 = -\frac{6}{5}(x - 3)$$
$$y - 0 = -\frac{6}{5}x + \frac{18}{5}$$
$$y = -\frac{6}{5}x + \frac{18}{5}$$

21. Graphing the linear inequality:

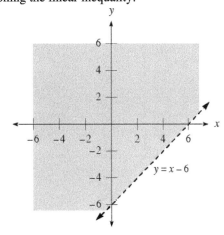

22. Graphing the linear inequality:

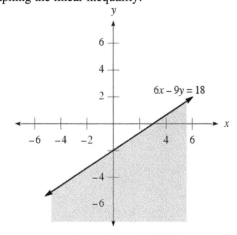

Chapter 3
Systems of Linear Equations

3.1 Solving Linear Systems by Graphing

1. Graphing both lines:

   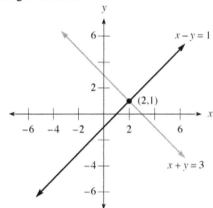

   The solution set is $\{(2,1)\}$.

3. Graphing both lines:

   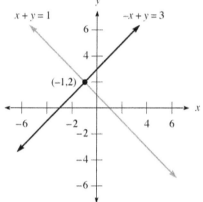

   The solution set is $\{(-1,2)\}$.

5. Graphing both lines:

   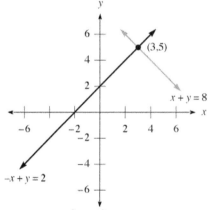

   The solution set is $\{(3,5)\}$.

7. Graphing both lines:

   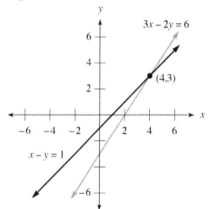

   The solution set is $\{(4,3)\}$.

9. Graphing both lines:

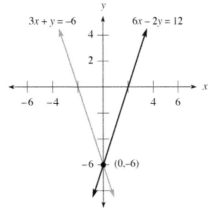

The solution set is $\{(0,-6)\}$.

11. Graphing both lines:

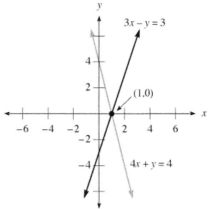

The solution set is $\{(1,0)\}$.

13. Graphing both lines:

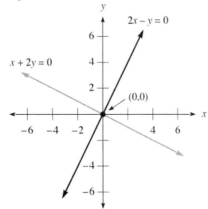

The solution set is $\{(0,0)\}$.

15. Graphing both lines:

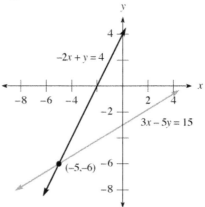

The solution set is $\{(-5,-6)\}$.

17. Graphing both lines:

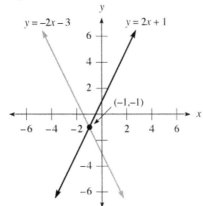

The solution set is $\{(-1,-1)\}$.

19. Graphing both lines:

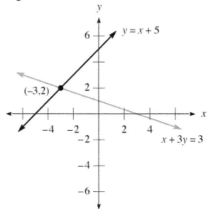

The solution set is $\{(-3,2)\}$.

21. Graphing both lines:

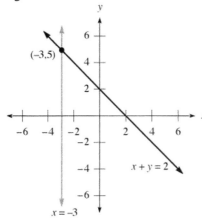

The solution set is $\{(-3,5)\}$.

23. Graphing both lines:

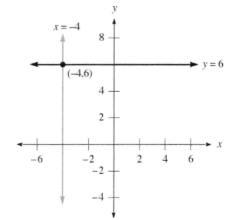

The solution set is $\{(-4,6)\}$.

25. Graphing both lines:

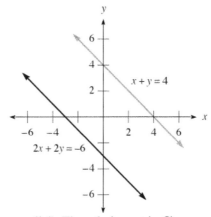

There is no intersection (the lines are parallel). The solution set is $\varnothing$.

27. Graphing both lines:

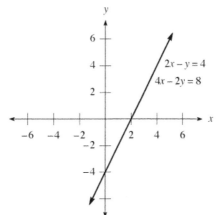

The system is dependent (both lines are the same, they coincide). The solution set is $\{(x,y) \mid 2x - y = 4\}$.

29. Graphing both lines:

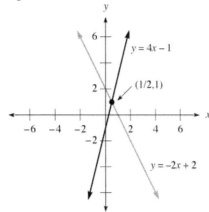

The intersection point is $\left(\dfrac{1}{2}, 1\right)$.

31. Graphing both lines:

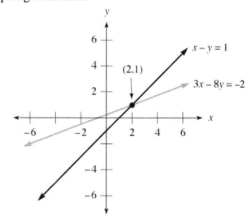

The intersection point is $(2, 1)$.

33.
 a. If Jane worked 25 hours, she would earn the same amount at each position.
 b. If Jane worked less than 20 hours, she should choose Gigi's since she earns more in that position.
 c. If Jane worked more than 30 hours, she should choose Marcy's since she earns more in that position.

35. Simplifying: $(x+y)+(x-y) = x+x+y-y = 2x$

37. Simplifying: $3(2x-y)+(x+3y) = 6x-3y+x+3y = 7x$

39. Simplifying: $-4(3x+5y)+5(5x+4y) = -12x-20y+25x+20y = 13x$

41. Simplifying: $6\left(\dfrac{1}{2}x - \dfrac{1}{3}y\right) = 6 \cdot \dfrac{1}{2}x - 6 \cdot \dfrac{1}{3}y = 3x - 2y$

43. Substituting $x = 3$:
$3 + y = 4$
$y = 1$

45. Substituting $x = 3$:
$3 + 3y = 3$
$3y = 0$
$y = 0$

47. Substituting $x = 6$:
$3(6) + 5y = -7$
$18 + 5y = -7$
$5y = -25$
$y = -5$

3.2 The Elimination Method

1. Adding the two equations:
   $2x = 4$
   $x = 2$
   Substituting into the first equation:
   $2 + y = 3$
   $y = 1$
   The solution is $(2,1)$.

3. Adding the two equations:
   $2y = 14$
   $y = 7$
   Substituting into the first equation:
   $x + 7 = 10$
   $x = 3$
   The solution is $(3,7)$.

5. Adding the two equations:
   $-2y = 10$
   $y = -5$
   Substituting into the first equation:
   $x - (-5) = 7$
   $x + 5 = 7$
   $x = 2$
   The solution is $(2,-5)$.

7. Adding the two equations:
   $4x = -4$
   $x = -1$
   Substituting into the first equation:
   $-1 + y = -1$
   $y = 0$
   The solution is $(-1,0)$.

9. Adding the two equations:
   $0 = 0$
   The lines coincide (the system is dependent). The solution set is $\{(x,y) \mid 3x + 2y = 1\}$.

11. Multiplying the first equation by 2:
    $6x - 2y = 8$
    $2x + 2y = 24$
    Adding the two equations:
    $8x = 32$
    $x = 4$
    Substituting into the first equation:
    $3(4) - y = 4$
    $12 - y = 4$
    $-y = -8$
    $y = 8$
    The solution is $(4,8)$.

13. Multiplying the second equation by -3:
    $5x - 3y = -2$
    $-30x + 3y = -3$
    Adding the two equations:
    $-25x = -5$
    $x = \frac{1}{5}$
    Substituting into the first equation:
    $5\left(\frac{1}{5}\right) - 3y = -2$
    $1 - 3y = -2$
    $-3y = -3$
    $y = 1$
    The solution is $\left(\frac{1}{5}, 1\right)$.

15. Multiplying the second equation by 4:
    $11x - 4y = 11$
    $20x + 4y = 20$
    Adding the two equations:
    $31x = 31$
    $x = 1$
    Substituting into the second equation:
    $5(1) + y = 5$
    $5 + y = 5$
    $y = 0$
    The solution is $(1,0)$.

17. Multiplying the second equation by 3:
    $3x - 5y = 7$
    $-3x + 3y = -3$
    Adding the two equations:
    $-2y = 4$
    $y = -2$
    Substituting into the second equation:
    $-x - 2 = -1$
    $-x = 1$
    $x = -1$
    The solution is $(-1,-2)$.

19. Multiplying the first equation by –2:
$$2x + 16y = 2$$
$$-2x + 4y = 13$$
Adding the two equations:
$$20y = 15$$
$$y = \frac{3}{4}$$
Substituting into the first equation:
$$-x - 8\left(\frac{3}{4}\right) = -1$$
$$-x - 6 = -1$$
$$-x = 5$$
$$x = -5$$
The solution is $\left(-5, \frac{3}{4}\right)$.

21. Multiplying the first equation by 2:
$$-6x - 2y = 14$$
$$6x + 7y = 11$$
Adding the two equations:
$$5y = 25$$
$$y = 5$$
Substituting into the first equation:
$$-3x - 5 = 7$$
$$-3x = 12$$
$$x = -4$$
The solution is $(-4, 5)$.

23. Adding the two equations:
$$8x = -24$$
$$x = -3$$
Substituting into the second equation:
$$2(-3) + y = -16$$
$$-6 + y = -16$$
$$y = -10$$
The solution is $(-3, -10)$.

25. Multiplying the second equation by 3:
$$x + 3y = 9$$
$$6x - 3y = 12$$
Adding the two equations:
$$7x = 21$$
$$x = 3$$
Substituting into the first equation:
$$3 + 3y = 9$$
$$3y = 6$$
$$y = 2$$
The solution is $(3, 2)$.

27. Multiplying the second equation by 2:
$$x - 6y = 3$$
$$8x + 6y = 42$$
Adding the two equations:
$$9x = 45$$
$$x = 5$$
Substituting into the second equation:
$$4(5) + 3y = 21$$
$$20 + 3y = 21$$
$$3y = 1$$
$$y = \frac{1}{3}$$
The solution is $\left(5, \frac{1}{3}\right)$.

29. Multiplying the second equation by –3:
$$2x + 9y = 2$$
$$-15x - 9y = 24$$
Adding the two equations:
$$-13x = 26$$
$$x = -2$$
Substituting into the first equation:
$$2(-2) + 9y = 2$$
$$-4 + 9y = 2$$
$$9y = 6$$
$$y = \frac{2}{3}$$
The solution is $\left(-2, \frac{2}{3}\right)$.

31. To clear each equation of fractions, multiply the first equation by 12 and the second equation by 6:

$$12\left(\frac{1}{3}x+\frac{1}{4}y\right)=12\left(\frac{7}{6}\right) \qquad 6\left(\frac{3}{2}x-\frac{1}{3}y\right)=6\left(\frac{7}{3}\right)$$
$$4x+3y=14 \qquad\qquad 9x-2y=14$$

The system of equations is:
$$4x+3y=14$$
$$9x-2y=14$$

Multiplying the first equation by 2 and the second equation by 3:
$$8x+6y=28$$
$$27x-6y=42$$

Adding the two equations:
$$35x=70$$
$$x=2$$

Substituting into $4x+3y=14$:
$$4(2)+3y=14$$
$$8+3y=14$$
$$3y=6$$
$$y=2$$

The solution is $(2,2)$.

33. Multiplying the first equation by –2:
$$-6x-4y=2$$
$$6x+4y=0$$

Adding the two equations:
$$0=2$$

Since this statement is false, there is no solution to the system. The system is inconsistent.

35. Multiplying the first equation by 2 and the second equation by 3:
$$22x+12y=34$$
$$15x-12y=3$$

Adding the two equations:
$$37x=37$$
$$x=1$$

Substituting into the second equation:
$$5(1)-4y=1$$
$$5-4y=1$$
$$-4y=-4$$
$$y=1$$

The solution is $(1,1)$.

37. To clear each equation of fractions, multiply the first equation by 6 and the second equation by 6:

$$6\left(\frac{1}{2}x+\frac{1}{6}y\right)=6\left(\frac{1}{3}\right) \qquad 6\left(-x-\frac{1}{3}y\right)=6\left(-\frac{1}{6}\right)$$
$$3x+y=2 \qquad\qquad -6x-2y=-1$$

The system of equations is:
$$3x+y=2$$
$$-6x-2y=-1$$

Multiplying the first equation by 2:
$$6x+2y=4$$
$$-6x-2y=-1$$

Adding the two equations:
$$0=3$$

Since this statement is false, there is no solution to the system. The system is inconsistent.

39. Multiplying the second equation by 100 (to eliminate decimals):
$$x + y = 22$$
$$5x + 10y = 170$$
Multiplying the first equation by –5:
$$-5x - 5y = -110$$
$$5x + 10y = 170$$
Adding the two equations:
$$5y = 60$$
$$y = 12$$
Substituting into the first equation:
$$x + 12 = 22$$
$$x = 10$$
The solution is $(10, 12)$.

41. Solving the equation:
$$x + (2x - 1) = 2$$
$$3x - 1 = 2$$
$$3x = 3$$
$$x = 1$$

43. Solving the equation:
$$2(3y - 1) - 3y = 4$$
$$6y - 2 - 3y = 4$$
$$3y - 2 = 4$$
$$3y = 6$$
$$y = 2$$

45. Solving the equation:
$$4x + 2(-2x + 4) = 8$$
$$4x - 4x + 8 = 8$$
$$8 = 8$$
The solution set is all real numbers.

47. Solving for x:
$$x - 3y = -1$$
$$x = 3y - 1$$

49. Substituting $x = 1$: $y = 2(1) - 1 = 2 - 1 = 1$

51. Substituting $y = 2$: $x = 3(2) - 1 = 6 - 1 = 5$

53. Substituting $x = 13$: $y = 1.5(13) + 15 = 19.5 + 15 = 34.5$

55. Substituting $x = 12$: $y = 0.75(12) + 24.95 = 9 + 24.95 = 33.95$

3.3 The Substitution Method

1. Substituting into the first equation:
$$x + (2x - 1) = 11$$
$$3x - 1 = 11$$
$$3x = 12$$
$$x = 4$$
$$y = 2(4) - 1 = 7$$
The solution is $(4, 7)$.

3. Substituting into the first equation:
$$x + (5x + 2) = 20$$
$$6x + 2 = 20$$
$$6x = 18$$
$$x = 3$$
$$y = 5(3) + 2 = 17$$
The solution is $(3, 17)$.

5. Substituting into the first equation:
$$-2x + (-4x + 8) = -1$$
$$-6x + 8 = -1$$
$$-6x = -9$$
$$x = \frac{3}{2}$$
$$y = -4\left(\frac{3}{2}\right) + 8 = -6 + 8 = 2$$
The solution is $\left(\frac{3}{2}, 2\right)$.

7. Substituting into the first equation:
$$3(-y + 6) - 2y = -2$$
$$-3y + 18 - 2y = -2$$
$$-5y + 18 = -2$$
$$-5y = -20$$
$$y = 4$$
$$x = -4 + 6 = 2$$
The solution is $(2, 4)$.

9. Substituting into the first equation:
$$5x - 4(4) = -16$$
$$5x - 16 = -16$$
$$5x = 0$$
$$x = 0$$
The solution is $(0, 4)$.

11. Substituting into the first equation:
$$5x + 4(-3x) = 7$$
$$5x - 12x = 7$$
$$-7x = 7$$
$$x = -1$$
$$y = -3(-1) = 3$$
The solution is $(-1, 3)$.

13. Solving the second equation for x:
$$x - 2y = -1$$
$$x = 2y - 1$$
Substituting into the first equation:
$$(2y - 1) + 3y = 4$$
$$5y - 1 = 4$$
$$5y = 5$$
$$y = 1$$
$$x = 2(1) - 1 = 1$$
The solution is $(1, 1)$.

15. Solving the second equation for x:
$$x - 5y = 17$$
$$x = 5y + 17$$
Substituting into the first equation:
$$2(5y + 17) + y = 1$$
$$10y + 34 + y = 1$$
$$11y + 34 = 1$$
$$11y = -33$$
$$y = -3$$
$$x = 5(-3) + 17 = 2$$
The solution is $(2, -3)$.

17. Solving the second equation for x:
$$x - 5y = -5$$
$$x = 5y - 5$$
Substituting into the first equation:
$$3(5y - 5) + 5y = -3$$
$$15y - 15 + 5y = -3$$
$$20y - 15 = -3$$
$$20y = 12$$
$$y = \tfrac{3}{5}$$
$$x = 5\left(\tfrac{3}{5}\right) - 5 = 3 - 5 = -2$$
The solution is $\left(-2, \tfrac{3}{5}\right)$.

19. Solving the second equation for x:
$$x - 3y = -18$$
$$x = 3y - 18$$
Substituting into the first equation:
$$5(3y - 18) + 3y = 0$$
$$15y - 90 + 3y = 0$$
$$18y - 90 = 0$$
$$18y = 90$$
$$y = 5$$
$$x = 3(5) - 18 = -3$$
The solution is $(-3, 5)$.

21. Solving the second equation for x:
$$x + 3y = 12$$
$$x = -3y + 12$$
Substituting into the first equation:
$$-3(-3y + 12) - 9y = 7$$
$$9y - 36 - 9y = 7$$
$$-36 = 7$$
Since this statement is false, there is no solution to the system. The system is inconsistent.

23. Substituting $y = 2x - 5$ into the first equation:

$$5x - 8(2x - 5) = 7$$
$$5x - 16x + 40 = 7$$
$$-11x + 40 = 7$$
$$-11x = -33$$
$$x = 3$$
$$y = 2(3) - 5 = 1$$

The solution is $(3,1)$.

25. Substituting $x = 2y - 1$ into the first equation:

$$7(2y - 1) - 6y = -1$$
$$14y - 7 - 6y = -1$$
$$8y - 7 = -1$$
$$8y = 6$$
$$y = \frac{3}{4}$$
$$x = 2\left(\frac{3}{4}\right) - 1 = \frac{3}{2} - 1 = \frac{1}{2}$$

The solution is $\left(\frac{1}{2}, \frac{3}{4}\right)$.

27. Substituting $y = 3x$ into the first equation:

$$-3x + 2(3x) = 6$$
$$-3x + 6x = 6$$
$$3x = 6$$
$$x = 2$$
$$y = 3(2) = 6$$

The solution is $(2,6)$.

29. Substituting $x = y$ into the first equation:

$$5(y) - 6y = -4$$
$$-y = -4$$
$$y = 4$$
$$x = 4$$

The solution is $(4,4)$.

31. Substituting $y = 2x - 12$ into the first equation:

$$3x + 3(2x - 12) = 9$$
$$3x + 6x - 36 = 9$$
$$9x = 45$$
$$x = 5$$
$$y = 2(5) - 12 = -2$$

The solution is $(5, -2)$.

33. Substituting $y = 10$ into the first equation:

$$7x - 11(10) = 16$$
$$7x - 110 = 16$$
$$7x = 126$$
$$x = 18$$
$$y = 10$$

The solution is $(18, 10)$.

35. Substituting $y = x - 2$ into the first equation:

$$-4(y + 2) + 4y = -8$$
$$-4y - 8 + 4y = -8$$
$$-8 = -8$$

Since this statement is true, the system is dependent. The two lines coincide. The solution is $\{(x,y) | y = x - 2\}$.

37. Substituting $y = 22 - x$ into the first equation:

$$0.05x + 0.10(22 - x) = 1.70$$
$$0.05x + 2.2 - 0.10x = 1.70$$
$$-0.05x + 2.2 = 1.7$$
$$-0.05x = -0.5$$
$$x = 10$$
$$y = 22 - 10 = 12$$

The solution is $(10, 12)$.

39. **a.** Setting the two equations equal:
$$\frac{1.40}{20}x+150 = \frac{1.40}{35}x+180$$
$$140\left(\frac{1.40}{20}x+150\right) = 140\left(\frac{1.40}{35}x+180\right)$$
$$9.8x+21{,}000 = 5.6x+25{,}200$$
$$4.2x+21{,}000 = 25{,}200$$
$$4.2x = 4{,}200$$
$$x = 1{,}000$$

At 1,000 miles the car and truck cost the same to operate.

b. If Daniel drives more than 1,200 miles, the car will be cheaper to operate.

c. If Daniel drives less than 800 miles, the truck will be cheaper to operate.

d. The graphs appear in the first quadrant only because all quantities are positive.

41. Let x and $5x + 8$ represent the two numbers. The equation is:
$$x+5x+8 = 26$$
$$6x+8 = 26$$
$$6x = 18$$
$$x = 3$$
$$5x+8 = 23$$

The two numbers are 3 and 23.

43. Let x represent the smaller number and $x + 9$ represent the larger number. The equation is:
$$x+9 = 2x-6$$
$$-x+9 = -6$$
$$-x = -15$$
$$x = 15$$
$$x+9 = 24$$

The two numbers are 15 and 24.

45. Let w represent the width and $3w + 5$ represent the length. Using the perimeter formula:
$$2(w)+2(3w+5) = 58$$
$$2w+6w+10 = 58$$
$$8w+10 = 58$$
$$8w = 48$$
$$w = 6$$
$$3w+5 = 23$$

The width is 6 inches and the length is 23 inches.

47. Completing the table:

	Nickels	Dimes
Number	$x+4$	x
Value (cents)	$5(x+4)$	$10x$

The equation is:
$$5(x+4)+10x = 170$$
$$5x+20+10x = 170$$
$$15x+20 = 170$$
$$15x = 150$$
$$x = 10$$
$$x+4 = 14$$

John has 14 nickels and 10 dimes.

3.4 Systems of Equations in Three Variables

1. Adding the first two equations and the first and third equations results in the system:
$$2x + 3z = 5$$
$$2x - 2z = 0$$
Solving the second equation yields $x = z$, now substituting:
$$2z + 3z = 5$$
$$5z = 5$$
$$z = 1$$
So $x = 1$, now substituting into the original first equation:
$$1 + y + 1 = 4$$
$$y + 2 = 4$$
$$y = 2$$
The solution is $(1, 2, 1)$.

3. Adding the first two equations and the first and third equations results in the system:
$$2x + 3z = 13$$
$$3x - 3z = -3$$
Adding yields:
$$5x = 10$$
$$x = 2$$
Substituting to find z:
$$2(2) + 3z = 13$$
$$4 + 3z = 13$$
$$3z = 9$$
$$z = 3$$
Substituting into the original first equation:
$$2 + y + 3 = 6$$
$$y + 5 = 6$$
$$y = 1$$
The solution is $(2, 1, 3)$.

5. Adding the second and third equations:
$$5x + z = 11$$
Multiplying the second equation by 2:
$$x + 2y + z = 3$$
$$4x - 2y + 4z = 12$$
Adding yields:
$$5x + 5z = 15$$
$$x + z = 3$$
So the system becomes:
$$5x + z = 11$$
$$x + z = 3$$
Multiply the second equation by -1:
$$5x + z = 11$$
$$-x - z = -3$$
Adding yields:
$$4x = 8$$
$$x = 2$$
Substituting to find z:
$$5(2) + z = 11$$
$$z + 10 = 11$$
$$z = 1$$

Substituting into the original first equation:
$$2 + 2y + 1 = 3$$
$$2y + 3 = 3$$
$$2y = 0$$
$$y = 0$$

The solution is $(2, 0, 1)$.

7. Multiply the second equation by –1 and add it to the first equation:
$$2x + 3y - 2z = 4$$
$$-x - 3y + 3z = -4$$

Adding results in the equation $x + z = 0$. Multiply the second equation by 2 and add it to the third equation:
$$2x + 6y - 6z = 8$$
$$3x - 6y + z = -3$$

Adding results in the equation:
$$5x - 5z = 5$$
$$x - z = 1$$

So the system becomes:
$$x - z = 1$$
$$x + z = 0$$

Adding yields:
$$2x = 1$$
$$x = \frac{1}{2}$$

Substituting to find z:
$$\frac{1}{2} + z = 0$$
$$z = -\frac{1}{2}$$

Substituting into the original first equation:
$$2\left(\frac{1}{2}\right) + 3y - 2\left(-\frac{1}{2}\right) = 4$$
$$1 + 3y + 1 = 4$$
$$3y + 2 = 4$$
$$3y = 2$$
$$y = \frac{2}{3}$$

The solution is $\left(\frac{1}{2}, \frac{2}{3}, -\frac{1}{2}\right)$.

9. Multiply the first equation by 2 and add it to the second equation:
$$-2x + 8y - 6z = 4$$
$$2x - 8y + 6z = 1$$

Adding yields 0 = 5, which is false. There is no solution (inconsistent system).

11. To clear the system of fractions, multiply the first equation by 2 and the second equation by 3:
$$x - 2y + 2z = 0$$
$$6x + y + 3z = 6$$
$$x + y + z = -4$$
Multiply the third equation by 2 and add it to the first equation:
$$x - 2y + 2z = 0$$
$$2x + 2y + 2z = -8$$
Adding yields the equation $3x + 4z = -8$. Multiply the third equation by –1 and add it to the second equation:
$$6x + y + 3z = 6$$
$$-x - y - z = 4$$
Adding yields the equation $5x + 2z = 10$. So the system becomes:
$$3x + 4z = -8$$
$$5x + 2z = 10$$
Multiply the second equation by –2:
$$3x + 4z = -8$$
$$-10x - 4z = -20$$
Adding yields:
$$-7x = -28$$
$$x = 4$$
Substituting to find z:
$$3(4) + 4z = -8$$
$$12 + 4z = -8$$
$$4z = -20$$
$$z = -5$$
Substituting into the original third equation:
$$4 + y - 5 = -4$$
$$y - 1 = -4$$
$$y = -3$$
The solution is $(4, -3, -5)$.

13. Multiply the first equation by –2 and add it to the third equation:
$$-4x + 2y + 6z = -2$$
$$4x - 2y - 6z = 2$$
Adding yields 0 = 0, which is true. Since there are now less equations than unknowns, there is no unique solution (dependent system).

15. Multiply the second equation by 3 and add it to the first equation:
$$2x - y + 3z = 4$$
$$3x + 6y - 3z = -9$$
Adding yields the equation $5x + 5y = -5$, or $x + y = -1$.

Multiply the second equation by 2 and add it to the third equation:
$$2x + 4y - 2z = -6$$
$$4x + 3y + 2z = -5$$
Adding yields the equation $6x + 7y = -11$. So the system becomes:
$$6x + 7y = -11$$
$$x + y = -1$$
Multiply the second equation by –6:
$$6x + 7y = -11$$
$$-6x - 6y = 6$$

Adding yields $y = -5$. Substituting to find x:
$$6x + 7(-5) = -11$$
$$6x - 35 = -11$$
$$6x = 24$$
$$x = 4$$
Substituting into the original first equation:
$$2(4) - (-5) + 3z = 4$$
$$13 + 3z = 4$$
$$3z = -9$$
$$z = -3$$
The solution is $(4, -5, -3)$.

17. Adding the second and third equations results in the equation $x + y = 9$. Since this is the same as the first equation, there are less equations than unknowns. There is no unique solution (dependent system).

19. Adding the second and third equations results in the equation $4x + y = 3$. So the system becomes:
$$4x + y = 3$$
$$2x + y = 2$$
Multiplying the second equation by -1:
$$4x + y = 3$$
$$-2x - y = -2$$
Adding yields:
$$2x = 1$$
$$x = \frac{1}{2}$$
Substituting to find y:
$$2\left(\frac{1}{2}\right) + y = 2$$
$$1 + y = 2$$
$$y = 1$$
Substituting into the original second equation:
$$1 + z = 3$$
$$z = 2$$
The solution is $\left(\frac{1}{2}, 1, 2\right)$.

21. Multiply the third equation by 2 and adding it to the second equation:
$$6y - 4z = 1$$
$$2x + 4z = 2$$
Adding yields the equation $2x + 6y = 3$. So the system becomes:
$$2x - 3y = 0$$
$$2x + 6y = 3$$
Multiply the first equation by 2:
$$4x - 6y = 0$$
$$2x + 6y = 3$$
Adding yields:
$$6x = 3$$
$$x = \frac{1}{2}$$

Substituting to find y:
$$2\left(\frac{1}{2}\right) + 6y = 3$$
$$1 + 6y = 3$$
$$6y = 2$$
$$y = \frac{1}{3}$$

Substituting into the original third equation to find z:
$$\frac{1}{2} + 2z = 1$$
$$2z = \frac{1}{2}$$
$$z = \frac{1}{4}$$

The solution is $\left(\frac{1}{2}, \frac{1}{3}, \frac{1}{4}\right)$.

23. Multiply the first equation by -2 and add it to the second equation:
$$-2x - 2y + 2z = -4$$
$$2x + y + 3z = 4$$

Adding yields $-y + 5z = 0$. Multiply the first equation by -1 and add it to the third equation:
$$-x - y + z = -2$$
$$x - 2y + 2z = 6$$

Adding yields $-3y + 3z = 4$. So the system becomes:
$$-y + 5z = 0$$
$$-3y + 3z = 4$$

Multiply the first equation by -3:
$$3y - 15z = 0$$
$$-3y + 3z = 4$$

Adding yields:
$$-12z = 4$$
$$z = -\frac{1}{3}$$

Substituting to find y:
$$-3y + 3\left(-\frac{1}{3}\right) = 4$$
$$-3y - 1 = 4$$
$$-3y = 5$$
$$y = -\frac{5}{3}$$

Substituting into the original first equation:
$$x - \frac{5}{3} + \frac{1}{3} = 2$$
$$x - \frac{4}{3} = 2$$
$$x = \frac{10}{3}$$

The solution is $\left(\frac{10}{3}, -\frac{5}{3}, -\frac{1}{3}\right)$.

25. Multiply the third equation by −1 and add it to the first equation:
$$2x + 3y = -\frac{1}{2}$$
$$-3y - 2z = \frac{3}{4}$$
Adding yields the equation $2x - 2z = \frac{1}{4}$. So the system becomes:
$$2x - 2z = \frac{1}{4}$$
$$4x + 8z = 2$$
Multiply the first equation by 4:
$$8x - 8z = 1$$
$$4x + 8z = 2$$
Adding yields:
$$12x = 3$$
$$x = \frac{1}{4}$$
Substituting to find z:
$$4\left(\frac{1}{4}\right) + 8z = 2$$
$$1 + 8z = 2$$
$$8z = 1$$
$$z = \frac{1}{8}$$
Substituting to find y:
$$2\left(\frac{1}{4}\right) + 3y = -\frac{1}{2}$$
$$\frac{1}{2} + 3y = -\frac{1}{2}$$
$$3y = -1$$
$$y = -\frac{1}{3}$$
The solution is $\left(\frac{1}{4}, -\frac{1}{3}, \frac{1}{8}\right)$.

27. To clear each equation of fractions, multiply the first equation by 6, the second equation by 4, and the third equation by 12:
$$2x + 3y - z = 24$$
$$x - 3y + 2z = 6$$
$$6x - 8y - 3z = -64$$
Multiply the first equation by 2 and add it to the second equation:
$$4x + 6y - 2z = 48$$
$$x - 3y + 2z = 6$$
Adding yields the equation $5x + 3y = 54$. Multiply the first equation by −3 and add it to the third equation:
$$-6x - 9y + 3z = -72$$
$$6x - 8y - 3z = -64$$

Adding yields:
$$-17y = -136$$
$$y = 8$$
Substituting to find x:
$$5x + 3(8) = 54$$
$$5x + 24 = 54$$
$$5x = 30$$
$$x = 6$$
Substituting to find z:
$$6 - 3(8) + 2z = 6$$
$$-18 + 2z = 6$$
$$2z = 24$$
$$z = 12$$
The solution is $(6, 8, 12)$.

29. To clear each equation of fractions, multiply the first equation by 6, the second equation by 6, and the third equation by 12:
$$6x - 3y - 2z = -8$$
$$2x - 3z = 30$$
$$-3x + 8y - 12z = -9$$
Multiply the first equation by 8 and the third equation by 3:
$$48x - 24y - 16z = -64$$
$$-9x + 24y - 36z = -27$$
Adding yields the equation:
$$39x - 52z = -91$$
$$3x - 4z = -7$$
So the system becomes:
$$2x - 3z = 30$$
$$3x - 4z = -7$$
Multiply the first equation by 3 and the second equation by –2:
$$6x - 9z = 90$$
$$-6x + 8z = 14$$
Adding yields:
$$-z = 104$$
$$z = -104$$
Substituting to find x:
$$2x - 3(-104) = 30$$
$$2x + 312 = 30$$
$$2x = -282$$
$$x = -141$$
Substituting to find y:
$$6(-141) - 3y - 2(-104) = -8$$
$$-846 - 3y + 208 = -8$$
$$-3y - 638 = -8$$
$$-3y = 630$$
$$y = -210$$
The solution is $(-141, -210, -104)$.

31. Divide the second equation by 5 and the third equation by 10 to produce the system:
$$x - y - z = 0$$
$$x + 4y = 16$$
$$2y - z = 5$$
Multiply the third equation by –1 and add it to the first equation:
$$x - y - z = 0$$
$$-2y + z = -5$$
Adding yields the equation $x - 3y = -5$. So the system becomes:
$$x + 4y = 16$$
$$x - 3y = -5$$
Multiply the second equation by –1:
$$x + 4y = 16$$
$$-x + 3y = 5$$
Adding yields:
$$7y = 21$$
$$y = 3$$
Substituting to find x:
$$x + 12 = 16$$
$$x = 4$$
Substituting to find z:
$$6 - z = 5$$
$$z = 1$$
The currents are 4 amps, 3 amps, and 1 amp.

33. Translating into symbols: $3x + 2$

35. Simplifying: $10(0.2x + 0.5y) = 2x + 5y$

37. Solving the equation:
$$x + (3x + 2) = 26$$
$$4x + 2 = 26$$
$$4x = 24$$
$$x = 6$$

39. Adding the two equations results in:
$$-9y = 9$$
$$y = -1$$
Substituting into the second equation:
$$-7(-1) + 4z = 27$$
$$7 + 4z = 27$$
$$4z = 20$$
$$z = 5$$
The solution is $(-1, 5)$.

41. Let x represent the smallest number, y represent the middle number, and z represent the larger number. The system of equations is:
$$x+y+z=20$$
$$z=x+y$$
$$y=2x+1$$

Substituting for y in equation 2: $z=x+(2x+1)=3x+1$

Substituting into equation 1:
$$x+(2x+1)+(3x+1)=20$$
$$6x+2=20$$
$$6x=18$$
$$x=3$$
$$y=2(3)+1=7$$
$$z=3+7=10$$

The numbers are 3, 7, and 10.

43. Let x represent the smallest piece, y represent the middle piece, and z represent the larger piece. The system of equations is:
$$x+y+z=20$$
$$z=x+y+2$$
$$y=2x$$

Substituting for y in equation 2: $z=x+(2x)+2=3x+2$

Substituting into equation 1:
$$x+(2x)+(3x+2)=20$$
$$6x+2=20$$
$$6x=18$$
$$x=3$$
$$y=2(3)=6$$
$$z=3+6+2=11$$

The pieces are 3 feet, 6 feet, and 11 feet in length.

3.5 Matrix Solutions to Linear Systems

1. Form the augmented matrix: $\begin{bmatrix} 1 & 1 & | & 5 \\ 3 & -1 & | & 3 \end{bmatrix}$

Add -3 times row 1 to row 2: $\begin{bmatrix} 1 & 1 & | & 5 \\ 0 & -4 & | & -12 \end{bmatrix}$

Dividing row 2 by -4: $\begin{bmatrix} 1 & 1 & | & 5 \\ 0 & 1 & | & 3 \end{bmatrix}$

So $y = 3$. Substituting to find x:
$$x+3=5$$
$$x=2$$

The solution is $(2,3)$.

3. Using the second equation as row 1, form the augmented matrix: $\begin{bmatrix} -1 & 1 & | & -1 \\ 3 & -5 & | & 7 \end{bmatrix}$

Add 3 times row 1 to row 2: $\begin{bmatrix} -1 & 1 & | & -1 \\ 0 & -2 & | & 4 \end{bmatrix}$

Dividing row 2 by –2 and row 1 by –1: $\begin{bmatrix} 1 & -1 & | & 1 \\ 0 & 1 & | & -2 \end{bmatrix}$

So $y = -2$. Substituting to find x:
$$x + 2 = 1$$
$$x = -1$$
The solution is $(-1, -2)$.

5. Form the augmented matrix: $\begin{bmatrix} 2 & -8 & | & 6 \\ 3 & -8 & | & 13 \end{bmatrix}$

Dividing row 1 by 2: $\begin{bmatrix} 1 & -4 & | & 3 \\ 3 & -8 & | & 13 \end{bmatrix}$

Add –3 times row 1 to row 2: $\begin{bmatrix} 1 & -4 & | & 3 \\ 0 & 4 & | & 4 \end{bmatrix}$

Dividing row 2 by 4: $\begin{bmatrix} 1 & -4 & | & 3 \\ 0 & 1 & | & 1 \end{bmatrix}$

So $y = 1$. Substituting to find x:
$$x - 4 = 3$$
$$x = 7$$
The solution is $(7, 1)$.

7. Form the augmented matrix: $\begin{bmatrix} 2 & -1 & | & -10 \\ 4 & 3 & | & 0 \end{bmatrix}$

Dividing row 1 by 2: $\begin{bmatrix} 1 & -\frac{1}{2} & | & -5 \\ 4 & 3 & | & 0 \end{bmatrix}$

Add –4 times row 1 to row 2: $\begin{bmatrix} 1 & -\frac{1}{2} & | & -5 \\ 0 & 5 & | & 20 \end{bmatrix}$

Divide row 2 by 5: $\begin{bmatrix} 1 & -\frac{1}{2} & | & -5 \\ 0 & 1 & | & 4 \end{bmatrix}$

So $y = 4$. Substituting to find x:
$$x - \frac{1}{2}(4) = -5$$
$$x - 2 = -5$$
$$x = -3$$
The solution is $(-3, 4)$.

9. Form the augmented matrix: $\begin{bmatrix} 5 & -3 & | & 27 \\ 6 & 2 & | & -18 \end{bmatrix}$

Dividing row 1 by 5: $\begin{bmatrix} 1 & -\frac{3}{5} & | & \frac{27}{5} \\ 6 & 2 & | & -18 \end{bmatrix}$

Add –6 times row 1 to row 2: $\begin{bmatrix} 1 & -\frac{3}{5} & | & \frac{27}{5} \\ 0 & \frac{28}{5} & | & -\frac{252}{5} \end{bmatrix}$

Multiply row 2 by $\frac{5}{28}$: $\begin{bmatrix} 1 & -\frac{3}{5} & | & \frac{27}{5} \\ 0 & 1 & | & -9 \end{bmatrix}$

So $y = -9$. Substituting to find x:
$$x - \frac{3}{5}(-9) = \frac{27}{5}$$
$$x + \frac{27}{5} = \frac{27}{5}$$
$$x = 0$$

The solution is $(0, -9)$.

11. First rewrite the system as:
$$5x + 2y = -14$$
$$-2x + y = 11$$

Form the augmented matrix: $\begin{bmatrix} 5 & 2 & | & -14 \\ -2 & 1 & | & 11 \end{bmatrix}$

Dividing row 1 by 5: $\begin{bmatrix} 1 & \frac{2}{5} & | & -\frac{14}{5} \\ -2 & 1 & | & 11 \end{bmatrix}$

Add 2 times row 1 to row 2: $\begin{bmatrix} 1 & \frac{2}{5} & | & -\frac{14}{5} \\ 0 & \frac{9}{5} & | & \frac{27}{5} \end{bmatrix}$

Multiply row 2 by $\frac{5}{9}$: $\begin{bmatrix} 1 & \frac{2}{5} & | & -\frac{14}{5} \\ 0 & 1 & | & 3 \end{bmatrix}$

So $y = 3$. Substituting to find x:
$$x + \frac{2}{5}(3) = -\frac{14}{5}$$
$$x + \frac{6}{5} = -\frac{14}{5}$$
$$x = -4$$

The solution is $(-4, 3)$.

13. Form the augmented matrix: $\begin{bmatrix} 2 & 3 & | & 11 \\ -1 & -1 & | & -2 \end{bmatrix}$

Dividing row 1 by 2: $\begin{bmatrix} 1 & \frac{3}{2} & | & \frac{11}{2} \\ -1 & -1 & | & -2 \end{bmatrix}$

Add row 1 to row 2: $\begin{bmatrix} 1 & \frac{3}{2} & | & \frac{11}{2} \\ 0 & \frac{1}{2} & | & \frac{7}{2} \end{bmatrix}$

Multiply row 2 by 2: $\begin{bmatrix} 1 & \frac{3}{2} & | & \frac{11}{2} \\ 0 & 1 & | & 7 \end{bmatrix}$

So $y = 7$. Substituting to find x:
$$x + \frac{3}{2}(7) = \frac{11}{2}$$
$$x + \frac{21}{2} = \frac{11}{2}$$
$$x = -5$$

The solution is $(-5, 7)$.

15. Form the augmented matrix: $\begin{bmatrix} 3 & -2 & | & 16 \\ 4 & 3 & | & -24 \end{bmatrix}$

Dividing row 1 by 3: $\begin{bmatrix} 1 & -\frac{2}{3} & | & \frac{16}{3} \\ 4 & 3 & | & -24 \end{bmatrix}$

Add –4 times row 1 to row 2: $\begin{bmatrix} 1 & -\frac{2}{3} & | & \frac{16}{3} \\ 0 & \frac{17}{3} & | & -\frac{136}{3} \end{bmatrix}$

Multiply row 2 by $\frac{3}{17}$: $\begin{bmatrix} 1 & -\frac{2}{3} & | & \frac{16}{3} \\ 0 & 1 & | & -8 \end{bmatrix}$

So $y = -8$. Substituting to find x:
$$x - \frac{2}{3}(-8) = \frac{16}{3}$$
$$x + \frac{16}{3} = \frac{16}{3}$$
$$x = 0$$

The solution is $(0, -8)$.

17. First rewrite the system as:
$$3x - 2y = 16$$
$$-2x + y = -12$$

Form the augmented matrix: $\begin{bmatrix} 3 & -2 & | & 16 \\ -2 & 1 & | & -12 \end{bmatrix}$

Divide row 1 by 3: $\begin{bmatrix} 1 & -\frac{2}{3} & | & \frac{16}{3} \\ -2 & 1 & | & -12 \end{bmatrix}$

Add 2 times row 1 to row 2: $\begin{bmatrix} 1 & -\frac{2}{3} & | & \frac{16}{3} \\ 0 & -\frac{1}{3} & | & -\frac{4}{3} \end{bmatrix}$

Multiplying row 2 by –3: $\begin{bmatrix} 1 & -\frac{2}{3} & | & \frac{16}{3} \\ 0 & 1 & | & 4 \end{bmatrix}$

So $y = 4$. Substituting to find x:
$$x - \frac{2}{3}(4) = \frac{16}{3}$$
$$x - \frac{8}{3} = \frac{16}{3}$$
$$x = 8$$

The solution is $(8, 4)$.

19. Form the augmented matrix: $\begin{bmatrix} 1 & 1 & 1 & | & 4 \\ 1 & -1 & 2 & | & 1 \\ 1 & -1 & -1 & | & -2 \end{bmatrix}$

Add –1 times row 1 to both row 2 and row 3: $\begin{bmatrix} 1 & 1 & 1 & | & 4 \\ 0 & -2 & 1 & | & -3 \\ 0 & -2 & -2 & | & -6 \end{bmatrix}$

Multiply row 2 by –1 and add it to row 3: $\begin{bmatrix} 1 & 1 & 1 & | & 4 \\ 0 & 2 & -1 & | & 3 \\ 0 & 0 & -3 & | & -3 \end{bmatrix}$

Divide row 3 by –3: $\begin{bmatrix} 1 & 1 & 1 & | & 4 \\ 0 & 2 & -1 & | & 3 \\ 0 & 0 & 1 & | & 1 \end{bmatrix}$

So $z = 1$. Substituting to find y:
$$2y - 1 = 3$$
$$2y = 4$$
$$y = 2$$

Substituting to find x:
$$x + 2 + 1 = 4$$
$$x = 1$$

The solution is $(1, 2, 1)$.

21. Form the augmented matrix: $\begin{bmatrix} 1 & 2 & 1 & | & 3 \\ 2 & -1 & 2 & | & 6 \\ 3 & 1 & -1 & | & 5 \end{bmatrix}$

Add –2 times row 1 to row 2 and –3 times row 1 to row 3: $\begin{bmatrix} 1 & 2 & 1 & | & 3 \\ 0 & -5 & 0 & | & 0 \\ 0 & -5 & -4 & | & -4 \end{bmatrix}$

Dividing row 2 by –5: $\begin{bmatrix} 1 & 2 & 1 & | & 3 \\ 0 & 1 & 0 & | & 0 \\ 0 & -5 & -4 & | & -4 \end{bmatrix}$

Adding 5 times row 2 to row 3: $\begin{bmatrix} 1 & 2 & 1 & | & 3 \\ 0 & 1 & 0 & | & 0 \\ 0 & 0 & -4 & | & -4 \end{bmatrix}$

Dividing row 3 by –4: $\begin{bmatrix} 1 & 2 & 1 & | & 3 \\ 0 & 1 & 0 & | & 0 \\ 0 & 0 & 1 & | & 1 \end{bmatrix}$

So $y = 0$ and $z = 1$. Substituting to find x:
$x + 0 + 1 = 3$
$x = 2$
The solution is $(2, 0, 1)$.

23. Form the augmented matrix: $\begin{bmatrix} 1 & -2 & 1 & | & -4 \\ 2 & 1 & -3 & | & 7 \\ 5 & -3 & 1 & | & -5 \end{bmatrix}$

Add –2 times row 1 to row 2 and –5 times row 1 to row 3: $\begin{bmatrix} 1 & -2 & 1 & | & -4 \\ 0 & 5 & -5 & | & 15 \\ 0 & 7 & -4 & | & 15 \end{bmatrix}$

Dividing row 2 by 5: $\begin{bmatrix} 1 & -2 & 1 & | & -4 \\ 0 & 1 & -1 & | & 3 \\ 0 & 7 & -4 & | & 15 \end{bmatrix}$

Add –7 times row 2 to row 3: $\begin{bmatrix} 1 & -2 & 1 & | & -4 \\ 0 & 1 & -1 & | & 3 \\ 0 & 0 & 3 & | & -6 \end{bmatrix}$

Dividing row 3 by 3: $\begin{bmatrix} 1 & -2 & 1 & | & -4 \\ 0 & 1 & -1 & | & 3 \\ 0 & 0 & 1 & | & -2 \end{bmatrix}$

So $z = -2$. Substituting to find y:
$y - (-2) = 3$
$y + 2 = 3$
$y = 1$

Substituting to find x:
$$x - 2(1) + (-2) = -4$$
$$x - 4 = -4$$
$$x = 0$$
The solution is $(0, 1, -2)$.

25. Form the augmented matrix (using the second equation as row 1): $\begin{bmatrix} 1 & -2 & -1 & | & 0 \\ 5 & -3 & 1 & | & 10 \\ 3 & -1 & 2 & | & 10 \end{bmatrix}$

Add -5 times row 1 to row 2 and -3 times row 1 to row 3: $\begin{bmatrix} 1 & -2 & -1 & | & 0 \\ 0 & 7 & 6 & | & 10 \\ 0 & 5 & 5 & | & 10 \end{bmatrix}$

Dividing row 5 by 5, and switching it with row 2: $\begin{bmatrix} 1 & -2 & -1 & | & 0 \\ 0 & 1 & 1 & | & 2 \\ 0 & 7 & 6 & | & 10 \end{bmatrix}$

Add -7 times row 2 to row 3: $\begin{bmatrix} 1 & -2 & -1 & | & 0 \\ 0 & 1 & 1 & | & 2 \\ 0 & 0 & -1 & | & -4 \end{bmatrix}$

Multiplying row 3 by -1: $\begin{bmatrix} 1 & -2 & -1 & | & 0 \\ 0 & 1 & 1 & | & 2 \\ 0 & 0 & 1 & | & 4 \end{bmatrix}$

So $z = 4$. Substituting to find y:
$$y + 4 = 2$$
$$y = -2$$
Substituting to find x:
$$x - 2(-2) - 4 = 0$$
$$x = 0$$
The solution is $(0, -2, 4)$.

27. Form the augmented matrix (using the third equation as row 1): $\begin{bmatrix} 1 & 1 & 2 & | & 5 \\ 2 & -5 & 3 & | & 2 \\ 3 & -7 & 1 & | & 0 \end{bmatrix}$

Add -2 times row 1 to row 2 and -3 times row 1 to row 3: $\begin{bmatrix} 1 & 1 & 2 & | & 5 \\ 0 & -7 & -1 & | & -8 \\ 0 & -10 & -5 & | & -15 \end{bmatrix}$

Dividing row 3 by -10 and switching it with row 2: $\begin{bmatrix} 1 & 1 & 2 & | & 5 \\ 0 & 1 & \frac{1}{2} & | & \frac{3}{2} \\ 0 & -7 & -1 & | & -8 \end{bmatrix}$

Add 7 times row 2 to row 3: $\begin{bmatrix} 1 & 1 & 2 & | & 5 \\ 0 & 1 & \frac{1}{2} & | & \frac{3}{2} \\ 0 & 0 & \frac{5}{2} & | & \frac{5}{2} \end{bmatrix}$

Multiplying row 3 by $\frac{2}{5}$: $\begin{bmatrix} 1 & 1 & 2 & | & 5 \\ 0 & 1 & \frac{1}{2} & | & \frac{3}{2} \\ 0 & 0 & 1 & | & 1 \end{bmatrix}$

So $z = 1$. Substituting to find y:
$$y + \frac{1}{2}(1) = \frac{3}{2}$$
$$y + \frac{1}{2} = \frac{3}{2}$$
$$y = 1$$

Substituting to find x:
$$x + 1(1) + 2(1) = 5$$
$$x + 3 = 5$$
$$x = 2$$

The solution is $(2,1,1)$.

29. Form the augmented matrix: $\begin{bmatrix} 1 & 2 & 0 & | & 3 \\ 0 & 1 & 1 & | & 3 \\ 4 & 0 & -1 & | & 2 \end{bmatrix}$

Add -4 times row 1 to row 3: $\begin{bmatrix} 1 & 2 & 0 & | & 3 \\ 0 & 1 & 1 & | & 3 \\ 0 & -8 & -1 & | & -10 \end{bmatrix}$

Add 8 times row 2 to row 3: $\begin{bmatrix} 1 & 2 & 0 & | & 3 \\ 0 & 1 & 1 & | & 3 \\ 0 & 0 & 7 & | & 14 \end{bmatrix}$

Dividing row 3 by 7: $\begin{bmatrix} 1 & 2 & 0 & | & 3 \\ 0 & 1 & 1 & | & 3 \\ 0 & 0 & 1 & | & 2 \end{bmatrix}$

So $z = 2$. Substituting to find y:
$$y + 2 = 3$$
$$y = 1$$

Substituting to find x:
$$x + 2 = 3$$
$$x = 1$$

The solution is $(1,1,2)$.

31. Form the augmented matrix: $\begin{bmatrix} 1 & 3 & 0 & | & 7 \\ 3 & 0 & -4 & | & -8 \\ 0 & 5 & -2 & | & -5 \end{bmatrix}$

Add –3 times row 1 to row 2: $\begin{bmatrix} 1 & 3 & 0 & | & 7 \\ 0 & -9 & -4 & | & -29 \\ 0 & 5 & -2 & | & -5 \end{bmatrix}$

Add 2 times row 3 to row 2: $\begin{bmatrix} 1 & 3 & 0 & | & 7 \\ 0 & 1 & -8 & | & -39 \\ 0 & 5 & -2 & | & -5 \end{bmatrix}$

Add –5 times row 2 to row 3: $\begin{bmatrix} 1 & 3 & 0 & | & 7 \\ 0 & 1 & -8 & | & -39 \\ 0 & 0 & 38 & | & 190 \end{bmatrix}$

Dividing row 3 by 38: $\begin{bmatrix} 1 & 3 & 0 & | & 7 \\ 0 & 1 & -8 & | & -39 \\ 0 & 0 & 1 & | & 5 \end{bmatrix}$

So $z = 5$. Substituting to find y:
$y - 40 = -39$
$y = 1$

Substituting to find x:
$x + 3 = 7$
$x = 4$

The solution is $(4, 1, 5)$.

33. Form the augmented matrix: $\begin{bmatrix} 1 & 4 & 0 & | & 13 \\ 2 & 0 & -5 & | & -3 \\ 0 & 4 & -3 & | & 9 \end{bmatrix}$

Add –2 times row 1 to row 2: $\begin{bmatrix} 1 & 4 & 0 & | & 13 \\ 0 & -8 & -5 & | & -29 \\ 0 & 4 & -3 & | & 9 \end{bmatrix}$

Dividing row 2 by –8: $\begin{bmatrix} 1 & 4 & 0 & | & 13 \\ 0 & 1 & \frac{5}{8} & | & \frac{29}{8} \\ 0 & 4 & -3 & | & 9 \end{bmatrix}$

Add –4 times row 2 to row 3: $\begin{bmatrix} 1 & 4 & 0 & | & 13 \\ 0 & 1 & \frac{5}{8} & | & \frac{29}{8} \\ 0 & 0 & -\frac{11}{2} & | & -\frac{11}{2} \end{bmatrix}$

Multiplying row 3 by $-\frac{2}{11}$: $\begin{bmatrix} 1 & 4 & 0 & | & 13 \\ 0 & 1 & \frac{5}{8} & | & \frac{29}{8} \\ 0 & 0 & 1 & | & 1 \end{bmatrix}$

So $z = 1$. Substituting to find y:
$$y + \frac{5}{8}(1) = \frac{29}{8}$$
$$y + \frac{5}{8} = \frac{29}{8}$$
$$y = \frac{24}{8} = 3$$

Substituting to find x:
$$x + 4(3) = 13$$
$$x + 12 = 13$$
$$x = 1$$

The solution is $(1, 3, 1)$.

35. Form the augmented matrix: $\begin{bmatrix} 1 & -2 & 1 & | & -5 \\ 2 & 3 & -2 & | & -9 \\ 2 & -1 & 2 & | & -1 \end{bmatrix}$

Add -2 times row 1 to row 2 and also to row 3: $\begin{bmatrix} 1 & -2 & 1 & | & -5 \\ 0 & 7 & -4 & | & 1 \\ 0 & 3 & 0 & | & 9 \end{bmatrix}$

Switching rows 2 and 3, and dividing the new row 2 by 3: $\begin{bmatrix} 1 & -2 & 1 & | & -5 \\ 0 & 1 & 0 & | & 3 \\ 0 & 7 & -4 & | & 1 \end{bmatrix}$

Add -7 times row 2 to row 3: $\begin{bmatrix} 1 & -2 & 1 & | & -5 \\ 0 & 1 & 0 & | & 3 \\ 0 & 0 & -4 & | & -20 \end{bmatrix}$

Dividing row 3 by -4: $\begin{bmatrix} 1 & -2 & 1 & | & -5 \\ 0 & 1 & 0 & | & 3 \\ 0 & 0 & 1 & | & 5 \end{bmatrix}$

So $z = 5$ and $y = 3$. Substituting to find x:
$$x - 2(3) + 1(5) = -5$$
$$x - 1 = -5$$
$$x = -4$$

The solution is $(-4, 3, 5)$.

37. Form the augmented matrix (using equation 2 as the first row): $\begin{bmatrix} 1 & 3 & -4 & | & 13 \\ 4 & -2 & -1 & | & -5 \\ 3 & -1 & -3 & | & 0 \end{bmatrix}$

Add −4 times row 1 to row 2 and −3 times row 1 to row 3: $\begin{bmatrix} 1 & 3 & -4 & | & 13 \\ 0 & -14 & 15 & | & -57 \\ 0 & -10 & 9 & | & -39 \end{bmatrix}$

Divide row 2 by −14: $\begin{bmatrix} 1 & 3 & -4 & | & 13 \\ 0 & 1 & -\frac{15}{14} & | & \frac{57}{14} \\ 0 & -10 & 9 & | & -39 \end{bmatrix}$

Add 10 times row 2 to row 3: $\begin{bmatrix} 1 & 3 & -4 & | & 13 \\ 0 & 1 & -\frac{15}{14} & | & \frac{57}{14} \\ 0 & 0 & -\frac{12}{7} & | & \frac{12}{7} \end{bmatrix}$

Multiplying row 3 by $-\frac{7}{12}$: $\begin{bmatrix} 1 & 3 & -4 & | & 13 \\ 0 & 1 & -\frac{15}{14} & | & \frac{57}{14} \\ 0 & 0 & 1 & | & -1 \end{bmatrix}$

So $z = -1$. Substituting to find y:

$$y - \frac{15}{14}(-1) = \frac{57}{14}$$

$$y + \frac{15}{14} = \frac{57}{14}$$

$$y = 3$$

Substituting to find x:

$$x + 3(3) - 4(-1) = 13$$

$$x + 13 = 13$$

$$x = 0$$

The solution is $(0, 3, -1)$.

39. Form the augmented matrix (using equation 2 as the first row): $\begin{bmatrix} 7 & -2 & 0 & | & 1 \\ 0 & 5 & 1 & | & 11 \\ 5 & 0 & 2 & | & -3 \end{bmatrix}$

Dividing row 1 by 7: $\begin{bmatrix} 1 & -\frac{2}{7} & 0 & | & \frac{1}{7} \\ 0 & 5 & 1 & | & 11 \\ 5 & 0 & 2 & | & -3 \end{bmatrix}$

Add -5 times row 1 to row 3: $\begin{bmatrix} 1 & -\frac{2}{7} & 0 & | & \frac{1}{7} \\ 0 & 5 & 1 & | & 11 \\ 0 & \frac{10}{7} & 2 & | & -\frac{26}{7} \end{bmatrix}$

Divide row 2 by 5: $\begin{bmatrix} 1 & -\frac{2}{7} & 0 & | & \frac{1}{7} \\ 0 & 1 & \frac{1}{5} & | & \frac{11}{5} \\ 0 & \frac{10}{7} & 2 & | & -\frac{26}{7} \end{bmatrix}$

Add $-\frac{10}{7}$ times row 2 to row 3: $\begin{bmatrix} 1 & -\frac{2}{7} & 0 & | & \frac{1}{7} \\ 0 & 1 & \frac{1}{5} & | & \frac{11}{5} \\ 0 & 0 & \frac{12}{7} & | & -\frac{48}{7} \end{bmatrix}$

Multiplying row 3 by $\frac{7}{12}$: $\begin{bmatrix} 1 & -\frac{2}{7} & 0 & | & \frac{1}{7} \\ 0 & 1 & \frac{1}{5} & | & \frac{11}{5} \\ 0 & 0 & 1 & | & -4 \end{bmatrix}$

So $z = -4$. Substituting to find y:

$$y + \frac{1}{5}(-4) = \frac{11}{5}$$
$$y - \frac{4}{5} = \frac{11}{5}$$
$$y = 3$$

Substituting to find x:

$$x - \frac{2}{7}(3) = \frac{1}{7}$$
$$x - \frac{6}{7} = \frac{1}{7}$$
$$x = 1$$

The solution is $(1, 3, -4)$.

41. Form the augmented matrix: $\begin{bmatrix} \frac{1}{3} & \frac{1}{5} & \Big| & 2 \\ \frac{1}{3} & -\frac{1}{2} & \Big| & -\frac{1}{3} \end{bmatrix}$

Multiplying row 1 by 15 and row 2 by 6: $\begin{bmatrix} 5 & 3 & | & 30 \\ 2 & -3 & | & -2 \end{bmatrix}$

Dividing row 2 by 2: $\begin{bmatrix} 5 & 3 & | & 30 \\ 1 & -\frac{3}{2} & | & -1 \end{bmatrix}$

Add -5 times row 2 to row 1: $\begin{bmatrix} 0 & \frac{21}{2} & | & 35 \\ 1 & -\frac{3}{2} & | & -1 \end{bmatrix}$

Multiplying row 1 by $\frac{2}{21}$: $\begin{bmatrix} 0 & 1 & | & \frac{10}{3} \\ 1 & -\frac{3}{2} & | & -1 \end{bmatrix}$

So $y = \frac{10}{3}$. Substituting to find x:

$$x - \frac{3}{2}\left(\frac{10}{3}\right) = -1$$
$$x - 5 = -1$$
$$x = 4$$

The solution is $\left(4, \frac{10}{3}\right)$.

43. Form the augmented matrix: $\begin{bmatrix} \frac{1}{3} & -\frac{1}{4} & | & 1 \\ \frac{1}{3} & \frac{1}{4} & | & 3 \end{bmatrix}$

Multiplying row 1 by 12 and row 2 by 12: $\begin{bmatrix} 4 & -3 & | & 12 \\ 4 & 3 & | & 36 \end{bmatrix}$

Dividing row 1 by 4: $\begin{bmatrix} 1 & -\frac{3}{4} & | & 3 \\ 4 & 3 & | & 36 \end{bmatrix}$

Add -4 times row 1 to row 2: $\begin{bmatrix} 1 & -\frac{3}{4} & | & 3 \\ 0 & 6 & | & 24 \end{bmatrix}$

Dividing row 2 by 6: $\begin{bmatrix} 1 & -\frac{3}{4} & | & 3 \\ 0 & 1 & | & 4 \end{bmatrix}$

So $y = 4$. Substituting to find x:

$$x - \frac{3}{4}(4) = 3$$
$$x - 3 = 3$$
$$x = 6$$

The solution is $(6, 4)$.

45. Form the augmented matrix: $\begin{bmatrix} 2 & -3 & | & 4 \\ 4 & -6 & | & 4 \end{bmatrix}$

Add –2 times row 1 to row 2: $\begin{bmatrix} 2 & -3 & | & 4 \\ 0 & 0 & | & -4 \end{bmatrix}$

The second row states that $0 = -4$, which is false. There is no solution.

47. Form the augmented matrix: $\begin{bmatrix} -6 & 4 & | & 8 \\ -3 & 2 & | & 4 \end{bmatrix}$

Divide row 1 by –2: $\begin{bmatrix} 3 & -2 & | & -4 \\ -3 & 2 & | & 4 \end{bmatrix}$

Adding row 1 to row 2: $\begin{bmatrix} 3 & -2 & | & -4 \\ 0 & 0 & | & 0 \end{bmatrix}$

The second row states that $0 = 0$, which is true. The system is dependent.

49. Graphing the line: 51. Graphing the line:

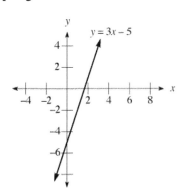
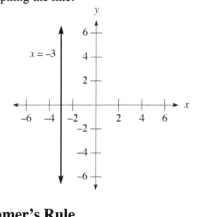

3.6 Determinants and Cramer's Rule

1. Evaluating the determinant: $\begin{vmatrix} 1 & 0 \\ 2 & 3 \end{vmatrix} = 1 \cdot 3 - 0 \cdot 2 = 3 - 0 = 3$

3. Evaluating the determinant: $\begin{vmatrix} 2 & 1 \\ 3 & 4 \end{vmatrix} = 2 \cdot 4 - 1 \cdot 3 = 8 - 3 = 5$

5. Evaluating the determinant: $\begin{vmatrix} 5 & 4 \\ 3 & 2 \end{vmatrix} = 5 \cdot 2 - 3 \cdot 4 = 10 - 12 = -2$

7. Evaluating the determinant: $\begin{vmatrix} 0 & 1 \\ 1 & 0 \end{vmatrix} = 0 \cdot 0 - 1 \cdot 1 = 0 - 1 = -1$

9. Evaluating the determinant: $\begin{vmatrix} -3 & 2 \\ 6 & -4 \end{vmatrix} = (-3) \cdot (-4) - 6 \cdot 2 = 12 - 12 = 0$

11. Evaluating the determinant: $\begin{vmatrix} -3 & -1 \\ 4 & -2 \end{vmatrix} = (-3) \cdot (-2) - (-1) \cdot 4 = 6 + 4 = 10$

13. Solving the equation:
$$\begin{vmatrix} 2x & 1 \\ x & 3 \end{vmatrix} = 10$$
$$6x - x = 10$$
$$5x = 10$$
$$x = 2$$

15. Solving the equation:
$$\begin{vmatrix} 1 & 2x \\ 2 & -3x \end{vmatrix} = 21$$
$$-3x - 4x = 21$$
$$-7x = 21$$
$$x = -3$$

17. Solving the equation:
$$\begin{vmatrix} 2x & -4 \\ x & 2 \end{vmatrix} = -16$$
$$2x(2) + 4x = -16$$
$$4x + 4x = -16$$
$$8x = -16$$
$$x = -2$$

19. Solving the equation:
$$\begin{vmatrix} 11x & -7x \\ 3 & -2 \end{vmatrix} = 3$$
$$11x(-2) + 7x(3) = 3$$
$$-22x + 21x = 3$$
$$-x = 3$$
$$x = -3$$

21. Solving the equation:
$$\begin{vmatrix} 2x & -4 \\ 2 & x \end{vmatrix} = -8x$$
$$2x(x) - 2(-4) = -8x$$
$$2x^2 + 8 = -8x$$
$$2x^2 + 8x + 8 = 0$$
$$x^2 + 4x + 4 = 0$$
$$(x+2)^2 = 0$$
$$x = -2$$

23. Solving the equation:
$$\begin{vmatrix} x^2 & 3 \\ x & 1 \end{vmatrix} = 10$$
$$x^2(1) - 3(x) = 10$$
$$x^2 - 3x = 10$$
$$x^2 - 3x - 10 = 0$$
$$(x+2)(x-5) = 0$$
$$x = -2, 5$$

25. Solving the equation:
$$\begin{vmatrix} x^2 & -4 \\ x & 1 \end{vmatrix} = 32$$
$$x^2(1) - (-4)(x) = 32$$
$$x^2 + 4x = 32$$
$$x^2 + 4x - 32 = 0$$
$$(x+8)(x-4) = 0$$
$$x = -8, 4$$

27. Solving the equation:
$$\begin{vmatrix} x & 5 \\ 1 & x \end{vmatrix} = 4$$
$$x(x) - (1)(5) = 4$$
$$x^2 - 5 = 4$$
$$x^2 - 9 = 0$$
$$(x+3)(x-3) = 0$$
$$x = -3, 3$$

29. First find the determinants:
$$D = \begin{vmatrix} 2 & -3 \\ 4 & -2 \end{vmatrix} = -4 + 12 = 8$$
$$D_x = \begin{vmatrix} 3 & -3 \\ 10 & -2 \end{vmatrix} = -6 + 30 = 24$$
$$D_y = \begin{vmatrix} 2 & 3 \\ 4 & 10 \end{vmatrix} = 20 - 12 = 8$$

Now use Cramer's rule:
$$x = \frac{D_x}{D} = \frac{24}{8} = 3 \qquad y = \frac{D_y}{D} = \frac{8}{8} = 1$$

The solution is $(3,1)$.

31. First find the determinants:
$$D = \begin{vmatrix} 5 & -2 \\ -10 & 4 \end{vmatrix} = 20 - 20 = 0$$
$$D_x = \begin{vmatrix} 4 & -2 \\ 1 & 4 \end{vmatrix} = 16 + 2 = 18$$
$$D_y = \begin{vmatrix} 5 & 4 \\ -10 & 1 \end{vmatrix} = 5 + 40 = 45$$

Since $D = 0$ and other determinants are nonzero, there is no solution, or $\emptyset$.

33. First find the determinants:
$$D = \begin{vmatrix} 4 & -7 \\ 5 & 2 \end{vmatrix} = 8 + 35 = 43$$
$$D_x = \begin{vmatrix} 3 & -7 \\ -3 & 2 \end{vmatrix} = 6 - 21 = -15$$
$$D_y = \begin{vmatrix} 4 & 3 \\ 5 & -3 \end{vmatrix} = -12 - 15 = -27$$

Now use Cramer's rule:
$$x = \frac{D_x}{D} = -\frac{15}{43} \qquad y = \frac{D_y}{D} = -\frac{27}{43}$$

The solution is $\left(-\frac{15}{43}, -\frac{27}{43}\right)$.

35. First find the determinants:
$$D = \begin{vmatrix} 9 & -8 \\ 2 & 3 \end{vmatrix} = 27 + 16 = 43$$
$$D_x = \begin{vmatrix} 4 & -8 \\ 6 & 3 \end{vmatrix} = 12 + 48 = 60$$
$$D_y = \begin{vmatrix} 9 & 4 \\ 2 & 6 \end{vmatrix} = 54 - 8 = 46$$

Now use Cramer's rule:
$$x = \frac{D_x}{D} = \frac{60}{43} \qquad y = \frac{D_y}{D} = \frac{46}{43}$$

The solution is $\left(\frac{60}{43}, \frac{46}{43}\right)$.

37. First find the determinants:
$$D = \begin{vmatrix} 3 & 2 \\ 4 & -5 \end{vmatrix} = -15 - 8 = -23$$
$$D_x = \begin{vmatrix} 6 & 2 \\ 8 & -5 \end{vmatrix} = -30 - 16 = -46$$
$$D_y = \begin{vmatrix} 3 & 6 \\ 4 & 8 \end{vmatrix} = 24 - 24 = 0$$

Now use Cramer's rule:
$$x = \frac{D_x}{D} = \frac{-46}{-23} = 2 \qquad y = \frac{D_y}{D} = \frac{0}{-23} = 0$$

The solution is $(2, 0)$.

39. First find the determinants:
$$D = \begin{vmatrix} 12 & -13 \\ 11 & 15 \end{vmatrix} = 180 + 143 = 323$$
$$D_x = \begin{vmatrix} 16 & -13 \\ 18 & 15 \end{vmatrix} = 240 + 234 = 474$$
$$D_y = \begin{vmatrix} 12 & 16 \\ 11 & 18 \end{vmatrix} = 216 - 176 = 40$$

Now use Cramer's rule:
$$x = \frac{D_x}{D} = \frac{474}{323} \qquad y = \frac{D_y}{D} = \frac{40}{323}$$

The solution is $\left(\frac{474}{323}, \frac{40}{323}\right)$.

41. Duplicating the first two columns:
$$\begin{vmatrix} 1 & 2 & 0 \\ 0 & 2 & 1 \\ 1 & 1 & 1 \end{vmatrix} \begin{matrix} 1 & 2 \\ 0 & 2 \\ 1 & 1 \end{matrix} = 1 \cdot 2 \cdot 1 + 2 \cdot 1 \cdot 1 + 0 \cdot 0 \cdot 1 - 1 \cdot 2 \cdot 0 - 1 \cdot 1 \cdot 1 - 1 \cdot 0 \cdot 2 = 2 + 2 + 0 - 0 - 1 - 0 = 3$$

43. Duplicating the first two columns:
$$\begin{vmatrix} 1 & 2 & 3 \\ 3 & 2 & 1 \\ 1 & 1 & 1 \end{vmatrix} \begin{matrix} 1 & 2 \\ 3 & 2 \\ 1 & 1 \end{matrix} = 1 \cdot 2 \cdot 1 + 2 \cdot 1 \cdot 1 + 3 \cdot 3 \cdot 1 - 1 \cdot 2 \cdot 3 - 1 \cdot 1 \cdot 1 - 1 \cdot 3 \cdot 2 = 2 + 2 + 9 - 6 - 1 - 6 = 0$$

45. Expanding across the first row:
$$\begin{vmatrix} 0 & 1 & 2 \\ 1 & 0 & 1 \\ -1 & 2 & 0 \end{vmatrix} = 0 \begin{vmatrix} 0 & 1 \\ 2 & 0 \end{vmatrix} - 1 \begin{vmatrix} 1 & 1 \\ -1 & 0 \end{vmatrix} + 2 \begin{vmatrix} 1 & 0 \\ -1 & 2 \end{vmatrix} = 0(0 - 2) - 1(0 + 1) + 2(2 - 0) = 0 - 1 + 4 = 3$$

47. Expanding across the first row:

$$\begin{vmatrix} 3 & 0 & 2 \\ 0 & -1 & -1 \\ 4 & 0 & 0 \end{vmatrix} = 3\begin{vmatrix} -1 & -1 \\ 0 & 0 \end{vmatrix} - 0\begin{vmatrix} 0 & -1 \\ 4 & 0 \end{vmatrix} + 2\begin{vmatrix} 0 & -1 \\ 4 & 0 \end{vmatrix} = 3(0-0) - 0(0+4) + 2(0+4) = 0 - 0 + 8 = 8$$

49. Expanding across the first row: $\begin{vmatrix} 2 & -1 & 0 \\ 1 & 0 & -2 \\ 0 & 1 & 2 \end{vmatrix} = 2\begin{vmatrix} 0 & -2 \\ 1 & 2 \end{vmatrix} + 1\begin{vmatrix} 1 & -2 \\ 0 & 2 \end{vmatrix} + 0\begin{vmatrix} 1 & 0 \\ 0 & 1 \end{vmatrix} = 2(0+2) + 1(2-0) + 0 = 4 + 2 = 6$

51. Expanding across the first row:

$$\begin{vmatrix} 1 & 3 & 7 \\ -2 & 6 & 4 \\ 3 & 7 & -1 \end{vmatrix} = 1\begin{vmatrix} 6 & 4 \\ 7 & -1 \end{vmatrix} - 3\begin{vmatrix} -2 & 4 \\ 3 & -1 \end{vmatrix} + 7\begin{vmatrix} -2 & 6 \\ 3 & 7 \end{vmatrix} = 1(-6-28) - 3(2-12) + 7(-14-18) = -34 + 30 - 224 = -228$$

53. Expanding across the first row:

$$\begin{vmatrix} -2 & 0 & 1 \\ 0 & 3 & 2 \\ 1 & 0 & -5 \end{vmatrix} = -2\begin{vmatrix} 3 & 2 \\ 0 & -5 \end{vmatrix} - 0\begin{vmatrix} 0 & 2 \\ 1 & -5 \end{vmatrix} + 1\begin{vmatrix} 0 & 3 \\ 1 & 0 \end{vmatrix} = 2(-15-0) - 0 + 1(0-3) = 30 - 3 = 27$$

55. Expanding across the first row:

$$\begin{vmatrix} -2 & 4 & -1 \\ 0 & 3 & 1 \\ -5 & -2 & 3 \end{vmatrix} = -2\begin{vmatrix} 3 & 1 \\ -2 & 3 \end{vmatrix} - 4\begin{vmatrix} 0 & 1 \\ -5 & 3 \end{vmatrix} - 1\begin{vmatrix} 0 & 3 \\ -5 & -2 \end{vmatrix} = -2(9+2) - 4(0+5) - 1(0+15) = -22 - 20 - 15 = -57$$

57. First find the determinants:

$$D = \begin{vmatrix} 1 & 1 & 1 \\ 1 & -1 & -1 \\ 2 & 2 & -1 \end{vmatrix} = 1\begin{vmatrix} -1 & -1 \\ 2 & -1 \end{vmatrix} - 1\begin{vmatrix} 1 & -1 \\ 2 & -1 \end{vmatrix} + 1\begin{vmatrix} 1 & -1 \\ 2 & 2 \end{vmatrix} = 3 - 1 + 4 = 6$$

$$D_x = \begin{vmatrix} 4 & 1 & 1 \\ 2 & -1 & -1 \\ 2 & 2 & -1 \end{vmatrix} = 4\begin{vmatrix} -1 & -1 \\ 2 & -1 \end{vmatrix} - 1\begin{vmatrix} 2 & -1 \\ 2 & -1 \end{vmatrix} + 1\begin{vmatrix} 2 & -1 \\ 2 & 2 \end{vmatrix} = 12 - 0 + 6 = 18$$

$$D_y = \begin{vmatrix} 1 & 4 & 1 \\ 1 & 2 & -1 \\ 2 & 2 & -1 \end{vmatrix} = 1\begin{vmatrix} 2 & -1 \\ 2 & -1 \end{vmatrix} - 4\begin{vmatrix} 1 & -1 \\ 2 & -1 \end{vmatrix} + 1\begin{vmatrix} 1 & 2 \\ 2 & 2 \end{vmatrix} = 0 - 4 - 2 = -6$$

$$D_z = \begin{vmatrix} 1 & 1 & 4 \\ 1 & -1 & 2 \\ 2 & 2 & 2 \end{vmatrix} = 1\begin{vmatrix} -1 & 2 \\ 2 & 2 \end{vmatrix} - 1\begin{vmatrix} 1 & 2 \\ 2 & 2 \end{vmatrix} + 4\begin{vmatrix} 1 & -1 \\ 2 & 2 \end{vmatrix} = -6 + 2 + 16 = 12$$

Now use Cramer's rule:

$$x = \frac{D_x}{D} = \frac{18}{6} = 3 \qquad y = \frac{D_y}{D} = \frac{-6}{6} = -1 \qquad z = \frac{D_z}{D} = \frac{12}{6} = 2$$

The solution is $(3, -1, 2)$.

59. First find the determinants:

$$D = \begin{vmatrix} 1 & 1 & -1 \\ -1 & 1 & 1 \\ 1 & 1 & 1 \end{vmatrix} = 1\begin{vmatrix} 1 & 1 \\ 1 & 1 \end{vmatrix} - 1\begin{vmatrix} -1 & 1 \\ 1 & 1 \end{vmatrix} - 1\begin{vmatrix} -1 & 1 \\ 1 & 1 \end{vmatrix} = 0 + 2 + 2 = 4$$

$$D_x = \begin{vmatrix} 2 & 1 & -1 \\ 3 & 1 & 1 \\ 4 & 1 & 1 \end{vmatrix} = 2\begin{vmatrix} 1 & 1 \\ 1 & 1 \end{vmatrix} - 1\begin{vmatrix} 3 & 1 \\ 4 & 1 \end{vmatrix} - 1\begin{vmatrix} 3 & 1 \\ 4 & 1 \end{vmatrix} = 0 + 1 + 1 = 2$$

$$D_y = \begin{vmatrix} 1 & 2 & -1 \\ -1 & 3 & 1 \\ 1 & 4 & 1 \end{vmatrix} = 1\begin{vmatrix} 3 & 1 \\ 4 & 1 \end{vmatrix} - 2\begin{vmatrix} -1 & 1 \\ 1 & 1 \end{vmatrix} - 1\begin{vmatrix} -1 & 3 \\ 1 & 4 \end{vmatrix} = -1 + 4 + 7 = 10$$

$$D_z = \begin{vmatrix} 1 & 1 & 2 \\ -1 & 1 & 3 \\ 1 & 1 & 4 \end{vmatrix} = 1\begin{vmatrix} 1 & 3 \\ 1 & 4 \end{vmatrix} - 1\begin{vmatrix} -1 & 3 \\ 1 & 4 \end{vmatrix} + 2\begin{vmatrix} -1 & 1 \\ 1 & 1 \end{vmatrix} = 1 + 7 - 4 = 4$$

Now use Cramer's rule:

$$x = \frac{D_x}{D} = \frac{2}{4} = \frac{1}{2} \qquad y = \frac{D_y}{D} = \frac{10}{4} = \frac{5}{2} \qquad z = \frac{D_z}{D} = \frac{4}{4} = 1$$

The solution is $\left(\frac{1}{2}, \frac{5}{2}, 1\right)$.

61. First find the determinants:

$$D = \begin{vmatrix} 3 & -1 & 2 \\ 6 & -2 & 4 \\ 1 & -5 & 2 \end{vmatrix} = 3\begin{vmatrix} -2 & 4 \\ -5 & 2 \end{vmatrix} + 1\begin{vmatrix} 6 & 4 \\ 1 & 2 \end{vmatrix} + 2\begin{vmatrix} 6 & -2 \\ 1 & -5 \end{vmatrix} = 48 + 8 - 56 = 0$$

$$D_x = \begin{vmatrix} 4 & -1 & 2 \\ 8 & -2 & 4 \\ 1 & -5 & 2 \end{vmatrix} = 4\begin{vmatrix} -2 & 4 \\ -5 & 2 \end{vmatrix} + 1\begin{vmatrix} 8 & 4 \\ 1 & 2 \end{vmatrix} + 2\begin{vmatrix} 8 & -2 \\ 1 & -5 \end{vmatrix} = 64 + 12 - 76 = 0$$

$$D_y = \begin{vmatrix} 3 & 4 & 2 \\ 6 & 8 & 4 \\ 1 & 1 & 2 \end{vmatrix} = 3\begin{vmatrix} 8 & 4 \\ 1 & 2 \end{vmatrix} - 4\begin{vmatrix} 6 & 4 \\ 1 & 2 \end{vmatrix} + 2\begin{vmatrix} 6 & 8 \\ 1 & 1 \end{vmatrix} = 36 - 32 - 4 = 0$$

$$D_z = \begin{vmatrix} 3 & -1 & 4 \\ 6 & -2 & 8 \\ 1 & -5 & 1 \end{vmatrix} = 3\begin{vmatrix} -2 & 8 \\ -5 & 1 \end{vmatrix} + 1\begin{vmatrix} 6 & 8 \\ 1 & 1 \end{vmatrix} + 4\begin{vmatrix} 6 & -2 \\ 1 & -5 \end{vmatrix} = 114 - 2 - 112 = 0$$

Since $D = 0$ and the other determinants are also 0, there is no unique solution (dependent).

63. First find the determinants:

$$D = \begin{vmatrix} 2 & -1 & 3 \\ 1 & -5 & -2 \\ -4 & -2 & 1 \end{vmatrix} = 2\begin{vmatrix} -5 & -2 \\ -2 & 1 \end{vmatrix} + 1\begin{vmatrix} 1 & -2 \\ -4 & 1 \end{vmatrix} + 3\begin{vmatrix} 1 & -5 \\ -4 & -2 \end{vmatrix} = -18 - 7 - 66 = -91$$

$$D_x = \begin{vmatrix} 4 & -1 & 3 \\ 1 & -5 & -2 \\ 3 & -2 & 1 \end{vmatrix} = 4\begin{vmatrix} -5 & -2 \\ -2 & 1 \end{vmatrix} + 1\begin{vmatrix} 1 & -2 \\ 3 & 1 \end{vmatrix} + 3\begin{vmatrix} 1 & -5 \\ 3 & -2 \end{vmatrix} = -36 + 7 + 39 = 10$$

$$D_y = \begin{vmatrix} 2 & 4 & 3 \\ 1 & 1 & -2 \\ -4 & 3 & 1 \end{vmatrix} = 2\begin{vmatrix} 1 & -2 \\ 3 & 1 \end{vmatrix} - 4\begin{vmatrix} 1 & -2 \\ -4 & 1 \end{vmatrix} + 3\begin{vmatrix} 1 & 1 \\ -4 & 3 \end{vmatrix} = 14 + 28 + 21 = 63$$

$$D_z = \begin{vmatrix} 2 & -1 & 4 \\ 1 & -5 & 1 \\ -4 & -2 & 3 \end{vmatrix} = 2\begin{vmatrix} -5 & 1 \\ -2 & 3 \end{vmatrix} + 1\begin{vmatrix} 1 & 1 \\ -4 & 3 \end{vmatrix} + 4\begin{vmatrix} 1 & -5 \\ -4 & -2 \end{vmatrix} = -26 + 7 - 88 = -107$$

Now use Cramer's rule:

$$x = \frac{D_x}{D} = -\frac{10}{91} \qquad y = \frac{D_y}{D} = -\frac{63}{91} = -\frac{9}{13} \qquad z = \frac{D_z}{D} = \frac{-107}{-91} = \frac{107}{91}$$

The solution is $\left(-\frac{10}{91}, -\frac{9}{13}, \frac{107}{91}\right)$.

65. First find the determinants:

$$D = \begin{vmatrix} 1 & 2 & -1 \\ 2 & 3 & 2 \\ 1 & -3 & 1 \end{vmatrix} = 1\begin{vmatrix} 3 & 2 \\ -3 & 1 \end{vmatrix} - 2\begin{vmatrix} 2 & 2 \\ 1 & 1 \end{vmatrix} - 1\begin{vmatrix} 2 & 3 \\ 1 & -3 \end{vmatrix} = 9 - 0 + 9 = 18$$

$$D_x = \begin{vmatrix} 4 & 2 & -1 \\ 5 & 3 & 2 \\ 6 & -3 & 1 \end{vmatrix} = 4\begin{vmatrix} 3 & 2 \\ -3 & 1 \end{vmatrix} - 2\begin{vmatrix} 5 & 2 \\ 6 & 1 \end{vmatrix} - 1\begin{vmatrix} 5 & 3 \\ 6 & -3 \end{vmatrix} = 36 + 14 + 33 = 83$$

$$D_y = \begin{vmatrix} 1 & 4 & -1 \\ 2 & 5 & 2 \\ 1 & 6 & 1 \end{vmatrix} = 1\begin{vmatrix} 5 & 2 \\ 6 & 1 \end{vmatrix} - 4\begin{vmatrix} 2 & 2 \\ 1 & 1 \end{vmatrix} - 1\begin{vmatrix} 2 & 5 \\ 1 & 6 \end{vmatrix} = -7 - 0 - 7 = -14$$

$$D_z = \begin{vmatrix} 1 & 2 & 4 \\ 2 & 3 & 5 \\ 1 & -3 & 6 \end{vmatrix} = 1\begin{vmatrix} 3 & 5 \\ -3 & 6 \end{vmatrix} - 2\begin{vmatrix} 2 & 5 \\ 1 & 6 \end{vmatrix} + 4\begin{vmatrix} 2 & 3 \\ 1 & -3 \end{vmatrix} = 33 - 14 - 36 = -17$$

Now use Cramer's rule:

$$x = \frac{D_x}{D} = \frac{83}{18} \qquad y = \frac{D_y}{D} = \frac{-14}{18} = -\frac{7}{9} \qquad z = \frac{D_z}{D} = -\frac{17}{18}$$

The solution is $\left(\frac{83}{18}, -\frac{7}{9}, -\frac{17}{18}\right)$.

67. First find the determinants:

$$D = \begin{vmatrix} 3 & -4 & 2 \\ 2 & -3 & 4 \\ 4 & 2 & -3 \end{vmatrix} = 3\begin{vmatrix} -3 & 4 \\ 2 & -3 \end{vmatrix} + 4\begin{vmatrix} 2 & 4 \\ 4 & -3 \end{vmatrix} + 2\begin{vmatrix} 2 & -3 \\ 4 & 2 \end{vmatrix} = 3 - 88 + 32 = -53$$

$$D_x = \begin{vmatrix} 5 & -4 & 2 \\ 7 & -3 & 4 \\ 6 & 2 & -3 \end{vmatrix} = 5\begin{vmatrix} -3 & 4 \\ 2 & -3 \end{vmatrix} + 4\begin{vmatrix} 7 & 4 \\ 6 & -3 \end{vmatrix} + 2\begin{vmatrix} 7 & -3 \\ 6 & 2 \end{vmatrix} = 5 - 180 + 64 = -111$$

$$D_y = \begin{vmatrix} 3 & 5 & 2 \\ 2 & 7 & 4 \\ 4 & 6 & -3 \end{vmatrix} = 3\begin{vmatrix} 7 & 4 \\ 6 & -3 \end{vmatrix} - 5\begin{vmatrix} 2 & 4 \\ 4 & -3 \end{vmatrix} + 2\begin{vmatrix} 2 & 7 \\ 4 & 6 \end{vmatrix} = -135 + 110 - 32 = -57$$

$$D_z = \begin{vmatrix} 3 & -4 & 5 \\ 2 & -3 & 7 \\ 4 & 2 & 6 \end{vmatrix} = 3\begin{vmatrix} -3 & 7 \\ 2 & 6 \end{vmatrix} + 4\begin{vmatrix} 2 & 7 \\ 4 & 6 \end{vmatrix} + 5\begin{vmatrix} 2 & -3 \\ 4 & 2 \end{vmatrix} = -96 - 64 + 80 = -80$$

Now use Cramer's rule:

$$x = \frac{D_x}{D} = \frac{-111}{-53} = \frac{111}{53} \qquad y = \frac{D_y}{D} = \frac{-57}{-53} = \frac{57}{53} \qquad z = \frac{D_z}{D} = \frac{-80}{-53} = \frac{80}{53}$$

The solution is $\left(\frac{111}{53}, \frac{57}{53}, \frac{80}{53}\right)$.

69. First find the determinants:

$$D = \begin{vmatrix} 1 & 0 & -3 \\ 0 & 1 & 2 \\ 1 & 0 & 1 \end{vmatrix} = 1\begin{vmatrix} 1 & 2 \\ 0 & 1 \end{vmatrix} - 0\begin{vmatrix} 0 & 2 \\ 1 & 1 \end{vmatrix} - 3\begin{vmatrix} 0 & 1 \\ 1 & 0 \end{vmatrix} = 1 - 0 + 3 = 4$$

$$D_x = \begin{vmatrix} 1 & 0 & -3 \\ 8 & 1 & 2 \\ 10 & 0 & 1 \end{vmatrix} = 1\begin{vmatrix} 1 & 2 \\ 0 & 1 \end{vmatrix} - 0\begin{vmatrix} 8 & 2 \\ 10 & 1 \end{vmatrix} - 3\begin{vmatrix} 8 & 1 \\ 10 & 0 \end{vmatrix} = 1 - 0 + 30 = 31$$

$$D_y = \begin{vmatrix} 1 & 1 & -3 \\ 0 & 8 & 2 \\ 1 & 10 & 1 \end{vmatrix} = 1\begin{vmatrix} 8 & 2 \\ 10 & 1 \end{vmatrix} - 1\begin{vmatrix} 0 & 2 \\ 1 & 1 \end{vmatrix} - 3\begin{vmatrix} 0 & 8 \\ 1 & 10 \end{vmatrix} = -12 + 2 + 24 = 14$$

$$D_z = \begin{vmatrix} 1 & 0 & 1 \\ 0 & 1 & 8 \\ 1 & 0 & 10 \end{vmatrix} = 1\begin{vmatrix} 1 & 8 \\ 0 & 10 \end{vmatrix} - 0\begin{vmatrix} 0 & 8 \\ 1 & 10 \end{vmatrix} + 1\begin{vmatrix} 0 & 1 \\ 1 & 0 \end{vmatrix} = 10 - 0 - 1 = 9$$

Now use Cramer's rule:

$$x = \frac{D_x}{D} = \frac{31}{4} \qquad y = \frac{D_y}{D} = \frac{14}{4} = \frac{7}{2} \qquad z = \frac{D_z}{D} = \frac{9}{4}$$

The solution is $\left(\frac{31}{4}, \frac{7}{2}, \frac{9}{4}\right)$.

71. First find the determinants:

$$D = \begin{vmatrix} -1 & -7 & 0 \\ 1 & 0 & 3 \\ 0 & 2 & 1 \end{vmatrix} = -1\begin{vmatrix} 0 & 3 \\ 2 & 1 \end{vmatrix} + 7\begin{vmatrix} 1 & 3 \\ 0 & 1 \end{vmatrix} + 0\begin{vmatrix} 1 & 0 \\ 0 & 2 \end{vmatrix} = 6+7+0 = 13$$

$$D_x = \begin{vmatrix} 1 & -7 & 0 \\ 11 & 0 & 3 \\ 0 & 2 & 1 \end{vmatrix} = 1\begin{vmatrix} 0 & 3 \\ 2 & 1 \end{vmatrix} + 7\begin{vmatrix} 11 & 3 \\ 0 & 1 \end{vmatrix} + 0\begin{vmatrix} 11 & 0 \\ 0 & 2 \end{vmatrix} = -6+77+0 = 71$$

$$D_y = \begin{vmatrix} -1 & 1 & 0 \\ 1 & 11 & 3 \\ 0 & 0 & 1 \end{vmatrix} = -1\begin{vmatrix} 11 & 3 \\ 0 & 1 \end{vmatrix} - 1\begin{vmatrix} 1 & 3 \\ 0 & 1 \end{vmatrix} + 0\begin{vmatrix} 1 & 11 \\ 0 & 0 \end{vmatrix} = -11-1+0 = -12$$

$$D_z = \begin{vmatrix} -1 & -7 & 1 \\ 1 & 0 & 11 \\ 0 & 2 & 0 \end{vmatrix} = -1\begin{vmatrix} 0 & 11 \\ 2 & 0 \end{vmatrix} + 7\begin{vmatrix} 1 & 11 \\ 0 & 0 \end{vmatrix} + 1\begin{vmatrix} 1 & 0 \\ 0 & 2 \end{vmatrix} = 22+0+2 = 24$$

Now use Cramer's rule:

$$x = \frac{D_x}{D} = \frac{71}{13} \qquad y = \frac{D_y}{D} = -\frac{12}{13} \qquad z = \frac{D_z}{D} = \frac{24}{13}$$

The solution is $\left(\frac{71}{13}, -\frac{12}{13}, \frac{24}{13}\right)$.

73. First find the determinants:

$$D = \begin{vmatrix} 1 & -1 & 0 \\ 3 & 0 & 1 \\ 0 & 1 & -2 \end{vmatrix} = 1\begin{vmatrix} 0 & 1 \\ 1 & -2 \end{vmatrix} + 1\begin{vmatrix} 3 & 1 \\ 0 & -2 \end{vmatrix} + 0\begin{vmatrix} 3 & 0 \\ 0 & 1 \end{vmatrix} = -1-6+0 = -7$$

$$D_x = \begin{vmatrix} 2 & -1 & 0 \\ 11 & 0 & 1 \\ -3 & 1 & -2 \end{vmatrix} = 2\begin{vmatrix} 0 & 1 \\ 1 & -2 \end{vmatrix} + 1\begin{vmatrix} 11 & 1 \\ -3 & -2 \end{vmatrix} + 0\begin{vmatrix} 11 & 0 \\ -3 & 1 \end{vmatrix} = -2-19+0 = -21$$

$$D_y = \begin{vmatrix} 1 & 2 & 0 \\ 3 & 11 & 1 \\ 0 & -3 & -2 \end{vmatrix} = 1\begin{vmatrix} 11 & 1 \\ -3 & -2 \end{vmatrix} - 2\begin{vmatrix} 3 & 1 \\ 0 & -2 \end{vmatrix} + 0\begin{vmatrix} 3 & 11 \\ 0 & -3 \end{vmatrix} = -19+12+0 = -7$$

$$D_z = \begin{vmatrix} 1 & -1 & 2 \\ 3 & 0 & 11 \\ 0 & 1 & -3 \end{vmatrix} = 1\begin{vmatrix} 0 & 11 \\ 1 & -3 \end{vmatrix} + 1\begin{vmatrix} 3 & 11 \\ 0 & -3 \end{vmatrix} + 2\begin{vmatrix} 3 & 0 \\ 0 & 1 \end{vmatrix} = -11-9+6 = -14$$

Now use Cramer's rule:

$$x = \frac{D_x}{D} = \frac{-21}{-7} = 3 \qquad y = \frac{D_y}{D} = \frac{-7}{-7} = 1 \qquad z = \frac{D_z}{D} = \frac{-14}{-7} = 2$$

The solution is $(3, 1, 2)$.

75. The determinant equation is:

$$\begin{vmatrix} y & x \\ m & 1 \end{vmatrix} = b$$

$$y - mx = b$$

$$y = mx + b$$

77. **a.** Writing the determinant equation:
$$\begin{vmatrix} x & -1.7 \\ 2 & 0.3 \end{vmatrix} = y$$
$$0.3x + 3.4 = y$$
$$y = 0.3x + 3.4$$

b. Substituting $x = 2$: $y = 0.3(2) + 3.4 = 0.6 + 3.4 = 4$ billion dollars

79. **a.** Writing the equation:
$$\begin{vmatrix} x & -3 \\ 7{,}121 & 767.5 \end{vmatrix} = I$$
$$767.5x + 3(7{,}121) = I$$
$$I = 767.5x + 21{,}363$$

b. Substituting $x = 4$: $I = 767.5(4) + 21{,}363 = \$24{,}433$

81. Substituting $x = 6$: $y = \begin{vmatrix} 0.1 & 6.9 \\ -2 & 6 \end{vmatrix} = 0.1(6) - (-2)(6.9) = 0.6 + 13.8 = 14.4$ million

83. Find the determinants:
$$D = \begin{vmatrix} -164.2 & 1 \\ 1 & 0 \end{vmatrix} = -164.2(0) - 1 = -1$$
$$D_H = \begin{vmatrix} -164.2 & 719 \\ 1 & 5 \end{vmatrix} = -164.2(5) - 719 = -1540$$

Now using Cramer's rule: $H = \dfrac{D_H}{D} = \dfrac{-1540}{-1} = 1{,}540$ heart transplants

85. First rewrite the system as:
$$-10x + y = 100$$
$$-12x + y = 0$$

Now find the determinants:
$$D = \begin{vmatrix} -10 & 1 \\ -12 & 1 \end{vmatrix} = -10 + 12 = 2$$
$$D_x = \begin{vmatrix} 100 & 1 \\ 0 & 1 \end{vmatrix} = 100 - 0 = 100$$
$$D_y = \begin{vmatrix} -10 & 100 \\ -12 & 0 \end{vmatrix} = 0 + 1200 = 1200$$

Now using Cramer's rule:
$$x = \dfrac{D_x}{D} = \dfrac{100}{2} = 50 \qquad y = \dfrac{D_y}{D} = \dfrac{1200}{2} = 600$$

The company must sell 50 items per week to break even.

87. First write the system in the correct form:
$$-0.98x + y = -1,915.8$$
$$y = 30$$
Find the determinants:
$$D = \begin{vmatrix} -0.98 & 1 \\ 0 & 1 \end{vmatrix} = -0.98(1) - 0 = -0.98$$
$$D_x = \begin{vmatrix} -1,915.8 & 1 \\ 30 & 1 \end{vmatrix} = -1,915.8(1) - 30 = -1,945.8$$

Now using Cramer's rule: $x = \dfrac{D_x}{D} = \dfrac{-1,945.8}{-0.98} \approx 1986$

In 1986 approximately 30 million residents were without health insurance.

89. Expanding across row 1:
$$\begin{vmatrix} 2 & 0 & 1 & -3 \\ -1 & 2 & 0 & 1 \\ -3 & 0 & 1 & 0 \\ 1 & 1 & 0 & 0 \end{vmatrix}$$

$$= 2\begin{vmatrix} 2 & 0 & 1 \\ 0 & 1 & 0 \\ 1 & 0 & 0 \end{vmatrix} - 0\begin{vmatrix} -1 & 0 & 1 \\ -3 & 1 & 0 \\ 1 & 0 & 0 \end{vmatrix} + 1\begin{vmatrix} -1 & 2 & 1 \\ -3 & 0 & 0 \\ 1 & 1 & 0 \end{vmatrix} + 3\begin{vmatrix} -1 & 2 & 0 \\ -3 & 0 & 1 \\ 1 & 1 & 0 \end{vmatrix}$$

$$= 2\left(2\begin{vmatrix} 1 & 0 \\ 0 & 0 \end{vmatrix} - 0 + 1\begin{vmatrix} 0 & 1 \\ 1 & 0 \end{vmatrix}\right) - 0 + 1\left(-1\begin{vmatrix} 0 & 0 \\ 1 & 0 \end{vmatrix} - 2\begin{vmatrix} -3 & 0 \\ 1 & 0 \end{vmatrix} + 1\begin{vmatrix} -3 & 0 \\ 1 & 1 \end{vmatrix}\right) + 3\left(-1\begin{vmatrix} 0 & 1 \\ 1 & 0 \end{vmatrix} - 2\begin{vmatrix} -3 & 1 \\ 1 & 0 \end{vmatrix}\right)$$

$$= 2(0 - 0 - 1) + 1(0 - 0 - 3) + 3(1 + 2)$$
$$= -2 - 3 + 9$$
$$= 4$$

91. Expanding down column 3:
$$\begin{vmatrix} 2 & 0 & 1 & -3 \\ -1 & 2 & 0 & 1 \\ -3 & 0 & 1 & 0 \\ 1 & 1 & 0 & 0 \end{vmatrix}$$

$$= 1\begin{vmatrix} -1 & 2 & 1 \\ -3 & 0 & 0 \\ 1 & 1 & 0 \end{vmatrix} - 0\begin{vmatrix} 2 & 0 & -3 \\ -3 & 0 & 0 \\ 1 & 1 & 0 \end{vmatrix} + 1\begin{vmatrix} 2 & 0 & -3 \\ -1 & 2 & 1 \\ 1 & 1 & 0 \end{vmatrix} - 0\begin{vmatrix} 2 & 0 & -3 \\ -1 & 2 & 1 \\ -3 & 0 & 0 \end{vmatrix}$$

$$= 1\left(-1\begin{vmatrix} 0 & 0 \\ 1 & 0 \end{vmatrix} - 2\begin{vmatrix} -3 & 0 \\ 1 & 0 \end{vmatrix} + 1\begin{vmatrix} -3 & 0 \\ 1 & 1 \end{vmatrix}\right) - 0 + 1\left(2\begin{vmatrix} 2 & 1 \\ 1 & 0 \end{vmatrix} - 0 - 3\begin{vmatrix} -1 & 2 \\ 1 & 1 \end{vmatrix}\right) - 0$$

$$= 1(0 - 0 - 3) + 1(-2 - 0 + 9)$$
$$= -3 + 7$$
$$= 4$$

93. Expanding the determinant:

$$\begin{vmatrix} 1 & 3 & 2 & -4 \\ 0 & 4 & 1 & 0 \\ -2 & 1 & 3 & 0 \\ 2 & 3 & 4 & -1 \end{vmatrix}$$

$$= 1\begin{vmatrix} 4 & 1 & 0 \\ 1 & 3 & 0 \\ 3 & 4 & -1 \end{vmatrix} - 3\begin{vmatrix} 0 & 1 & 0 \\ -2 & 3 & 0 \\ 2 & 4 & -1 \end{vmatrix} + 2\begin{vmatrix} 0 & 4 & 0 \\ -2 & 1 & 0 \\ 2 & 3 & -1 \end{vmatrix} + 4\begin{vmatrix} 0 & 4 & 1 \\ -2 & 1 & 3 \\ 2 & 3 & 4 \end{vmatrix}$$

$$= 4\begin{vmatrix} 3 & 0 \\ 4 & -1 \end{vmatrix} - 1\begin{vmatrix} 1 & 0 \\ 3 & -1 \end{vmatrix} - 3\left(-1\begin{vmatrix} -2 & 0 \\ 2 & -1 \end{vmatrix}\right) + 2\left(-4\begin{vmatrix} -2 & 0 \\ 2 & -1 \end{vmatrix}\right) + 4\left(-4\begin{vmatrix} -2 & 3 \\ 2 & 4 \end{vmatrix} + 1\begin{vmatrix} -2 & 1 \\ 2 & 3 \end{vmatrix}\right)$$

$$= 4(-3) - 1(-1) - 3(-2) - 8(2) - 16(-14) + 4(-8)$$

$$= -12 + 1 + 6 - 16 + 224 - 32$$

$$= 171$$

95. Find the determinants:

$$D = \begin{vmatrix} 1 & 2 & -1 & 3 \\ 2 & 1 & 2 & -2 \\ 1 & -3 & 1 & -1 \\ -2 & 1 & -1 & 3 \end{vmatrix}$$

$$= 1\begin{vmatrix} 1 & 2 & -2 \\ -3 & 1 & -1 \\ 1 & -1 & 3 \end{vmatrix} - 2\begin{vmatrix} 2 & 2 & -2 \\ 1 & 1 & -1 \\ -2 & -1 & 3 \end{vmatrix} - 1\begin{vmatrix} 2 & 1 & -2 \\ 1 & -3 & -1 \\ -2 & 1 & 3 \end{vmatrix} - 3\begin{vmatrix} 2 & 1 & 2 \\ 1 & -3 & 1 \\ -2 & 1 & -1 \end{vmatrix}$$

$$= 1\left(1\begin{vmatrix} 1 & -1 \\ -1 & 3 \end{vmatrix} - 2\begin{vmatrix} -3 & -1 \\ 1 & 3 \end{vmatrix} - 2\begin{vmatrix} -3 & 1 \\ 1 & -1 \end{vmatrix}\right) - 2\left(2\begin{vmatrix} 1 & -1 \\ -1 & 3 \end{vmatrix} - 2\begin{vmatrix} 1 & -1 \\ -2 & 3 \end{vmatrix} - 2\begin{vmatrix} 1 & 1 \\ -2 & -1 \end{vmatrix}\right)$$

$$-1\left(2\begin{vmatrix} -3 & -1 \\ 1 & 3 \end{vmatrix} - 1\begin{vmatrix} 1 & -1 \\ -2 & 3 \end{vmatrix} - 2\begin{vmatrix} 1 & -3 \\ -2 & 1 \end{vmatrix}\right) - 3\left(2\begin{vmatrix} -3 & 1 \\ 1 & -1 \end{vmatrix} - 1\begin{vmatrix} 1 & 1 \\ -2 & -1 \end{vmatrix} + 2\begin{vmatrix} 1 & -3 \\ -2 & 1 \end{vmatrix}\right)$$

$$= 1[1(2) - 2(-8) - 2(2)] - 2[2(2) - 2(1) - 2(1)] - 1[2(-8) - 1(1) - 2(-5)] - 3[2(2) - 1(1) + 2(-5)]$$

$$= 1(2 + 16 - 4) - 2(4 - 2 - 2) - 1(-16 - 1 + 10) - 3(4 - 1 - 10)$$

$$= 1(14) - 2(0) - 1(-7) - 3(-7)$$

$$= 14 - 0 + 7 + 21$$

$$= 42$$

$$D_x = \begin{vmatrix} 4 & 2 & -1 & 3 \\ 9 & 1 & 2 & -2 \\ 1 & -3 & 1 & -1 \\ -3 & 1 & -1 & 3 \end{vmatrix}$$

$$= 4\begin{vmatrix} 1 & 2 & -2 \\ -3 & 1 & -1 \\ 1 & -1 & 3 \end{vmatrix} - 2\begin{vmatrix} 9 & 2 & -2 \\ 1 & 1 & -1 \\ -3 & -1 & 3 \end{vmatrix} - 1\begin{vmatrix} 9 & 1 & -2 \\ 1 & -3 & -1 \\ -3 & 1 & 3 \end{vmatrix} - 3\begin{vmatrix} 9 & 1 & 2 \\ 1 & -3 & 1 \\ -3 & 1 & -1 \end{vmatrix}$$

$$= 4\left(1\begin{vmatrix} 1 & -1 \\ -1 & 3 \end{vmatrix} - 2\begin{vmatrix} -3 & -1 \\ 1 & 3 \end{vmatrix} - 2\begin{vmatrix} -3 & 1 \\ 1 & -1 \end{vmatrix}\right) - 2\left(9\begin{vmatrix} 1 & -1 \\ -1 & 3 \end{vmatrix} - 2\begin{vmatrix} 1 & -1 \\ -3 & 3 \end{vmatrix} - 2\begin{vmatrix} 1 & 1 \\ -3 & -1 \end{vmatrix}\right)$$

$$-1\left(9\begin{vmatrix} -3 & -1 \\ 1 & 3 \end{vmatrix} - 1\begin{vmatrix} 1 & -1 \\ -3 & 3 \end{vmatrix} - 2\begin{vmatrix} 1 & -3 \\ -3 & 1 \end{vmatrix}\right) - 3\left(9\begin{vmatrix} -3 & 1 \\ 1 & -1 \end{vmatrix} - 1\begin{vmatrix} 1 & 1 \\ -3 & -1 \end{vmatrix} + 2\begin{vmatrix} 1 & -3 \\ -3 & 1 \end{vmatrix}\right)$$

$$= 4[1(2) - 2(-8) - 2(2)] - 2[9(2) - 2(0) - 2(2)] - 1[9(-8) - 1(0) - 2(-8)] - 3[9(2) - 1(2) + 2(-8)]$$

$$= 4(2 + 16 - 4) - 2(18 - 0 - 4) - 1(-72 - 0 + 16) - 3(18 - 2 - 16)$$

$$= 4(14) - 2(14) - 1(-56) - 3(0)$$

$$= 56 - 28 + 56 - 0$$

$$= 84$$

$$D_y = \begin{vmatrix} 1 & 4 & -1 & 3 \\ 2 & 9 & 2 & -2 \\ 1 & 1 & 1 & -1 \\ -2 & -3 & -1 & 3 \end{vmatrix}$$

$$= 1\begin{vmatrix} 9 & 2 & -2 \\ 1 & 1 & -1 \\ -3 & -1 & 3 \end{vmatrix} - 4\begin{vmatrix} 2 & 2 & -2 \\ 1 & 1 & -1 \\ -2 & -1 & 3 \end{vmatrix} - 1\begin{vmatrix} 2 & 9 & -2 \\ 1 & 1 & -1 \\ -2 & -3 & 3 \end{vmatrix} - 3\begin{vmatrix} 2 & 9 & 2 \\ 1 & 1 & 1 \\ -2 & -3 & -1 \end{vmatrix}$$

$$= 1\left(9\begin{vmatrix} 1 & -1 \\ -1 & 3 \end{vmatrix} - 2\begin{vmatrix} 1 & -1 \\ -3 & 3 \end{vmatrix} - 2\begin{vmatrix} 1 & 1 \\ -3 & -1 \end{vmatrix}\right) - 4\left(2\begin{vmatrix} 1 & -1 \\ -1 & 3 \end{vmatrix} - 2\begin{vmatrix} 1 & -1 \\ -2 & 3 \end{vmatrix} - 2\begin{vmatrix} 1 & 1 \\ -2 & -1 \end{vmatrix}\right)$$

$$-1\left(2\begin{vmatrix} 1 & -1 \\ -3 & 3 \end{vmatrix} - 9\begin{vmatrix} 1 & -1 \\ -2 & 3 \end{vmatrix} - 2\begin{vmatrix} 1 & 1 \\ -2 & -3 \end{vmatrix}\right) - 3\left(2\begin{vmatrix} 1 & 1 \\ -3 & -1 \end{vmatrix} - 9\begin{vmatrix} 1 & 1 \\ -2 & -1 \end{vmatrix} + 2\begin{vmatrix} 1 & 1 \\ -2 & -3 \end{vmatrix}\right)$$

$$= 1[9(2) - 2(0) - 2(2)] - 4[2(2) - 2(1) - 2(1)] - 1[2(0) - 9(1) - 2(-1)] - 3[2(2) - 9(1) + 2(-1)]$$

$$= 1(18 - 0 - 4) - 4(4 - 2 - 2) - 1(0 - 9 + 2) - 3(4 - 9 - 2)$$

$$= 1(14) - 4(0) - 1(-7) - 3(-7)$$

$$= 14 - 0 + 7 + 21$$

$$= 42$$

$$D_z = \begin{vmatrix} 1 & 2 & 4 & 3 \\ 2 & 1 & 9 & -2 \\ 1 & -3 & 1 & -1 \\ -2 & 1 & -3 & 3 \end{vmatrix}$$

$$= 1\begin{vmatrix} 1 & 9 & -2 \\ -3 & 1 & -1 \\ 1 & -3 & 3 \end{vmatrix} - 2\begin{vmatrix} 2 & 9 & -2 \\ 1 & 1 & -1 \\ -2 & -3 & 3 \end{vmatrix} + 4\begin{vmatrix} 2 & 1 & -2 \\ 1 & -3 & -1 \\ -2 & 1 & 3 \end{vmatrix} - 3\begin{vmatrix} 2 & 1 & 9 \\ 1 & -3 & 1 \\ -2 & 1 & -3 \end{vmatrix}$$

$$= 1\left(1\begin{vmatrix} 1 & -1 \\ -3 & 3 \end{vmatrix} - 9\begin{vmatrix} -3 & -1 \\ 1 & 3 \end{vmatrix} - 2\begin{vmatrix} -3 & 1 \\ 1 & -3 \end{vmatrix}\right) - 2\left(2\begin{vmatrix} 1 & -1 \\ -3 & 3 \end{vmatrix} - 9\begin{vmatrix} 1 & -1 \\ -2 & 3 \end{vmatrix} - 2\begin{vmatrix} 1 & 1 \\ -2 & -3 \end{vmatrix}\right)$$

$$+ 4\left(2\begin{vmatrix} -3 & -1 \\ 1 & 3 \end{vmatrix} - 1\begin{vmatrix} 1 & -1 \\ -2 & 3 \end{vmatrix} - 2\begin{vmatrix} 1 & -3 \\ -2 & 1 \end{vmatrix}\right) - 3\left(2\begin{vmatrix} -3 & 1 \\ 1 & -3 \end{vmatrix} - 1\begin{vmatrix} 1 & 1 \\ -2 & -3 \end{vmatrix} + 9\begin{vmatrix} 1 & -3 \\ -2 & 1 \end{vmatrix}\right)$$

$$= 1\bigl[1(0) - 9(-8) - 2(8)\bigr] - 2\bigl[2(0) - 9(1) - 2(-1)\bigr] + 4\bigl[2(-8) - 1(1) - 2(-5)\bigr] - 3\bigl[2(8) - 1(-1) + 9(-5)\bigr]$$

$$= 1(0 + 72 - 16) - 2(0 - 9 + 2) + 4(-16 - 1 + 10) - 3(16 + 1 - 45)$$

$$= 1(56) - 2(-7) + 4(-7) - 3(-28)$$

$$= 56 + 14 - 28 + 84$$

$$= 126$$

$$D_w = \begin{vmatrix} 1 & 2 & -1 & 4 \\ 2 & 1 & 2 & 9 \\ 1 & -3 & 1 & 1 \\ -2 & 1 & -1 & -3 \end{vmatrix}$$

$$= 1\begin{vmatrix} 1 & 2 & 9 \\ -3 & 1 & 1 \\ 1 & -1 & -3 \end{vmatrix} - 2\begin{vmatrix} 2 & 2 & 9 \\ 1 & 1 & 1 \\ -2 & -1 & 3 \end{vmatrix} - 1\begin{vmatrix} 2 & 1 & 9 \\ 1 & -3 & 1 \\ -2 & 1 & -3 \end{vmatrix} - 4\begin{vmatrix} 2 & 1 & 2 \\ 1 & -3 & 1 \\ -2 & 1 & -1 \end{vmatrix}$$

$$= 1\left(1\begin{vmatrix} 1 & -1 \\ -1 & -3 \end{vmatrix} - 2\begin{vmatrix} -3 & 1 \\ 1 & -3 \end{vmatrix} + 9\begin{vmatrix} -3 & 1 \\ 1 & -1 \end{vmatrix}\right) - 2\left(2\begin{vmatrix} 1 & 1 \\ -1 & 3 \end{vmatrix} - 2\begin{vmatrix} 1 & 1 \\ -2 & 3 \end{vmatrix} + 9\begin{vmatrix} 1 & 1 \\ -2 & -1 \end{vmatrix}\right)$$

$$- 1\left(2\begin{vmatrix} -3 & 1 \\ 1 & -3 \end{vmatrix} - 1\begin{vmatrix} 1 & 1 \\ -2 & -3 \end{vmatrix} + 9\begin{vmatrix} 1 & -3 \\ -2 & 1 \end{vmatrix}\right) - 4\left(2\begin{vmatrix} -3 & 1 \\ 1 & -1 \end{vmatrix} - 1\begin{vmatrix} 1 & 1 \\ -2 & -1 \end{vmatrix} + 2\begin{vmatrix} 1 & -3 \\ -2 & 1 \end{vmatrix}\right)$$

$$= 1\bigl[1(-4) - 2(8) + 9(2)\bigr] - 2\bigl[2(4) - 2(5) + 9(1)\bigr] - 1\bigl[2(8) - 1(-1) + 9(-5)\bigr] - 4\bigl[2(2) - 1(1) + 2(-5)\bigr]$$

$$= 1(-4 - 16 + 18) - 2(8 - 10 + 9) - 1(16 + 1 - 45) - 4(4 - 1 - 10)$$

$$= 1(-2) - 2(7) - 1(-28) - 4(-7)$$

$$= -2 - 14 + 28 + 28$$

$$= 42$$

Now use Cramer's rule:

$$x = \frac{D_x}{D} = \frac{84}{42} = 2 \qquad y = \frac{D_y}{D} = \frac{42}{42} = 1 \qquad z = \frac{D_z}{D} = \frac{126}{42} = 3 \qquad w = \frac{D_w}{D} = \frac{42}{42} = 1$$

The solution is $(2, 1, 3, 1)$.

97. First find the determinants:

$$D = \begin{vmatrix} a & 1 & 1 \\ 1 & a & 1 \\ 1 & 1 & a \end{vmatrix}$$

$$= a\begin{vmatrix} a & 1 \\ 1 & a \end{vmatrix} - 1\begin{vmatrix} 1 & 1 \\ 1 & a \end{vmatrix} + 1\begin{vmatrix} 1 & a \\ 1 & 1 \end{vmatrix}$$

$$= a(a^2 - 1) - 1(a - 1) + 1(1 - a)$$
$$= a(a+1)(a-1) - 1(a-1) - 1(a-1)$$
$$= (a-1)(a^2 + a - 2)$$
$$= (a-1)(a-1)(a+2)$$

$$D_x = \begin{vmatrix} 1 & 1 & 1 \\ 1 & a & 1 \\ 1 & 1 & a \end{vmatrix}$$

$$= 1\begin{vmatrix} a & 1 \\ 1 & a \end{vmatrix} - 1\begin{vmatrix} 1 & 1 \\ 1 & a \end{vmatrix} + 1\begin{vmatrix} 1 & a \\ 1 & 1 \end{vmatrix}$$

$$= 1(a^2 - 1) - 1(a - 1) + 1(1 - a)$$
$$= (a+1)(a-1) - 1(a-1) - 1(a-1)$$
$$= (a-1)(a+1-2)$$
$$= (a-1)(a-1)$$

$$D_y = \begin{vmatrix} a & 1 & 1 \\ 1 & 1 & 1 \\ 1 & 1 & a \end{vmatrix}$$

$$= a\begin{vmatrix} 1 & 1 \\ 1 & a \end{vmatrix} - 1\begin{vmatrix} 1 & 1 \\ 1 & a \end{vmatrix} + 1\begin{vmatrix} 1 & 1 \\ 1 & 1 \end{vmatrix}$$

$$= a(a-1) - 1(a-1) + 1(0)$$
$$= a(a-1) - 1(a-1)$$
$$= (a-1)(a-1)$$

$$D_z = \begin{vmatrix} a & 1 & 1 \\ 1 & a & 1 \\ 1 & 1 & 1 \end{vmatrix}$$

$$= a\begin{vmatrix} a & 1 \\ 1 & 1 \end{vmatrix} - 1\begin{vmatrix} 1 & 1 \\ 1 & 1 \end{vmatrix} + 1\begin{vmatrix} 1 & a \\ 1 & 1 \end{vmatrix}$$

$$= a(a-1) - 1(0) + 1(1 - a)$$
$$= a(a-1) - 1(a-1)$$
$$= (a-1)(a-1)$$

Now use Cramer's rule:

$$x = \frac{D_x}{D} = \frac{(a-1)(a-1)}{(a-1)(a-1)(a+2)} = \frac{1}{a+2}$$

$$y = \frac{D_y}{D} = \frac{(a-1)(a-1)}{(a-1)(a-1)(a+2)} = \frac{1}{a+2}$$

$$z = \frac{D_z}{D} = \frac{(a-1)(a-1)}{(a-1)(a-1)(a+2)} = \frac{1}{a+2}$$

The solution is $\left(\frac{1}{a+2}, \frac{1}{a+2}, \frac{1}{a+2}\right)$.

99. Translating into symbols: $3x + 2$

101. Simplifying: $25 - \frac{385}{9} = \frac{225}{9} - \frac{385}{9} = -\frac{160}{9}$

103. Simplifying: $0.08(4,000) = 320$

105. Simplifying: $10(0.2x + 0.5y) = 2x + 5y$

107. Solving the equation:
$$x + (3x + 2) = 26$$
$$4x + 2 = 26$$
$$4x = 24$$
$$x = 6$$

109. Multiply the first equation by −20:
$$-60y - 20z = -340$$
$$5y + 20z = 65$$
Adding yields:
$$-55y = -275$$
$$y = 5$$
Substituting into the first equation:
$$3(5) + z = 17$$
$$15 + z = 17$$
$$z = 2$$
The solution is $y = 5, z = 2$.

3.7 Applications of Systems of Equations

1. Let x and y represent the two numbers. The system of equations is:
$$y = 2x + 3$$
$$x + y = 18$$
Substituting into the second equation:
$$x + 2x + 3 = 18$$
$$3x = 15$$
$$x = 5$$
$$y = 2(5) + 3 = 13$$
The two numbers are 5 and 13.

3. Let x and y represent the two numbers. The system of equations is:
$$y - x = 6$$
$$2x = 4 + y$$
The second equation is $y = 2x - 4$. Substituting into the first equation:
$$2x - 4 - x = 6$$
$$x = 10$$
$$y = 2(10) - 4 = 16$$
The two numbers are 10 and 16.

5. Let x, y, and z represent the three numbers. The system of equations is:
$$x + y + z = 8$$
$$2x = z - 2$$
$$x + z = 5$$
The third equation is $z = 5 - x$. Substituting into the second equation:
$$2x = 5 - x - 2$$
$$3x = 3$$
$$x = 1$$
$$z = 5 - 1 = 4$$
Substituting into the first equation:
$$1 + y + 4 = 8$$
$$y = 3$$
The three numbers are 1, 3, and 4.

7. Let x and y represent the two numbers. The system of equations is:
$$y = 5x + 8$$
$$x + y = 26$$
Substituting into the second equation:
$$x + 5x + 8 = 26$$
$$6x + 8 = 26$$
$$6x = 18$$
$$x = 3$$
$$y = 5(3) + 8 = 23$$
The two numbers are 3 and 23.

9. Let x and y represent the two numbers. The system of equations is:
$$y - x = 9$$
$$y = 2x - 6$$
Substituting into the first equation:
$$2x - 6 - x = 9$$
$$x - 6 = 9$$
$$x = 15$$
$$y = 2(15) - 6 = 24$$
The two numbers are 15 and 24.

11. Let a represent the number of adult tickets and c represent the number of children's tickets. The system of equations is:
$$a + c = 925$$
$$2a + c = 1150$$
Multiply the first equation by -1:
$$-a - c = -925$$
$$2a + c = 1150$$
Adding yields:
$$a = 225$$
$$c = 700$$
Linda sold 225 adult tickets and 700 children's tickets.

13. Let x represent the amount charged at 6% and y represent the amount charged at 7%. The system of equations is:
$$x + y = 20{,}000$$
$$0.06x + 0.07y = 1{,}280$$
Multiplying the first equation by -0.06:
$$-0.06x - 0.06y = -1{,}200$$
$$0.06x + 0.07y = 1{,}280$$
Adding yields:
$$0.01y = 80$$
$$y = 8{,}000$$
$$x = 12{,}000$$
Mr. Jones charged $12,000 at 6% and $8,000 at 7%.

15. Let x represent the amount charged at 6% and $2x$ represent the amount charged at 7.5%. The equation is:
$$0.075(2x) + 0.06(x) = 840$$
$$0.21x = 840$$
$$x = 4{,}000$$
$$2x = 8{,}000$$
Susan charged $4,000 at 6% and $8,000 at 7.5%.

17. Let x, y and z represent the amounts withdrawn in the three accounts. The system of equations is:
$$x+y+z=2,200$$
$$z=3x$$
$$0.06x+0.08y+0.09z=178$$
Substituting into the first equation:
$$x+y+3x=2,200$$
$$4x+y=2,200$$
Substituting into the third equation:
$$0.06x+0.08y+0.09(3x)=178$$
$$0.33x+0.08y=178$$
The system of equations becomes:
$$4x+y=2,200$$
$$0.33x+0.08y=178$$
Multiply the first equation by –0.08:
$$-0.32x-0.08y=-176$$
$$0.33x+0.08y=178$$
Adding yields:
$$0.01x=2$$
$$x=200$$
$$z=3(200)=600$$
$$y=2,200-4(200)=1,400$$
William withdrew $200 at 6%, $1,400 at 8%, and $600 at 9%.

19. Let x represent the amount of 20% alcohol and y represent the amount of 50% alcohol. The system of equations is:
$$x+y=9$$
$$0.20x+0.50y=0.30(9)$$
Multiplying the first equation by –0.2:
$$-0.20x-0.20y=-1.8$$
$$0.20x+0.50y=2.7$$
Adding yields:
$$0.30y=0.9$$
$$y=3$$
$$x=6$$
The mixture contains 3 gallons of 50% alcohol and 6 gallons of 20% alcohol.

21. Let x represent the amount of 20% disinfectant and y represent the amount of 14% disinfectant. The system of equations is:
$$x+y=15$$
$$0.20x+0.14y=0.16(15)$$
Multiplying the first equation by –0.14:
$$-0.14x-0.14y=-2.1$$
$$0.20x+0.14y=2.4$$
Adding yields:
$$0.06x=0.3$$
$$x=5$$
$$y=10$$
The mixture contains 5 gallons of 20% disinfectant and 10 gallons of 14% disinfectant.

23. Let x represent the amount of nuts and y represent the amount of oats. The system of equations is:
$$x + y = 25$$
$$1.55x + 1.35y = 1.45(25)$$
Multiplying the first equation by -1.35:
$$-1.35x - 1.35y = -33.75$$
$$1.55x + 1.35y = 36.25$$
Adding yields:
$$0.20x = 2.5$$
$$x = 12.5$$
$$y = 12.5$$
The mixture contains 12.5 pounds of oats and 12.5 pounds of nuts.

25. Let b represent the rate of the boat and c represent the rate of the current. The system of equations is:
$$2(b+c) = 24$$
$$3(b-c) = 18$$
The system of equations simplifies to:
$$b + c = 12$$
$$b - c = 6$$
Adding yields:
$$2b = 18$$
$$b = 9$$
$$c = 3$$
The rate of the boat is 9 mph and the rate of the current is 3 mph.

27. Let a represent the rate of the airplane and w represent the rate of the wind. The system of equations is:
$$2(a+w) = 600$$
$$\frac{5}{2}(a-w) = 600$$
The system of equations simplifies to:
$$a + w = 300$$
$$a - w = 240$$
Adding yields:
$$2a = 540$$
$$a = 270$$
$$w = 30$$
The rate of the airplane is 270 mph and the rate of the wind is 30 mph.

29. Let n represent the number of nickels and d represent the number of dimes. The system of equations is:
$$n + d = 20$$
$$0.05n + 0.10d = 1.40$$
Multiplying the first equation by -0.05:
$$-0.05n - 0.05d = -1$$
$$0.05n + 0.10d = 1.40$$
Adding yields:
$$0.05d = 0.40$$
$$d = 8$$
$$n = 12$$
Bob has 12 nickels and 8 dimes.

31. Let n, d, and q represent the number of nickels, dimes, and quarters. The system of equations is:
$$n + d + q = 9$$
$$0.05n + 0.10d + 0.25q = 1.20$$
$$d = n$$
Substituting into the first equation:
$$n + n + q = 9$$
$$2n + q = 9$$
Substituting into the second equation:
$$0.05n + 0.10n + 0.25q = 1.20$$
$$0.15n + 0.25q = 1.20$$
The system of equations becomes:
$$2n + q = 9$$
$$0.15n + 0.25q = 1.20$$
Multiplying the first equation by -0.25:
$$-0.50n - 0.25q = -2.25$$
$$0.15n + 0.25q = 1.20$$
Adding yields:
$$-0.35n = -1.05$$
$$n = 3$$
$$d = 3$$
$$q = 9 - 2(3) = 3$$
The collection contains 3 nickels, 3 dimes, and 3 quarters.

33. Let n, d, and q represent the number of nickels, dimes, and quarters. The system of equations is:
$$n + d + q = 140$$
$$0.05n + 0.10d + 0.25q = 10.00$$
$$d = 2q$$
Substituting into the first equation:
$$n + 2q + q = 140$$
$$n + 3q = 140$$
Substituting into the second equation:
$$0.05n + 0.10(2q) + 0.25q = 10.00$$
$$0.05n + 0.45q = 10.00$$
The system of equations becomes:
$$n + 3q = 140$$
$$0.05n + 0.45q = 10.00$$
Multiplying the first equation by -0.05:
$$-0.05n - 0.15q = -7$$
$$0.05n + 0.45q = 10$$
Adding yields:
$$0.30q = 3$$
$$q = 10$$
$$d = 2(10) = 20$$
$$n = 140 - 3(10) = 110$$
Kaela has 110 nickels in the collection.

35. Let n represent the number of nickels and d represent the number of dimes. The system of equations is:
$$0.05n + 0.10d = 1.70$$
$$n = d + 4$$
Substituting into the first equation:
$$0.05(d+4) + 0.10d = 1.70$$
$$0.05d + 0.2 + 0.10d = 1.70$$
$$0.15d + 0.2 = 1.7$$
$$0.15d = 1.5$$
$$d = 10$$
$$n = 10 + 4 = 14$$
John has 14 nickels and 10 dimes.

37. Let $x = mp + b$ represent the relationship. Using the points (2,300) and (1.5,400) results in the system:
$$300 = 2m + b$$
$$400 = 1.5m + b$$
Multiplying the second equation by –1:
$$300 = 2m + b$$
$$-400 = -1.5m - b$$
Adding yields:
$$-100 = 0.5m$$
$$m = -200$$
$$b = 300 - 2(-200) = 700$$
The equation is $x = -200p + 700$. Substituting $p = 3$: $x = -200(3) + 700 = 100$ items

39. Let w represent the width and l represent the length. The system of equations is:
$$l = 3w + 5$$
$$2w + 2l = 58$$
Substituting into the second equation:
$$2w + 2(3w + 5) = 58$$
$$2w + 6w + 10 = 58$$
$$8w + 10 = 58$$
$$8w = 48$$
$$w = 6$$
$$l = 3(6) + 5 = 23$$
The width is 6 inches and the length is 23 inches.

41. The system of equations is:
$$a + b + c = 128$$
$$9a + 3b + c = 128$$
$$25a + 5b + c = 0$$
Multiply the first equation by –1 and add it to the second equation:
$$-a - b - c = -128$$
$$9a + 3b + c = 128$$
Adding yields:
$$8a + 2b = 0$$
$$4a + b = 0$$
Multiply the first equation by –1 and add it to the third equation:
$$-a - b - c = -128$$
$$25a + 5b + c = 0$$
Adding yields:
$$24a + 4b = -128$$
$$6a + b = -32$$
The system simplifies to:
$$4a + b = 0$$
$$6a + b = -32$$

Multiplying the first equation by –1:
$$-4a - b = 0$$
$$6a + b = -32$$
Adding yields:
$$2a = -32$$
$$a = -16$$
Substituting to find b:
$$4(-16) + b = 0$$
$$b = 64$$
Substituting to find c:
$$-16 + 64 + c = 128$$
$$c = 80$$
The equation for the height is $h = -16t^2 + 64t + 80$.

43. Let p represent the pre-registration tickets sold and o represent the on-site registration tickets sold. The system of equations is:
$$p + o = 29$$
$$28p + 38o = 922$$
Multiply the first equation by –28:
$$-28p - 28o = -812$$
$$28p + 38o = 922$$
Adding yields:
$$10o = 110$$
$$o = 11$$
Substituting into the first equation:
$$p + 11 = 29$$
$$p = 18$$
There were 18 pre-registration tickets sold and 11 on-site registration tickets sold.

45. No, the graph does not include the boundary line.

47. Substituting $x = 4 - y$ into the second equation:
$$(4 - y) - 2y = 4$$
$$4 - 3y = 4$$
$$-3y = 0$$
$$y = 0$$
The solution is $(4, 0)$.

49. Solving the inequality:
$$20x + 9,300 > 18,000$$
$$20x > 8,700$$
$$x > 435$$

Chapter 3 Test

1. The solution is $(0,6)$.

2. Graphing the two equations:

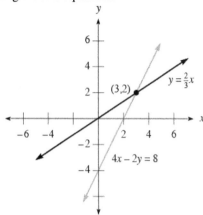

3. Graphing the two equations:

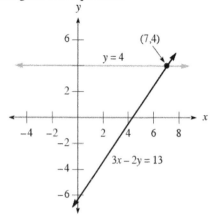

The intersection point is $(3,2)$.

The intersection point is $(7,4)$.

4. Graphing both lines:

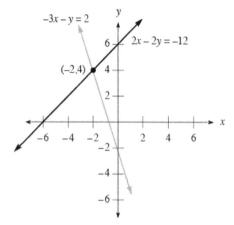

The intersection point is $(-2,4)$.

5. Multiplying the first equation by 3:
$$3x - 3y = -27$$
$$2x + 3y = 7$$
Adding the two equations:
$$5x = -20$$
$$x = -4$$
Substituting into the second equation:
$$2(-4) + 3y = 7$$
$$-8 + 3y = 7$$
$$3y = 15$$
$$y = 5$$
The solution is $(-4, 5)$.

6. Multiplying the first equation by –1:
$$-3x + y = -1$$
$$5x - y = 3$$
Adding the two equations:
$$2x = 2$$
$$x = 1$$
Substituting into the first equation:
$$3(1) - y = 1$$
$$3 - y = 1$$
$$-y = -2$$
$$y = 2$$
The solution is $(1, 2)$.

7. Multiplying the second equation by –2:
$$2x + 3y = -3$$
$$-2x - 12y = -24$$
Adding the two equations:
$$-9y = -27$$
$$y = 3$$
Substituting into the second equation:
$$x + 6(3) = 12$$
$$x + 18 = 12$$
$$x = -6$$
The solution is $(-6, 3)$.

8. Multiplying the first equation by –2:
$$-4x - 6y = -8$$
$$4x + 6y = 8$$
Adding the two equations: $0 = 0$
Since this statement is true, the system is dependent. The two lines coincide. The solution set is $\{(x,y) \mid 2x + 3y = 4\}$.

9. Substituting into the first equation:
$$3x - (2x - 8) = 12$$
$$3x - 2x + 8 = 12$$
$$x + 8 = 12$$
$$x = 4$$
Substituting into the second equation: $y = 2(4) - 8 = 8 - 8 = 0$. The solution is $(4, 0)$.

10. Substituting into the first equation:
$$3(4y - 17) - 6y = 3$$
$$12y - 51 - 6y = 3$$
$$6y - 51 = 3$$
$$6y = 54$$
$$y = 9$$
Substituting into the second equation: $x = 4(9) - 17 = 36 - 17 = 19$. The solution is $(19, 9)$.

11. Solving the second equation for y:
$$3x + y = -5$$
$$y = -3x - 5$$
Substituting into the first equation:
$$2x - 3(-3x - 5) = -18$$
$$2x + 9x + 15 = -18$$
$$11x + 15 = -18$$
$$11x = -33$$
$$x = -3$$
Substituting into the second equation: $y = -3(-3) - 5 = 9 - 5 = 4$. The solution is $(-3, 4)$.

12. Solving the second equation for x:
$$x - 4y = -1$$
$$x = 4y - 1$$
Substituting into the first equation:
$$2(4y - 1) - 3y = 13$$
$$8y - 2 - 3y = 13$$
$$5y - 2 = 13$$
$$5y = 15$$
$$y = 3$$
Substituting into the second equation: $x = 4(3) - 1 = 12 - 1 = 11$. The solution is $(11, 3)$.

13. Adding the first and third equations:
$$5x + 3y = 7$$
Multiplying the first equation by 2 and adding to the second equation:
$$2x + 8y + 4z = 10$$
$$x - 2y - 4z = -3$$
Adding yields:
$$3x + 6y = 7$$
So we have the system:
$$5x + 3y = 7$$
$$3x + 6y = 7$$
Multiplying the first equation by -2:
$$-10x - 6y = -14$$
$$3x + 6y = 7$$
Adding the two equations:
$$-7x = -7$$
$$x = 1$$
Substituting to find y:
$$5(1) + 3y = 7$$
$$5 + 3y = 7$$
$$3y = 2$$
$$y = \frac{2}{3}$$

Substituting into the original first equation:
$$1+4\left(\frac{2}{3}\right)+2z=5$$
$$\frac{11}{3}+2z=5$$
$$2z=\frac{4}{3}$$
$$z=\frac{2}{3}$$
The solution is $\left(1,\frac{2}{3},\frac{2}{3}\right)$.

14. Form the augmented matrix using the third equation as row 1: $\begin{bmatrix} -1 & 5 & 1 & | & 13 \\ 4 & -1 & -2 & | & -12 \\ -3 & -6 & 1 & | & -5 \end{bmatrix}$

Multiply row 1 by –1: $\begin{bmatrix} 1 & -5 & -1 & | & -13 \\ 4 & -1 & -2 & | & -12 \\ -3 & -6 & 1 & | & -5 \end{bmatrix}$

Add –4 times row 1 to row 2 and 3 times row 1 to row 3: $\begin{bmatrix} 1 & -5 & -1 & | & -13 \\ 0 & 19 & 2 & | & 40 \\ 0 & -21 & -2 & | & -44 \end{bmatrix}$

Add row 2 to row 3: $\begin{bmatrix} 1 & -5 & -1 & | & -13 \\ 0 & 19 & 2 & | & 40 \\ 0 & -2 & 0 & | & -4 \end{bmatrix}$

Divide row 3 by –2 and switch with row 2: $\begin{bmatrix} 1 & -5 & -1 & | & -13 \\ 0 & 1 & 0 & | & 2 \\ 0 & 19 & 2 & | & 40 \end{bmatrix}$

Add –19 times row 2 to row 3: $\begin{bmatrix} 1 & -5 & -1 & | & -13 \\ 0 & 1 & 0 & | & 2 \\ 0 & 0 & 2 & | & 2 \end{bmatrix}$

Divide row 3 by 2: $\begin{bmatrix} 1 & -5 & -1 & | & -13 \\ 0 & 1 & 0 & | & 2 \\ 0 & 0 & 1 & | & 1 \end{bmatrix}$

So $z = 1$ and $y = 2$. Substituting to find x:
$$x-5(2)-1(1)=-13$$
$$x-11=-13$$
$$x=-2$$
The solution is $(-2,2,1)$.

15. Evaluating the determinant: $\begin{vmatrix} 3 & -5 \\ -2 & 4 \end{vmatrix} = 3 \cdot 4 - (-5) \cdot (-2) = 12 - 10 = 2$

16. Expanding across the first row:
$\begin{vmatrix} 1 & 2 & 3 \\ 4 & 5 & 6 \\ 7 & 8 & 9 \end{vmatrix} = 1\begin{vmatrix} 5 & 6 \\ 8 & 9 \end{vmatrix} - 2\begin{vmatrix} 4 & 6 \\ 7 & 9 \end{vmatrix} + 3\begin{vmatrix} 4 & 5 \\ 7 & 8 \end{vmatrix} = 1(45-48) - 2(36-42) + 3(32-35) = -3 + 12 - 9 = 0$

17. First find the determinants:
$$D = \begin{vmatrix} 4 & -2 \\ 3 & 5 \end{vmatrix} = 20 + 6 = 26$$
$$D_x = \begin{vmatrix} 5 & -2 \\ 11 & 5 \end{vmatrix} = 25 + 22 = 47$$
$$D_y = \begin{vmatrix} 4 & 5 \\ 3 & 11 \end{vmatrix} = 44 - 15 = 29$$

Now use Cramer's rule:
$$x = \frac{D_x}{D} = \frac{47}{26} \qquad y = \frac{D_y}{D} = \frac{29}{26}$$

The solution is $\left(\frac{47}{26}, \frac{29}{26}\right)$.

18. Let x and y represent the two numbers. The system of equations is:
$$y = \frac{1}{2}x + 2$$
$$x + y = 8$$

Substituting into the second equation:
$$x + \frac{1}{2}x + 2 = 8$$
$$\frac{3}{2}x + 2 = 8$$
$$\frac{3}{2}x = 6$$
$$x = 4$$
$$y = \frac{1}{2}(4) + 2 = 4$$

The numbers are 4 and 4.

19. Let x represent the amount Ralph owes at 13% and $3x$ represent the amount he owes at 17%. The equation is:
$$0.13(x) + 0.17(3x) = 768$$
$$0.64x = 768$$
$$x = 1{,}200$$
$$3x = 3{,}600$$

Ralph owes $1,200 at 13% and $3,600 at 17%.

20. Let a represent the adult tickets and c represent the children's tickets. The system of equations is:
$$a + c = 890$$
$$15a + 5c = 11{,}750$$

Multiply the first equation by –5:
$$-5a - 5c = -4{,}450$$
$$15a + 5c = 11{,}750$$

Adding yields:
$$10a = 7{,}300$$
$$a = 730$$
$$c = 890 - 730 = 160$$

There were 730 adult tickets and 160 children's tickets sold.

21. Let b and c represent the rates of the boat and current. The system of equations is:
$$4(b+c) = 36$$
$$11(b-c) = 33$$
Simplifying the system by dividing the first equation by 4 and the second equation by 11:
$$b+c = 9$$
$$b-c = 3$$
Adding yields:
$$2b = 12$$
$$b = 6$$
$$c = 9-6 = 3$$
The boat's speed is 6 mph and the current's speed is 3 mph.

22. Let n, d, and q represent the number of nickels, dimes, and quarters. The system of equations is:
$$n+d+q = 8$$
$$0.05n + 0.10d + 0.25q = 1.05$$
$$n = d+q$$
Substituting into the first equation:
$$d+q+d+q = 8$$
$$2d+2q = 8$$
$$d+q = 4$$
Substituting into the second equation:
$$0.05(d+q) + 0.10d + 0.25q = 1.05$$
$$0.15d + 0.30q = 1.05$$
$$d+2q = 7$$
The system of equations becomes:
$$d+2q = 7$$
$$d+q = 4$$
Subtracting:
$$q = 3$$
$$d = 4-3 = 1$$
$$n = 1+3 = 4$$
The collection consists of 4 nickels, 1 dime, and 3 quarters.

Chapter 4
Exponents and Polynomials

4.1 Multiplication with Exponents

1. The base is 4 and the exponent is 2. Evaluating the expression: $4^2 = 4 \cdot 4 = 16$
3. The base is 0.3 and the exponent is 2. Evaluating the expression: $(0.3)^2 = 0.3 \cdot 0.3 = 0.09$
5. The base is 4 and the exponent is 3. Evaluating the expression: $4^3 = 4 \cdot 4 \cdot 4 = 64$
7. The base is -5 and the exponent is 2. Evaluating the expression: $(-5)^2 = (-5) \cdot (-5) = 25$
9. The base is 2 and the exponent is 3. Evaluating the expression: $-2^3 = -2 \cdot 2 \cdot 2 = -8$
11. The base is 3 and the exponent is 4. Evaluating the expression: $3^4 = 3 \cdot 3 \cdot 3 \cdot 3 = 81$
13. The base is $\frac{2}{3}$ and the exponent is 2. Evaluating the expression: $\left(\frac{2}{3}\right)^2 = \left(\frac{2}{3}\right) \cdot \left(\frac{2}{3}\right) = \frac{4}{9}$
15. The base is $\frac{1}{2}$ and the exponent is 4. Evaluating the expression: $\left(\frac{1}{2}\right)^4 = \left(\frac{1}{2}\right) \cdot \left(\frac{1}{2}\right) \cdot \left(\frac{1}{2}\right) \cdot \left(\frac{1}{2}\right) = \frac{1}{16}$

17. a. Completing the table:

Number (x)	1	2	3	4	5	6	7
Square (x^2)	1	4	9	16	25	36	49

   b. For numbers larger than 1, the square of the number is larger than the number.

19. Simplifying the expression: $x^4 \cdot x^5 = x^{4+5} = x^9$
21. Simplifying the expression: $y^{10} \cdot y^{20} = y^{10+20} = y^{30}$
23. Simplifying the expression: $2^5 \cdot 2^4 \cdot 2^3 = 2^{5+4+3} = 2^{12}$
25. Simplifying the expression: $x^4 \cdot x^6 \cdot x^8 \cdot x^{10} = x^{4+6+8+10} = x^{28}$
27. Simplifying the expression: $(x^2)^5 = x^{2 \cdot 5} = x^{10}$
29. Simplifying the expression: $(5^4)^3 = 5^{4 \cdot 3} = 5^{12}$
31. Simplifying the expression: $(y^3)^3 = y^{3 \cdot 3} = y^9$
33. Simplifying the expression: $(2^5)^{10} = 2^{5 \cdot 10} = 2^{50}$
35. Simplifying the expression: $(a^3)^x = a^{3x}$
37. Simplifying the expression: $(b^x)^y = b^{xy}$
39. Simplifying the expression: $(4x)^2 = 4^2 \cdot x^2 = 16x^2$
41. Simplifying the expression: $(2y)^5 = 2^5 \cdot y^5 = 32y^5$
43. Simplifying the expression: $(-3x)^4 = (-3)^4 \cdot x^4 = 81x^4$
45. Simplifying the expression: $(0.5ab)^2 = (0.5)^2 \cdot a^2 b^2 = 0.25 a^2 b^2$
47. Simplifying the expression: $(4xyz)^3 = 4^3 \cdot x^3 y^3 z^3 = 64 x^3 y^3 z^3$
49. Simplifying using properties of exponents: $(2x^4)^3 = 2^3 (x^4)^3 = 8x^{12}$
51. Simplifying using properties of exponents: $(4a^3)^2 = 4^2 (a^3)^2 = 16 a^6$
53. Simplifying using properties of exponents: $(x^2)^3 (x^4)^2 = x^6 \cdot x^8 = x^{14}$

55. Simplifying using properties of exponents: $(a^3)^1(a^2)^4 = a^3 \cdot a^8 = a^{11}$

57. Simplifying using properties of exponents: $(2x)^3(2x)^4 = (2x)^7 = 2^7 x^7 = 128x^7$

59. Simplifying using properties of exponents: $(3x^2)^3(2x)^4 = 3^3 x^6 \cdot 2^4 x^4 = 27x^6 \cdot 16x^4 = 432x^{10}$

61. Simplifying using properties of exponents: $(4x^2y^3)^2 = 4^2 x^4 y^6 = 16x^4 y^6$

63. Simplifying using properties of exponents: $\left(\frac{2}{3}a^4b^5\right)^3 = \left(\frac{2}{3}\right)^3 a^{12}b^{15} = \frac{8}{27}a^{12}b^{15}$

65. Completing the table:

Number (x)	−3	−2	−1	0	1	2	3
Square (x^2)	9	4	1	0	1	4	9

Constructing a line graph:

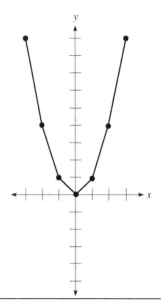

67. Completing the table:

Number (x)	−2.5	−1.5	−0.5	0	0.5	1.5	2.5
Square (x^2)	6.25	2.25	0.25	0	0.25	2.25	6.25

69. Writing in scientific notation: $43{,}200 = 4.32 \times 10^4$

71. Writing in scientific notation: $570 = 5.7 \times 10^2$

73. Writing in scientific notation: $238{,}000 = 2.38 \times 10^5$

75. Writing in expanded form: $2.49 \times 10^3 = 2{,}490$

77. Writing in expanded form: $3.52 \times 10^2 = 352$

79. Writing in expanded form: $2.8 \times 10^4 = 28{,}000$

81. The volume is given by: $V = (3 \text{ in.})^3 = 27 \text{ inches}^3$

83. The volume is given by: $V = (2.5 \text{ in.})^3 \approx 15.6 \text{ inches}^3$

85. The volume is given by: $V = (8 \text{ in.})(4.5 \text{ in.})(1 \text{ in.}) = 36 \text{ inches}^3$

87. Possibly, it depends on the actual dimensions of the box.

89. Writing in scientific notation: $650{,}000{,}000 \text{ seconds} = 6.5 \times 10^8 \text{ seconds}$

91. Writing in expanded form: $3.78 \times 10^7 \text{ heartbeats} = 37{,}800{,}000 \text{ heartbeats}$

93. Writing in expanded form: $3.27 \times 10^3 \text{ dollars} = \$3{,}270$

95. Substitute $c = 8$, $b = 3.32$, and $s = 3.66$: $d = \pi \cdot 3.66 \cdot 8 \cdot \left(\frac{1}{2} \cdot 3.32\right)^2 \approx 253 \text{ inches}^3$

97. Substitute $c = 6$, $b = 3.94$, and $s = 3.01$: $d = \pi \cdot 3.01 \cdot 6 \cdot \left(\frac{1}{2} \cdot 3.94\right)^2 \approx 220 \text{ inches}^3$

99. Subtracting: $-4-7=-4+(-7)=-11$

101. Subtracting: $-4-(-7)=-4+7=3$

103. Subtracting: $15-(-20)=15+20=35$

105. Subtracting: $-15-20=-15+(-20)=-35$

107. Simplifying: $5(3)-10=15-10=5$

109. Simplifying: $-8-2(3)=-8-6=-14$

111. Simplifying: $2(3)-4-3(-4)=6-4+12=14$

113. Simplifying: $2(3)+4(5)-5(2)=6+20-10=16$

4.2 Division with Exponents

1. Writing with positive exponents: $3^{-2}=\dfrac{1}{3^2}=\dfrac{1}{9}$

3. Writing with positive exponents: $6^{-2}=\dfrac{1}{6^2}=\dfrac{1}{36}$

5. Writing with positive exponents: $8^{-2}=\dfrac{1}{8^2}=\dfrac{1}{64}$

7. Writing with positive exponents: $5^{-3}=\dfrac{1}{5^3}=\dfrac{1}{125}$

9. Writing with positive exponents: $2x^{-3}=2\cdot\dfrac{1}{x^3}=\dfrac{2}{x^3}$

11. Writing with positive exponents: $(2x)^{-3}=\dfrac{1}{(2x)^3}=\dfrac{1}{8x^3}$

13. Writing with positive exponents: $(5y)^{-2}=\dfrac{1}{(5y)^2}=\dfrac{1}{25y^2}$

15. Writing with positive exponents: $10^{-2}=\dfrac{1}{10^2}=\dfrac{1}{100}$

17. Completing the table:

Number (x)	Square (x^2)	Power of 2 (2^x)
-3	9	$\tfrac{1}{8}$
-2	4	$\tfrac{1}{4}$
-1	1	$\tfrac{1}{2}$
0	0	1
1	1	2
2	4	4
3	9	8

19. Simplifying: $\dfrac{5^1}{5^3}=5^{1-3}=5^{-2}=\dfrac{1}{5^2}=\dfrac{1}{25}$

21. Simplifying: $\dfrac{x^{10}}{x^4}=x^{10-4}=x^6$

23. Simplifying: $\dfrac{4^3}{4^0}=4^{3-0}=4^3=64$

25. Simplifying: $\dfrac{(2x)^7}{(2x)^4}=(2x)^{7-4}=(2x)^3=2^3x^3=8x^3$

27. Simplifying: $\dfrac{6^{11}}{6}=\dfrac{6^{11}}{6^1}=6^{11-1}=6^{10}\ (=60,466,176)$

29. Simplifying: $\dfrac{6}{6^{11}}=\dfrac{6^1}{6^{11}}=6^{1-11}=6^{-10}=\dfrac{1}{6^{10}}\ \left(=\dfrac{1}{60,466,176}\right)$

31. Simplifying: $\dfrac{2^{-5}}{2^3}=2^{-5-3}=2^{-8}=\dfrac{1}{2^8}=\dfrac{1}{256}$

33. Simplifying: $\dfrac{2^5}{2^{-3}}=2^{5-(-3)}=2^{5+3}=2^8=256$

35. Simplifying: $\dfrac{(3x)^{-5}}{(3x)^{-8}}=(3x)^{-5-(-8)}=(3x)^{-5+8}=(3x)^3=3^3x^3=27x^3$

37. Simplifying: $(3xy)^4=3^4x^4y^4=81x^4y^4$

39. Simplifying: $10^0=1$

41. Simplifying: $(2a^2b)^1=2a^2b$

43. Simplifying: $(7y^3)^{-2}=\dfrac{1}{(7y^3)^2}=\dfrac{1}{49y^6}$

45. Simplifying: $x^{-3}\cdot x^{-5}=x^{-3-5}=x^{-8}=\dfrac{1}{x^8}$

47. Simplifying: $y^7\cdot y^{-10}=y^{7-10}=y^{-3}=\dfrac{1}{y^3}$

49. Simplifying: $\dfrac{(x^2)^3}{x^4} = \dfrac{x^6}{x^4} = x^{6-4} = x^2$

51. Simplifying: $\dfrac{(a^4)^3}{(a^3)^2} = \dfrac{a^{12}}{a^6} = a^{12-6} = a^6$

53. Simplifying: $\dfrac{y^7}{(y^2)^8} = \dfrac{y^7}{y^{16}} = y^{7-16} = y^{-9} = \dfrac{1}{y^9}$

55. Simplifying: $\left(\dfrac{y^7}{y^2}\right)^8 = (y^{7-2})^8 = (y^5)^8 = y^{40}$

57. Simplifying: $\dfrac{(x^{-2})^3}{x^{-5}} = \dfrac{x^{-6}}{x^{-5}} = x^{-6-(-5)} = x^{-6+5} = x^{-1} = \dfrac{1}{x}$

59. Simplifying: $\left(\dfrac{x^{-2}}{x^{-5}}\right)^3 = (x^{-2+5})^3 = (x^3)^3 = x^9$

61. Simplifying: $\dfrac{(a^3)^2(a^4)^5}{(a^5)^2} = \dfrac{a^6 \cdot a^{20}}{a^{10}} = \dfrac{a^{26}}{a^{10}} = a^{26-10} = a^{16}$

63. Simplifying: $\dfrac{(a^{-2})^3(a^4)^2}{(a^{-3})^{-2}} = \dfrac{a^{-6} \cdot a^8}{a^6} = \dfrac{a^2}{a^6} = a^{2-6} = a^{-4} = \dfrac{1}{a^4}$

65. Completing the table:

Number (x)	Power of 2 (2^x)
−3	1/8
−2	1/4
−1	1/2
0	1
1	2
2	4
3	8

Constructing the line graph:

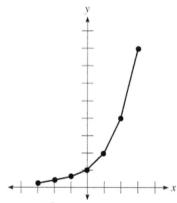

67. Writing in scientific notation: $0.0048 = 4.8 \times 10^{-3}$

69. Writing in scientific notation: $25 = 2.5 \times 10^1$

71. Writing in scientific notation: $0.000009 = 9 \times 10^{-6}$

73. Completing the table:

Expanded Form	Scientific Notation $(n \times 10^r)$
0.000357	3.57×10^{-4}
0.00357	3.57×10^{-3}
0.0357	3.57×10^{-2}
0.357	3.57×10^{-1}
3.57	3.57×10^{0}
35.7	3.57×10^{1}
357	3.57×10^{2}
3,570	3.57×10^{3}
35,700	3.57×10^{4}

75. Writing in expanded form: $4.23 \times 10^{-3} = 0.00423$

77. Writing in expanded form: $8 \times 10^{-5} = 0.00008$

79. Writing in expanded form: $4.2 \times 10^{0} = 4.2$

81. Writing in expanded form: 2×10^{-3} seconds = 0.002 seconds

83. Writing each number in scientific notation:
    Craven/Busch: $0.002 = 2 \times 10^{-3}$ Earnhardt/Irvan: $0.005 = 5 \times 10^{-3}$
    Harvick/Gordon: $0.006 = 6 \times 10^{-3}$ Kahne/Kenseth: $0.01 = 1 \times 10^{-2}$
    Kenseth/Kahne: $0.01 = 1 \times 10^{-2}$

85. Writing in scientific notation: $25 \times 10^3 = 2.5 \times 10^4$

87. Writing in scientific notation: $23.5 \times 10^4 = 2.35 \times 10^5$

89. Writing in scientific notation: $0.82 \times 10^{-3} = 8.2 \times 10^{-4}$

91. The area of the smaller square is $(10 \text{ in.})^2 = 100 \text{ inches}^2$, while the area of the larger square is $(20 \text{ in.})^2 = 400 \text{ inches}^2$. It would take 4 smaller squares to cover the larger square.

93. The area of the smaller square is x^2, while the area of the larger square is $(2x)^2 = 4x^2$. It would take 4 smaller squares to cover the larger square.

95. The volume of the smaller box is $(6 \text{ in.})^3 = 216 \text{ inches}^3$, while the volume of the larger box is $(12 \text{ in.})^3 = 1{,}728 \text{ inches}^3$. Thus 8 smaller boxes will fit inside the larger box ($8 \cdot 216 = 1{,}728$).

97. The volume of the smaller box is x^3, while the volume of the larger box is $(2x)^3 = 8x^3$. Thus 8 smaller boxes will fit inside the larger box.

99. Simplifying: $3(4.5) = 13.5$

101. Simplifying: $\frac{4}{5}(10) = \frac{40}{5} = 8$

103. Simplifying: $6.8(3.9) = 26.52$

105. Simplifying: $-3 + 15 = 12$

107. Simplifying: $x^5 \cdot x^3 = x^{5+3} = x^8$

109. Simplifying: $\frac{x^3}{x^2} = x^{3-2} = x$

111. Simplifying: $\frac{y^3}{y^5} = y^{3-5} = y^{-2} = \frac{1}{y^2}$

113. Writing in expanded form: $3.4 \times 10^2 = 340$

4.3 Operations with Monomials

1. Multiplying the monomials: $(3x^4)(4x^3) = 12x^{4+3} = 12x^7$

3. Multiplying the monomials: $(-2y^4)(8y^7) = -16y^{4+7} = -16y^{11}$

5. Multiplying the monomials: $(8x)(4x) = 32x^{1+1} = 32x^2$

7. Multiplying the monomials: $(10a^3)(10a)(2a^2) = 200a^{3+1+2} = 200a^6$

9. Multiplying the monomials: $(6ab^2)(-4a^2b) = -24a^{1+2}b^{2+1} = -24a^3b^3$

11. Multiplying the monomials: $(4x^2y)(3x^3y^3)(2xy^4) = 24x^{2+3+1}y^{1+3+4} = 24x^6y^8$

13. Dividing the monomials: $\dfrac{15x^3}{5x^2} = \dfrac{15}{5} \cdot \dfrac{x^3}{x^2} = 3x$

15. Dividing the monomials: $\dfrac{18y^9}{3y^{12}} = \dfrac{18}{3} \cdot \dfrac{y^9}{y^{12}} = 6 \cdot \dfrac{1}{y^3} = \dfrac{6}{y^3}$

17. Dividing the monomials: $\dfrac{32a^3}{64a^4} = \dfrac{32}{64} \cdot \dfrac{a^3}{a^4} = \dfrac{1}{2} \cdot \dfrac{1}{a} = \dfrac{1}{2a}$

19. Dividing the monomials: $\dfrac{21a^2b^3}{-7ab^5} = \dfrac{21}{-7} \cdot \dfrac{a^2}{a} \cdot \dfrac{b^3}{b^5} = -3 \cdot a \cdot \dfrac{1}{b^2} = -\dfrac{3a}{b^2}$

21. Dividing the monomials: $\dfrac{3x^3y^2z}{27xy^2z^3} = \dfrac{3}{27} \cdot \dfrac{x^3}{x} \cdot \dfrac{y^2}{y^2} \cdot \dfrac{z}{z^3} = \dfrac{1}{9} \cdot x^2 \cdot \dfrac{1}{z^2} = \dfrac{x^2}{9z^2}$

23. Completing the table:

a	b	ab	$\dfrac{a}{b}$	$\dfrac{b}{a}$
10	$5x$	$50x$	$\dfrac{2}{x}$	$\dfrac{x}{2}$
$20x^3$	$6x^2$	$120x^5$	$\dfrac{10x}{3}$	$\dfrac{3}{10x}$
$25x^5$	$5x^4$	$125x^9$	$5x$	$\dfrac{1}{5x}$
$3x^{-2}$	$3x^2$	9	$\dfrac{1}{x^4}$	x^4
$-2y^4$	$8y^7$	$-16y^{11}$	$-\dfrac{1}{4y^3}$	$-4y^3$

25. Finding the product: $(3 \times 10^3)(2 \times 10^5) = 6 \times 10^8$

27. Finding the product: $(3.5 \times 10^4)(5 \times 10^{-6}) = 17.5 \times 10^{-2} = 1.75 \times 10^{-1}$

29. Finding the product: $(5.5 \times 10^{-3})(2.2 \times 10^{-4}) = 12.1 \times 10^{-7} = 1.21 \times 10^{-6}$

31. Finding the quotient: $\dfrac{8.4 \times 10^5}{2 \times 10^2} = 4.2 \times 10^3$

33. Finding the quotient: $\dfrac{6 \times 10^8}{2 \times 10^{-2}} = 3 \times 10^{10}$

35. Finding the quotient: $\dfrac{2.5 \times 10^{-6}}{5 \times 10^{-4}} = 0.5 \times 10^{-2} = 5 \times 10^{-3}$

37. Combining the monomials: $3x^2 + 5x^2 = (3+5)x^2 = 8x^2$

39. Combining the monomials: $8x^5 - 19x^5 = (8-19)x^5 = -11x^5$

41. Combining the monomials: $2a + a - 3a = (2+1-3)a = 0a = 0$

43. Combining the monomials: $10x^3 - 8x^3 + 2x^3 = (10 - 8 + 2)x^3 = 4x^3$

45. Combining the monomials: $20ab^2 - 19ab^2 + 30ab^2 = (20 - 19 + 30)ab^2 = 31ab^2$

47. Completing the table:

a	b	ab	$a+b$
$5x$	$3x$	$15x^2$	$8x$
$4x^2$	$2x^2$	$8x^4$	$6x^2$
$3x^3$	$6x^3$	$18x^6$	$9x^3$
$2x^4$	$-3x^4$	$-6x^8$	$-x^4$
x^5	$7x^5$	$7x^{10}$	$8x^5$

49. Simplifying the expression: $\dfrac{(3x^2)(8x^5)}{6x^4} = \dfrac{24x^7}{6x^4} = \dfrac{24}{6} \cdot \dfrac{x^7}{x^4} = 4x^3$

51. Simplifying the expression: $\dfrac{(9a^2b)(2a^3b^4)}{18a^5b^7} = \dfrac{18a^5b^5}{18a^5b^7} = \dfrac{18}{18} \cdot \dfrac{a^5}{a^5} \cdot \dfrac{b^5}{b^7} = 1 \cdot \dfrac{1}{b^2} = \dfrac{1}{b^2}$

53. Simplifying the expression: $\dfrac{(4x^3y^2)(9x^4y^{10})}{(3x^5y)(2x^6y)} = \dfrac{36x^7y^{12}}{6x^{11}y^2} = \dfrac{36}{6} \cdot \dfrac{x^7}{x^{11}} \cdot \dfrac{y^{12}}{y^2} = 6 \cdot \dfrac{1}{x^4} \cdot y^{10} = \dfrac{6y^{10}}{x^4}$

55. Applying the distributive property: $xy\left(x + \dfrac{1}{y}\right) = xy \cdot x + xy \cdot \dfrac{1}{y} = x^2y + x$

57. Applying the distributive property: $xy\left(\dfrac{1}{y} + \dfrac{1}{x}\right) = xy \cdot \dfrac{1}{y} + xy \cdot \dfrac{1}{x} = x + y$

59. Applying the distributive property: $x^2\left(1 - \dfrac{4}{x^2}\right) = x^2 \cdot 1 - x^2 \cdot \dfrac{4}{x^2} = x^2 - 4$

61. Applying the distributive property: $x^2\left(1 - \dfrac{1}{x} - \dfrac{6}{x^2}\right) = x^2 \cdot 1 - x^2 \cdot \dfrac{1}{x} - x^2 \cdot \dfrac{6}{x^2} = x^2 - x - 6$

63. Applying the distributive property: $x^2\left(1 - \dfrac{5}{x}\right) = x^2 \cdot 1 - x^2 \cdot \dfrac{5}{x} = x^2 - 5x$

65. Applying the distributive property: $x^2\left(1 - \dfrac{8}{x}\right) = x^2 \cdot 1 - x^2 \cdot \dfrac{8}{x} = x^2 - 8x$

67. Simplifying the expression: $\dfrac{(6 \times 10^8)(3 \times 10^5)}{9 \times 10^7} = \dfrac{18 \times 10^{13}}{9 \times 10^7} = 2 \times 10^6$

69. Simplifying the expression: $\dfrac{(5 \times 10^3)(4 \times 10^{-5})}{2 \times 10^{-2}} = \dfrac{20 \times 10^{-2}}{2 \times 10^{-2}} = 10 = 1 \times 10^1$

71. Simplifying the expression: $\dfrac{(2.8 \times 10^{-7})(3.6 \times 10^4)}{2.4 \times 10^3} = \dfrac{10.08 \times 10^{-3}}{2.4 \times 10^3} = 4.2 \times 10^{-6}$

73. Simplifying the expression: $\dfrac{18x^4}{3x} + \dfrac{21x^7}{7x^4} = 6x^3 + 3x^3 = 9x^3$

75. Simplifying the expression: $\dfrac{45a^6}{9a^4} - \dfrac{50a^8}{2a^6} = 5a^2 - 25a^2 = -20a^2$

77. Simplifying the expression: $\dfrac{6x^7y^4}{3x^2y^2} + \dfrac{8x^5y^8}{2y^6} = 2x^5y^2 + 4x^5y^2 = 6x^5y^2$

79. Simplifying: $3 - 8 = -5$

81. Simplifying: $-1 + 7 = 6$

83. Simplifying: $3(5)^2 + 1 = 3(25) + 1 = 75 + 1 = 76$

85. Simplifying: $2x^2 + 4x^2 = 6x^2$

87. Simplifying: $-5x + 7x = 2x$

89. Simplifying: $-(2x+9) = -2x - 9$

91. Substituting $x = 4$: $2x + 3 = 2(4) + 3 = 8 + 3 = 11$

4.4 Addition and Subtraction of Polynomials

1. This is a trinomial of degree 3.
3. This is a trinomial of degree 3.
5. This is a binomial of degree 1.
7. This is a binomial of degree 2.
9. This is a monomial of degree 2.
11. This is a monomial of degree 0.

13. Combining the polynomials: $(2x^2 + 3x + 4) + (3x^2 + 2x + 5) = (2x^2 + 3x^2) + (3x + 2x) + (4 + 5) = 5x^2 + 5x + 9$

15. Combining the polynomials: $(3a^2 - 4a + 1) + (2a^2 - 5a + 6) = (3a^2 + 2a^2) + (-4a - 5a) + (1 + 6) = 5a^2 - 9a + 7$

17. Combining the polynomials: $x^2 + 4x + 2x + 8 = x^2 + (4x + 2x) + 8 = x^2 + 6x + 8$

19. Combining the polynomials: $6x^2 - 3x - 10x + 5 = 6x^2 + (-3x - 10x) + 5 = 6x^2 - 13x + 5$

21. Combining the polynomials: $x^2 - 3x + 3x - 9 = x^2 + (-3x + 3x) - 9 = x^2 - 9$

23. Combining the polynomials: $3y^2 - 5y - 6y + 10 = 3y^2 + (-5y - 6y) + 10 = 3y^2 - 11y + 10$

25. Combining the polynomials:
$$(6x^3 - 4x^2 + 2x) + (9x^2 - 6x + 3) = 6x^3 + (-4x^2 + 9x^2) + (2x - 6x) + 3 = 6x^3 + 5x^2 - 4x + 3$$

27. Combining the polynomials:
$$\left(\frac{2}{3}x^2 - \frac{1}{5}x - \frac{3}{4}\right) + \left(\frac{4}{3}x^2 - \frac{4}{5}x + \frac{7}{4}\right) = \left(\frac{2}{3}x^2 + \frac{4}{3}x^2\right) + \left(-\frac{1}{5}x - \frac{4}{5}x\right) + \left(-\frac{3}{4} + \frac{7}{4}\right) = 2x^2 - x + 1$$

29. Combining the polynomials: $(a^2 - a - 1) - (-a^2 + a + 1) = a^2 - a - 1 + a^2 - a - 1 = 2a^2 - 2a - 2$

31. Combining the polynomials:
$$\left(\frac{5}{9}x^3 + \frac{1}{3}x^2 - 2x + 1\right) - \left(\frac{2}{3}x^3 + x^2 + \frac{1}{2}x - \frac{3}{4}\right) = \frac{5}{9}x^3 + \frac{1}{3}x^2 - 2x + 1 - \frac{2}{3}x^3 - x^2 - \frac{1}{2}x + \frac{3}{4}$$
$$= -\frac{1}{9}x^3 - \frac{2}{3}x^2 - \frac{5}{2}x + \frac{7}{4}$$

33. Combining the polynomials:
$$(4y^2 - 3y + 2) + (5y^2 + 12y - 4) - (13y^2 - 6y + 20) = 4y^2 - 3y + 2 + 5y^2 + 12y - 4 - 13y^2 + 6y - 20$$
$$= (4y^2 + 5y^2 - 13y^2) + (-3y + 12y + 6y) + (2 - 4 - 20)$$
$$= -4y^2 + 15y - 22$$

35. Performing the subtraction:
$$(11x^2 - 10x + 13) - (10x^2 + 23x - 50) = 11x^2 - 10x + 13 - 10x^2 - 23x + 50$$
$$= (11x^2 - 10x^2) + (-10x - 23x) + (13 + 50)$$
$$= x^2 - 33x + 63$$

37. Performing the subtraction:
$$(11y^2 + 11y + 11) - (3y^2 + 7y - 15) = 11y^2 + 11y + 11 - 3y^2 - 7y + 15$$
$$= (11y^2 - 3y^2) + (11y - 7y) + (11 + 15)$$
$$= 8y^2 + 4y + 26$$

39. Performing the addition:
$$(25x^2 - 50x + 75) + (50x^2 - 100x - 150) = (25x^2 + 50x^2) + (-50x - 100x) + (75 - 150) = 75x^2 - 150x - 75$$

41. Performing the operations:
$(3x-2)+(11x+5)-(2x+1) = 3x-2+11x+5-2x-1 = (3x+11x-2x)+(-2+5-1) = 12x+2$

43. Evaluating when $x = 3$: $x^2 - 2x + 1 = (3)^2 - 2(3) + 1 = 9 - 6 + 1 = 4$

45. Finding the volume of the cylinder and sphere:
$$V_{\text{cylinder}} = \pi(3^2)(6) = 54\pi \qquad V_{\text{sphere}} = \frac{4}{3}\pi(3^3) = 36\pi$$
Subtracting to find the amount of space to pack: $V = 54\pi - 36\pi = 18\pi \approx 56.52 \text{ inches}^3$

47. Simplifying: $(-5)(-1) = 5$
49. Simplifying: $(-1)(6) = -6$
51. Simplifying: $(5x)(-4x) = -20x^2$
53. Simplifying: $3x(-7) = -21x$
55. Simplifying: $5x + (-3x) = 2x$
57. Multiplying: $3(2x-6) = 6x-18$

4.5 Multiplication with Polynomials

1. Using the distributive property: $2x(3x+1) = 2x(3x) + 2x(1) = 6x^2 + 2x$

3. Using the distributive property: $2x^2(3x^2 - 2x + 1) = 2x^2(3x^2) - 2x^2(2x) + 2x^2(1) = 6x^4 - 4x^3 + 2x^2$

5. Using the distributive property: $2ab(a^2 - ab + 1) = 2ab(a^2) - 2ab(ab) + 2ab(1) = 2a^3b - 2a^2b^2 + 2ab$

7. Using the distributive property: $y^2(3y^2 + 9y + 12) = y^2(3y^2) + y^2(9y) + y^2(12) = 3y^4 + 9y^3 + 12y^2$

9. Using the distributive property:
$4x^2y(2x^3y + 3x^2y^2 + 8y^3) = 4x^2y(2x^3y) + 4x^2y(3x^2y^2) + 4x^2y(8y^3) = 8x^5y^2 + 12x^4y^3 + 32x^2y^4$

11. Multiplying using the FOIL method: $(x+3)(x+4) = x^2 + 3x + 4x + 12 = x^2 + 7x + 12$

13. Multiplying using the FOIL method: $(x+6)(x+1) = x^2 + 6x + 1x + 6 = x^2 + 7x + 6$

15. Multiplying using the FOIL method: $\left(x + \frac{1}{2}\right)\left(x + \frac{3}{2}\right) = x^2 + \frac{1}{2}x + \frac{3}{2}x + \frac{3}{4} = x^2 + 2x + \frac{3}{4}$

17. Multiplying using the FOIL method: $(a+5)(a-3) = a^2 + 5a - 3a - 15 = a^2 + 2a - 15$

19. Multiplying using the FOIL method: $(x-a)(y+b) = xy + bx - ay - ab$

21. Multiplying using the FOIL method: $(x+6)(x-6) = x^2 - 6x + 6x - 36 = x^2 - 36$

23. Multiplying using the FOIL method: $\left(y + \frac{5}{6}\right)\left(y - \frac{5}{6}\right) = y^2 - \frac{5}{6}y + \frac{5}{6}y - \frac{25}{36} = y^2 - \frac{25}{36}$

25. Multiplying using the FOIL method: $(2x-3)(x-4) = 2x^2 - 8x - 3x + 12 = 2x^2 - 11x + 12$

27. Multiplying using the FOIL method: $(a+2)(2a-1) = 2a^2 - a + 4a - 2 = 2a^2 + 3a - 2$

29. Multiplying using the FOIL method: $(2x-5)(3x-2) = 6x^2 - 4x - 15x + 10 = 6x^2 - 19x + 10$

31. Multiplying using the FOIL method: $(2x+3)(a+4) = 2ax + 8x + 3a + 12$

33. Multiplying using the FOIL method: $(5x-4)(5x+4) = 25x^2 + 20x - 20x - 16 = 25x^2 - 16$

35. Multiplying using the FOIL method: $\left(2x - \frac{1}{2}\right)\left(x + \frac{3}{2}\right) = 2x^2 + 3x - \frac{1}{2}x - \frac{3}{4} = 2x^2 + \frac{5}{2}x - \frac{3}{4}$

37. Multiplying using the FOIL method: $(1-2a)(3-4a) = 3 - 4a - 6a + 8a^2 = 3 - 10a + 8a^2$

39. The product is $(x+2)(x+3) = x^2 + 5x + 6$:

	x	3
x	x^2	$3x$
2	$2x$	6

41. The product is $(x+1)(2x+2) = 2x^2 + 4x + 2$:

	x	x	2
x	x^2	x^2	$2x$
1	x	x	2

43. Multiplying using the column method:
$$\begin{array}{r} a^2 - 3a + 2 \\ a - 3 \\ \hline a^3 - 3a^2 + 2a \\ -3a^2 + 9a - 6 \\ \hline a^3 - 6a^2 + 11a - 6 \end{array}$$

45. Multiplying using the column method:
$$\begin{array}{r} x^2 - 2x + 4 \\ x + 2 \\ \hline x^3 - 2x^2 + 4x \\ 2x^2 - 4x + 8 \\ \hline x^3 + 8 \end{array}$$

47. Multiplying using the column method:
$$\begin{array}{r} x^2 + 8x + 9 \\ 2x + 1 \\ \hline 2x^3 + 16x^2 + 18x \\ x^2 + 8x + 9 \\ \hline 2x^3 + 17x^2 + 26x + 9 \end{array}$$

49. Multiplying using the column method:
$$\begin{array}{r} 5x^2 + 2x + 1 \\ x^2 - 3x + 5 \\ \hline 5x^4 + 2x^3 + x^2 \\ -15x^3 - 6x^2 - 3x \\ 25x^2 + 10x + 5 \\ \hline 5x^4 - 13x^3 + 20x^2 + 7x + 5 \end{array}$$

51. Multiplying using the FOIL method: $(x^2 + 3)(2x^2 - 5) = 2x^4 - 5x^2 + 6x^2 - 15 = 2x^4 + x^2 - 15$

53. Multiplying using the FOIL method: $(3a^4 + 2)(2a^2 + 5) = 6a^6 + 15a^4 + 4a^2 + 10$

55. First multiply two polynomials using the FOIL method: $(x + 3)(x + 4) = x^2 + 3x + 4x + 12 = x^2 + 7x + 12$
Now using the column method:
$$\begin{array}{r} x^2 + 7x + 12 \\ x + 5 \\ \hline x^3 + 7x^2 + 12x \\ 5x^2 + 35x + 60 \\ \hline x^3 + 12x^2 + 47x + 60 \end{array}$$

57. Simplifying: $(x - 3)(x - 2) + 2 = x^2 - 3x - 2x + 6 + 2 = x^2 - 5x + 8$
59. Simplifying: $(2x - 3)(4x + 3) + 4 = 8x^2 + 6x - 12x - 9 + 4 = 8x^2 - 6x - 5$
61. Simplifying: $(x + 4)(x - 5) + (-5)(2) = x^2 - 5x + 4x - 20 - 10 = x^2 - x - 30$
63. Simplifying: $2(x - 3) + x(x + 2) = 2x - 6 + x^2 + 2x = x^2 + 4x - 6$
65. Simplifying: $3x(x + 1) - 2x(x - 5) = 3x^2 + 3x - 2x^2 + 10x = x^2 + 13x$
67. Simplifying: $x(x + 2) - 3 = x^2 + 2x - 3$
69. Simplifying: $a(a - 3) + 6 = a^2 - 3a + 6$
71. Let x represent the width and $2x + 5$ represent the length. The area is given by: $A = x(2x + 5) = 2x^2 + 5x$
73. Let x and $x + 1$ represent the width and length, respectively. The area is given by: $A = x(x + 1) = x^2 + x$
75. Simplifying: $13 \cdot 13 = 169$
77. Simplifying: $2(x)(-5) = -10x$
79. Simplifying: $6x + (-6x) = 0$
81. Simplifying: $(2x)(-3) + (2x)(3) = -6x + 6x = 0$
83. Multiplying: $-4(3x - 4) = -12x + 16$
85. Multiplying: $(x - 1)(x + 2) = x^2 + 2x - x - 2 = x^2 + x - 2$
87. Multiplying: $(x + 3)(x + 3) = x^2 + 3x + 3x + 9 = x^2 + 6x + 9$

4.6 Binomial Squares and Other Special Products

1. Multiplying using the FOIL method: $(x-2)^2 = (x-2)(x-2) = x^2 - 2x - 2x + 4 = x^2 - 4x + 4$

3. Multiplying using the FOIL method: $(a+3)^2 = (a+3)(a+3) = a^2 + 3a + 3a + 9 = a^2 + 6a + 9$

5. Multiplying using the FOIL method: $(x-5)^2 = (x-5)(x-5) = x^2 - 5x - 5x + 25 = x^2 - 10x + 25$

7. Multiplying using the FOIL method: $\left(a - \frac{1}{2}\right)^2 = \left(a - \frac{1}{2}\right)\left(a - \frac{1}{2}\right) = a^2 - \frac{1}{2}a - \frac{1}{2}a + \frac{1}{4} = a^2 - a + \frac{1}{4}$

9. Multiplying using the FOIL method: $(x+10)^2 = (x+10)(x+10) = x^2 + 10x + 10x + 100 = x^2 + 20x + 100$

11. Multiplying using the square of binomial formula: $(a+0.8)^2 = a^2 + 2(a)(0.8) + (0.8)^2 = a^2 + 1.6a + 0.64$

13. Multiplying using the square of binomial formula: $(2x-1)^2 = (2x)^2 - 2(2x)(1) + (1)^2 = 4x^2 - 4x + 1$

15. Multiplying using the square of binomial formula: $(4a+5)^2 = (4a)^2 + 2(4a)(5) + (5)^2 = 16a^2 + 40a + 25$

17. Multiplying using the square of binomial formula: $(3x-2)^2 = (3x)^2 - 2(3x)(2) + (2)^2 = 9x^2 - 12x + 4$

19. Multiplying using the square of binomial formula: $(3a+5b)^2 = (3a)^2 + 2(3a)(5b) + (5b)^2 = 9a^2 + 30ab + 25b^2$

21. Multiplying using the square of binomial formula: $(4x-5y)^2 = (4x)^2 - 2(4x)(5y) + (5y)^2 = 16x^2 - 40xy + 25y^2$

23. Multiplying using the square of binomial formula: $(7m+2n)^2 = (7m)^2 + 2(7m)(2n) + (2n)^2 = 49m^2 + 28mn + 4n^2$

25. Multiplying using the square of binomial formula: $(6x-10y)^2 = (6x)^2 - 2(6x)(10y) + (10y)^2 = 36x^2 - 120xy + 100y^2$

27. Multiplying using the square of binomial formula: $(x^2+5)^2 = (x^2)^2 + 2(x^2)(5) + (5)^2 = x^4 + 10x^2 + 25$

29. Multiplying using the square of binomial formula: $(a^2+1)^2 = (a^2)^2 + 2(a^2)(1) + (1)^2 = a^4 + 2a^2 + 1$

31. Completing the table:

x	$(x+3)^2$	x^2+9	x^2+6x+9
1	16	10	16
2	25	13	25
3	36	18	36
4	49	25	49

33. Completing the table:

a	1	3	3	4
b	1	5	4	5
$(a+b)^2$	4	64	49	81
a^2+b^2	2	34	25	41
a^2+ab+b^2	3	49	37	61
$a^2+2ab+b^2$	4	64	49	81

35. Multiplying using the FOIL method: $(a+5)(a-5) = a^2 - 5a + 5a - 25 = a^2 - 25$

37. Multiplying using the FOIL method: $(y-1)(y+1) = y^2 + y - y - 1 = y^2 - 1$

39. Multiplying using the difference of squares formula: $(9+x)(9-x) = (9)^2 - (x)^2 = 81 - x^2$

41. Multiplying using the difference of squares formula: $(2x+5)(2x-5) = (2x)^2 - (5)^2 = 4x^2 - 25$

43. Multiplying using the difference of squares formula: $\left(4x+\frac{1}{3}\right)\left(4x-\frac{1}{3}\right) = (4x)^2 - \left(\frac{1}{3}\right)^2 = 16x^2 - \frac{1}{9}$

45. Multiplying using the difference of squares formula: $(2a+7)(2a-7) = (2a)^2 - (7)^2 = 4a^2 - 49$

47. Multiplying using the difference of squares formula: $(6-7x)(6+7x) = (6)^2 - (7x)^2 = 36 - 49x^2$

49. Multiplying using the difference of squares formula: $(x^2+3)(x^2-3) = (x^2)^2 - (3)^2 = x^4 - 9$

51. Multiplying using the difference of squares formula: $(a^2+4)(a^2-4) = (a^2)^2 - (4)^2 = a^4 - 16$

53. Multiplying using the difference of squares formula: $(5y^4-8)(5y^4+8) = (5y^4)^2 - (8)^2 = 25y^8 - 64$

55. Multiplying and simplifying: $(x+3)(x-3)+(x+5)(x-5) = (x^2-9)+(x^2-25) = 2x^2 - 34$

57. Multiplying and simplifying:
$(2x+3)^2 - (4x-1)^2 = (4x^2+12x+9) - (16x^2-8x+1) = 4x^2+12x+9-16x^2+8x-1 = -12x^2+20x+8$

59. Multiplying and simplifying:
$$(a+1)^2 - (a+2)^2 + (a+3)^2 = (a^2+2a+1) - (a^2+4a+4) + (a^2+6a+9)$$
$$= a^2+2a+1-a^2-4a-4+a^2+6a+9$$
$$= a^2+4a+6$$

61. Multiplying and simplifying:
$$(2x+3)^3 = (2x+3)(2x+3)^2$$
$$= (2x+3)(4x^2+12x+9)$$
$$= 8x^3+24x^2+18x+12x^2+36x+27$$
$$= 8x^3+36x^2+54x+27$$

63. Multiplying: $(49)(51) = (50-1)(50+1) = 50^2 - 1^2 = 2{,}500 - 1 = 2{,}499$

65. Evaluating when $x = 2$:
$$(x+3)^2 = (2+3)^2 = (5)^2 = 25$$
$$x^2+6x+9 = (2)^2+6(2)+9 = 4+12+9 = 25$$
Both expressions are equal to 25.

67. Let x and $x+1$ represent the two integers. The expression can be written as:
$$(x)^2+(x+1)^2 = x^2+(x^2+2x+1) = 2x^2+2x+1$$

69. Let x, $x+1$, and $x+2$ represent the three integers. The expression can be written as:
$$(x)^2+(x+1)^2+(x+2)^2 = x^2+(x^2+2x+1)+(x^2+4x+4) = 3x^2+6x+5$$

71. Verifying the areas: $(a+b)^2 = a^2+ab+ab+b^2 = a^2+2ab+b^2$

73. Simplifying: $\dfrac{10x^3}{5x} = 2x^{3-1} = 2x^2$

75. Simplifying: $\dfrac{3x^2}{3} = x^2$

77. Simplifying: $\dfrac{9x^2}{3x} = 3x^{2-1} = 3x$

79. Simplifying: $\dfrac{24x^3y^2}{8x^2y} = 3x^{3-2}y^{2-1} = 3xy$

4.7 Dividing a Polynomial by a Monomial

1. Performing the division: $\dfrac{5x^2-10x}{5x} = \dfrac{5x^2}{5x} - \dfrac{10x}{5x} = x-2$

3. Performing the division: $\dfrac{15x-10x^3}{5x} = \dfrac{15x}{5x} - \dfrac{10x^3}{5x} = 3-2x^2$

5. Performing the division: $\dfrac{25x^2y-10xy}{5x} = \dfrac{25x^2y}{5x} - \dfrac{10xy}{5x} = 5xy-2y$

7. Performing the division: $\dfrac{35x^5-30x^4+25x^3}{5x} = \dfrac{35x^5}{5x} - \dfrac{30x^4}{5x} + \dfrac{25x^3}{5x} = 7x^4-6x^3+5x^2$

9. Performing the division: $\dfrac{50x^5-25x^3+5x}{5x} = \dfrac{50x^5}{5x} - \dfrac{25x^3}{5x} + \dfrac{5x}{5x} = 10x^4-5x^2+1$

11. Performing the division: $\dfrac{8a^2-4a}{-2a} = \dfrac{8a^2}{-2a} + \dfrac{-4a}{-2a} = -4a+2$

13. Performing the division: $\dfrac{16a^5+24a^4}{-2a} = \dfrac{16a^5}{-2a} + \dfrac{24a^4}{-2a} = -8a^4-12a^3$

15. Performing the division: $\dfrac{8ab+10a^2}{-2a} = \dfrac{8ab}{-2a} + \dfrac{10a^2}{-2a} = -4b-5a$

17. Performing the division: $\dfrac{12a^3b-6a^2b^2+14ab^3}{-2a} = \dfrac{12a^3b}{-2a} + \dfrac{-6a^2b^2}{-2a} + \dfrac{14ab^3}{-2a} = -6a^2b+3ab^2-7b^3$

19. Performing the division: $\dfrac{a^2+2ab+b^2}{-2a} = \dfrac{a^2}{-2a} + \dfrac{2ab}{-2a} + \dfrac{b^2}{-2a} = -\dfrac{a}{2}-b-\dfrac{b^2}{2a}$

21. Performing the division: $\dfrac{6x+8y}{2} = \dfrac{6x}{2} + \dfrac{8y}{2} = 3x+4y$

23. Performing the division: $\dfrac{7y-21}{-7} = \dfrac{7y}{-7} + \dfrac{-21}{-7} = -y+3$

25. Performing the division: $\dfrac{10xy-8x}{2x} = \dfrac{10xy}{2x} - \dfrac{8x}{2x} = 5y-4$

27. Performing the division: $\dfrac{x^2y-x^3y^2}{x} = \dfrac{x^2y}{x} - \dfrac{x^3y^2}{x} = xy-x^2y^2$

29. Performing the division: $\dfrac{x^2y-x^3y^2}{-x^2y} = \dfrac{x^2y}{-x^2y} + \dfrac{-x^3y^2}{-x^2y} = -1+xy$

31. Performing the division: $\dfrac{a^2b^2-ab^2}{-ab^2} = \dfrac{a^2b^2}{-ab^2} + \dfrac{-ab^2}{-ab^2} = -a+1$

33. Performing the division: $\dfrac{x^3-3x^2y+xy^2}{x} = \dfrac{x^3}{x} - \dfrac{3x^2y}{x} + \dfrac{xy^2}{x} = x^2-3xy+y^2$

35. Performing the division: $\dfrac{10a^2-15a^2b+25a^2b^2}{5a^2} = \dfrac{10a^2}{5a^2} - \dfrac{15a^2b}{5a^2} + \dfrac{25a^2b^2}{5a^2} = 2-3b+5b^2$

37. Performing the division: $\dfrac{26x^2y^2-13xy}{-13xy} = \dfrac{26x^2y^2}{-13xy} + \dfrac{-13xy}{-13xy} = -2xy+1$

39. Performing the division: $\dfrac{4x^2y^2-2xy}{4xy} = \dfrac{4x^2y^2}{4xy} - \dfrac{2xy}{4xy} = xy-\dfrac{1}{2}$

41. Performing the division: $\dfrac{5a^2x - 10ax^2 + 15a^2x^2}{20a^2x^2} = \dfrac{5a^2x}{20a^2x^2} - \dfrac{10ax^2}{20a^2x^2} + \dfrac{15a^2x^2}{20a^2x^2} = \dfrac{1}{4x} - \dfrac{1}{2a} + \dfrac{3}{4}$

43. Performing the division: $\dfrac{16x^5 + 8x^2 + 12x}{12x^3} = \dfrac{16x^5}{12x^3} + \dfrac{8x^2}{12x^3} + \dfrac{12x}{12x^3} = \dfrac{4x^2}{3} + \dfrac{2}{3x} + \dfrac{1}{x^2}$

45. Performing the division: $\dfrac{9a^{5m} - 27a^{3m}}{3a^{2m}} = \dfrac{9a^{5m}}{3a^{2m}} - \dfrac{27a^{3m}}{3a^{2m}} = 3a^{5m-2m} - 9a^{3m-2m} = 3a^{3m} - 9a^m$

47. Performing the division:
$\dfrac{10x^{5m} - 25x^{3m} + 35x^m}{5x^m} = \dfrac{10x^{5m}}{5x^m} - \dfrac{25x^{3m}}{5x^m} + \dfrac{35x^m}{5x^m} = 2x^{5m-m} - 5x^{3m-m} + 7x^{m-m} = 2x^{4m} - 5x^{2m} + 7$

49. Simplifying and then dividing:
$$\dfrac{2x^3(3x+2) - 3x^2(2x-4)}{2x^2} = \dfrac{6x^4 + 4x^3 - 6x^3 + 12x^2}{2x^2}$$
$$= \dfrac{6x^4 - 2x^3 + 12x^2}{2x^2}$$
$$= \dfrac{6x^4}{2x^2} - \dfrac{2x^3}{2x^2} + \dfrac{12x^2}{2x^2}$$
$$= 3x^2 - x + 6$$

51. Simplifying and then dividing:
$\dfrac{(x+2)^2 - (x-2)^2}{2x} = \dfrac{(x^2+4x+4)-(x^2-4x+4)}{2x} = \dfrac{x^2+4x+4-x^2+4x-4}{2x} = \dfrac{8x}{2x} = 4$

53. Simplifying and then dividing:
$\dfrac{(x+5)^2 + (x+5)(x-5)}{2x} = \dfrac{(x^2+10x+25)+(x^2-25)}{2x} = \dfrac{2x^2+10x}{2x} = \dfrac{2x^2}{2x} + \dfrac{10x}{2x} = x+5$

55. Evaluating when $x = 2$: $2x + 3 = 2(2) + 3 = 4 + 3 = 7$

Evaluating when $x = 2$: $\dfrac{10x+15}{5} = \dfrac{10(2)+15}{5} = \dfrac{20+15}{5} = \dfrac{35}{5} = 7$

57. Evaluating when $x = 10$: $\dfrac{3x+8}{2} = \dfrac{3(10)+8}{2} = \dfrac{30+8}{2} = \dfrac{38}{2} = 19$

Evaluating when $x = 10$: $3x + 4 = 3(10) + 4 = 30 + 4 = 34$

59. Dividing:

$$\begin{array}{r} 146 \\ 27\overline{)3962} \\ \underline{27} \\ 126 \\ \underline{108} \\ 182 \\ \underline{162} \\ 20 \end{array}$$

The quotient is $146\dfrac{20}{27}$.

61. Dividing: $\dfrac{2x^2+5x}{x} = \dfrac{2x^2}{x} + \dfrac{5x}{x} = 2x+5$

63. Multiplying: $(x-3)x = x(x-3) = x^2 - 3x$

65. Multiplying: $2x^2(x-5) = 2x^3 - 10x^2$

67. Subtracting: $(x^2-5x)-(x^2-3x) = x^2-5x-x^2+3x = -2x$

69. Subtracting: $(-2x+8)-(-2x+6) = -2x+8+2x-6 = 2$

4.8 Dividing a Polynomial by a Polynomial

1. Using long division:

$$\begin{array}{r} x-2 \\ x-3\overline{\smash{)}x^2-5x+6} \\ \underline{x^2-3x} \\ -2x+6 \\ \underline{-2x+6} \\ 0 \end{array}$$

The quotient is $x-2$.

3. Using long division:

$$\begin{array}{r} a+4 \\ a+5\overline{\smash{)}a^2+9a+20} \\ \underline{a^2+5a} \\ 4a+20 \\ \underline{4a+20} \\ 0 \end{array}$$

The quotient is $a+4$.

5. Using long division:

$$\begin{array}{r} x-3 \\ x-3\overline{\smash{)}x^2-6x+9} \\ \underline{x^2-3x} \\ -3x+9 \\ \underline{-3x+9} \\ 0 \end{array}$$

The quotient is $x-3$.

7. Using long division:

$$\begin{array}{r} x+3 \\ 2x-1\overline{\smash{)}2x^2+5x-3} \\ \underline{2x^2-x} \\ 6x-3 \\ \underline{6x-3} \\ 0 \end{array}$$

The quotient is $x+3$.

9. Using long division:

$$\begin{array}{r} a-5 \\ 2a+1\overline{\smash{)}2a^2-9a-5} \\ \underline{2a^2+a} \\ -10a-5 \\ \underline{-10a-5} \\ 0 \end{array}$$

The quotient is $a-5$.

11. Using long division:

$$\begin{array}{r} x+2 \\ x+3\overline{\smash{)}x^2+5x+8} \\ \underline{x^2+3x} \\ 2x+8 \\ \underline{2x+6} \\ 2 \end{array}$$

The quotient is $x+2+\dfrac{2}{x+3}$.

13. Using long division:

$$\begin{array}{r} a-2 \\ a+5\overline{\smash{)}a^2+3a+2} \\ \underline{a^2+5a} \\ -2a+2 \\ \underline{-2a-10} \\ 12 \end{array}$$

The quotient is $a-2+\dfrac{12}{a+5}$.

15. Using long division:

$$\begin{array}{r} x+4 \\ x-2\overline{\smash{)}x^2+2x+1} \\ \underline{x^2-2x} \\ 4x+1 \\ \underline{4x-8} \\ 9 \end{array}$$

The quotient is $x+4+\dfrac{9}{x-2}$.

17. Using long division:
$$\begin{array}{r} x+4 \\ x+1{\overline{\smash{\big)}\,x^2+5x-6}} \\ \underline{x^2+x} \\ 4x-6 \\ \underline{4x+4} \\ -10 \end{array}$$

The quotient is $x+4+\dfrac{-10}{x+1}$.

19. Using long division:
$$\begin{array}{r} a+1 \\ a+2{\overline{\smash{\big)}\,a^2+3a+1}} \\ \underline{a^2+2a} \\ a+1 \\ \underline{a+2} \\ -1 \end{array}$$

The quotient is $a+1+\dfrac{-1}{a+2}$.

21. Using long division:
$$\begin{array}{r} x-3 \\ 2x+4{\overline{\smash{\big)}\,2x^2-2x+5}} \\ \underline{2x^2+4x} \\ -6x+5 \\ \underline{-6x-12} \\ 17 \end{array}$$

The quotient is $x-3+\dfrac{17}{2x+4}$.

23. Using long division:
$$\begin{array}{r} 3a-2 \\ 2a+3{\overline{\smash{\big)}\,6a^2+5a+1}} \\ \underline{6a^2+9a} \\ -4a+1 \\ \underline{-4a-6} \\ 7 \end{array}$$

The quotient is $3a-2+\dfrac{7}{2a+3}$.

25. Using long division:
$$\begin{array}{r} 2a^2-a-3 \\ 3a-5{\overline{\smash{\big)}\,6a^3-13a^2-4a+15}} \\ \underline{6a^3-10a^2} \\ -3a^2-4a \\ \underline{-3a^2+5a} \\ -9a+15 \\ \underline{-9a+15} \\ 0 \end{array}$$

The quotient is $2a^2-a-3$.

27. Using long division:
$$\begin{array}{r} x^2-x+5 \\ x+1{\overline{\smash{\big)}\,x^3+0x^2+4x+5}} \\ \underline{x^3+x^2} \\ -x^2+4x \\ \underline{-x^2-x} \\ 5x+5 \\ \underline{5x+5} \\ 0 \end{array}$$

The quotient is x^2-x+5.

29. Using long division:
$$\begin{array}{r} x^2+x+1 \\ x-1{\overline{\smash{\big)}\,x^3+0x^2+0x-1}} \\ \underline{x^3-x^2} \\ x^2+0x \\ \underline{x^2-x} \\ x-1 \\ \underline{x-1} \\ 0 \end{array}$$

The quotient is x^2+x+1.

31. Using long division:
$$\begin{array}{r} x^2+2x+4 \\ x-2{\overline{\smash{\big)}\,x^3+0x^2+0x-8}} \\ \underline{x^3-2x^2} \\ 2x^2+0x \\ \underline{2x^2-4x} \\ 4x-8 \\ \underline{4x-8} \\ 0 \end{array}$$

The quotient is x^2+2x+4.

33. Dividing: $\dfrac{\$5{,}894}{12} \approx \$491.17$ per month

35. Dividing: $\dfrac{\$3{,}977}{12} \approx \$331.42$ per month

37. Evaluating: $6(3+4)+5 = 6(7)+5 = 42+5 = 47$

39. Evaluating: $1^2+2^2+3^2 = 1+4+9 = 14$

41. Evaluating: $5(6+3\cdot 2)+4+3\cdot 2 = 5(6+6)+4+3\cdot 2 = 5(12)+4+6 = 60+4+6 = 70$

43. Evaluating: $(1^3+2^3)+[(2\cdot 3)+(4\cdot 5)] = (1+8)+(6+20) = 9+26 = 35$

45. Evaluating: $(2 \cdot 3 + 4 + 5) \div 3 = (6 + 4 + 5) \div 3 = 15 \div 3 = 5$

47. Evaluating: $6 \cdot 10^3 + 5 \cdot 10^2 + 4 \cdot 10^1 = 6,000 + 500 + 40 = 6,540$

49. Evaluating: $1 \cdot 10^3 + 7 \cdot 10^2 + 6 \cdot 10^1 + 0 = 1,000 + 700 + 60 + 0 = 1,760$

51. Evaluating: $4 \cdot 2 - 1 + 5 \cdot 3 - 2 = 8 - 1 + 15 - 2 = 20$

53. Evaluating: $(2^3 + 3^2) \cdot 4 - 5 = (8 + 9) \cdot 4 - 5 = 17 \cdot 4 - 5 = 68 - 5 = 63$

55. Evaluating: $2(2^2 + 3^2) + 3(3^2) = 2(4 + 9) + 3(9) = 2(13) + 3(9) = 26 + 27 = 53$

Chapter 4 Test

1. Simplifying the expression: $(-2)^5 = (-2)(-2)(-2)(-2)(-2) = -32$

2. Simplifying the expression: $\left(\frac{2}{3}\right)^3 = \left(\frac{2}{3}\right) \cdot \left(\frac{2}{3}\right) \cdot \left(\frac{2}{3}\right) = \frac{8}{27}$

3. Simplifying the expression: $(4x^2)^2 (2x^3)^3 = 16x^4 \cdot 8x^9 = 128x^{13}$

4. Simplifying the expression: $4^{-2} = \frac{1}{4^2} = \frac{1}{16}$

5. Simplifying the expression: $(4a^5 b^3)^0 = 1$

6. Simplifying the expression: $\frac{x^{-4}}{x^{-7}} = x^{-4-(-7)} = x^{-4+7} = x^3$

7. Simplifying the expression: $\frac{(x^{-3})^2 (x^{-5})^{-3}}{(x^{-3})^{-4}} = \frac{x^{-6} x^{15}}{x^{12}} = \frac{x^9}{x^{12}} = x^{9-12} = x^{-3} = \frac{1}{x^3}$

8. Writing in scientific notation: $0.04307 = 4.307 \times 10^{-2}$

9. Writing in expanded form: $7.63 \times 10^6 = 7,630,000$

10. Simplifying the expression: $\frac{17x^2 y^5 z^3}{51x^4 y^2 z} = \frac{17}{51} \cdot \frac{x^2}{x^4} \cdot \frac{y^5}{y^2} \cdot \frac{z^3}{z} = \frac{1}{3} \cdot \frac{1}{x^2} \cdot y^3 \cdot z^2 = \frac{y^3 z^2}{3x^2}$

11. Simplifying the expression: $\frac{(3a^3 b)(4a^2 b^5)}{24a^2 b^4} = \frac{12a^5 b^6}{24a^2 b^4} = \frac{a^3 b^2}{2}$

12. Simplifying the expression: $\frac{28x^4}{4x} + \frac{30x^7}{6x^4} = 7x^3 + 5x^3 = 12x^3$

13. Simplifying the expression: $\frac{(1.1 \times 10^5)(3 \times 10^{-2})}{4.4 \times 10^{-5}} = \frac{3.3 \times 10^3}{4.4 \times 10^{-5}} = 0.75 \times 10^8 = 7.5 \times 10^7$

14. Performing the operations: $(9x^2 - 2x) + (7x + 4) = 9x^2 - 2x + 7x + 4 = 9x^2 + 5x + 4$

15. Performing the operations: $(4x^2 + 5x - 6) - (2x^2 - x - 4) = 4x^2 + 5x - 6 - 2x^2 + x + 4 = 2x^2 + 6x - 2$

16. Performing the operations: $(7x + 3) - (2x + 7) = 7x + 3 - 2x - 7 = 5x - 4$

17. Evaluating when $a = -3$: $3a^2 + 4a + 6 = 3(-3)^2 + 4(-3) + 6 = 27 - 12 + 6 = 21$

18. Multiplying using the distributive property: $3x^2(5x^2 - 2x + 4) = 3x^2(5x^2) - 3x^2(2x) + 3x^2(4) = 15x^4 - 6x^3 + 12x^2$

19. Multiplying using the FOIL method: $\left(x + \frac{1}{4}\right)\left(x - \frac{1}{3}\right) = x^2 - \frac{1}{3}x + \frac{1}{4}x - \frac{1}{12} = x^2 - \frac{1}{12}x - \frac{1}{12}$

20. Multiplying using the FOIL method: $(2x - 3)(5x + 6) = 10x^2 + 12x - 15x - 18 = 10x^2 - 3x - 18$

21. Multiplying using the column method:

$$\begin{array}{r} x^2 - 4x + 16 \\ \underline{x + 4} \\ x^3 - 4x^2 + 16x \\ \underline{4x^2 - 16x + 64} \\ x^3 + 64 \end{array}$$

22. Multiplying using the square of binomial formula: $(x-6)^2 = (x)^2 - 2(x)(6) + (6)^2 = x^2 - 12x + 36$

23. Multiplying using the square of binomial formula: $(2a+4b)^2 = (2a)^2 + 2(2a)(4b) + (4b)^2 = 4a^2 + 16ab + 16b^2$

24. Multiplying using the difference of squares formula: $(3x-6)(3x+6) = (3x)^2 - (6)^2 = 9x^2 - 36$

25. Multiplying using the difference of squares formula: $(x^2-4)(x^2+4) = (x^2)^2 - (4)^2 = x^4 - 16$

26. Dividing the monomial: $\dfrac{18x^3 - 36x^2 + 6x}{6x} = \dfrac{18x^3}{6x} - \dfrac{36x^2}{6x} + \dfrac{6x}{6x} = 3x^2 - 6x + 1$

27. Using long division:

$$\begin{array}{r} 3x - 1 \\ 3x-1{\overline{\smash{\big)}\,9x^2 - 6x - 4}} \\ \underline{9x^2 - 3x} \\ -3x - 4 \\ \underline{-3x + 1} \\ -5 \end{array}$$

The quotient is $3x - 1 + \dfrac{-5}{3x-1}$.

28. Using long division:

$$\begin{array}{r} 4x + 11 \\ x-4{\overline{\smash{\big)}\,4x^2 - 5x + 6}} \\ \underline{4x^2 - 16x} \\ 11x + 6 \\ \underline{11x - 44} \\ 50 \end{array}$$

The quotient is $4x + 11 + \dfrac{50}{x-4}$.

29. Using the volume formula: $V = (3.2 \text{ in.})^3 \approx 32.77 \text{ inches}^3$

30. Let w represent the width, $3w$ represent the length, and $\dfrac{1}{3}w$ represent the height. The volume is given by:

$$V = (w)(3w)\left(\dfrac{1}{3}w\right) = w^3$$

Chapter 5
Factoring

5.1 The Greatest Common Factor and Factoring by Grouping

1. Factoring out the greatest common factor: $15x + 25 = 5(3x + 5)$
3. Factoring out the greatest common factor: $6a + 9 = 3(2a + 3)$
5. Factoring out the greatest common factor: $4x - 8y = 4(x - 2y)$
7. Factoring out the greatest common factor: $3x^2 - 6x + 9 = 3(x^2 - 2x + 3)$
9. Factoring out the greatest common factor: $3a^2 - 3a - 60 = 3(a^2 - a - 20)$
11. Factoring out the greatest common factor: $24y^2 - 52y + 24 = 4(6y^2 - 13y + 6)$
13. Factoring out the greatest common factor: $9x^2 - 8x^3 = x^2(9 - 8x)$
15. Factoring out the greatest common factor: $13a^2 - 26a^3 = 13a^2(1 - 2a)$
17. Factoring out the greatest common factor: $21x^2y - 28xy^2 = 7xy(3x - 4y)$
19. Factoring out the greatest common factor: $22a^2b^2 - 11ab^2 = 11ab^2(2a - 1)$
21. Factoring out the greatest common factor: $7x^3 + 21x^2 - 28x = 7x(x^2 + 3x - 4)$
23. Factoring out the greatest common factor: $121y^4 - 11x^4 = 11(11y^4 - x^4)$
25. Factoring out the greatest common factor: $100x^4 - 50x^3 + 25x^2 = 25x^2(4x^2 - 2x + 1)$
27. Factoring out the greatest common factor: $8a^2 + 16b^2 + 32c^2 = 8(a^2 + 2b^2 + 4c^2)$
29. Factoring out the greatest common factor: $4a^2b - 16ab^2 + 32a^2b^2 = 4ab(a - 4b + 8ab)$
31. Factoring out the greatest common factor: $121a^3b^2 - 22a^2b^3 + 33a^3b^3 = 11a^2b^2(11a - 2b + 3ab)$
33. Factoring out the greatest common factor: $12x^2y^3 - 72x^5y^3 - 36x^4y^4 = 12x^2y^3(1 - 6x^3 - 3x^2y)$
35. Factoring by grouping: $xy + 5x + 3y + 15 = x(y + 5) + 3(y + 5) = (y + 5)(x + 3)$
37. Factoring by grouping: $xy + 6x + 2y + 12 = x(y + 6) + 2(y + 6) = (y + 6)(x + 2)$
39. Factoring by grouping: $ab + 7a - 3b - 21 = a(b + 7) - 3(b + 7) = (b + 7)(a - 3)$
41. Factoring by grouping: $ax - bx + ay - by = x(a - b) + y(a - b) = (a - b)(x + y)$
43. Factoring by grouping: $2ax + 6x - 5a - 15 = 2x(a + 3) - 5(a + 3) = (a + 3)(2x - 5)$
45. Factoring by grouping: $3xb - 4b - 6x + 8 = b(3x - 4) - 2(3x - 4) = (3x - 4)(b - 2)$
47. Factoring by grouping: $x^2 + 2a + 2x + ax = x^2 + ax + 2x + 2a = x(x + a) + 2(x + a) = (x + a)(x + 2)$
49. Factoring by grouping: $x^2 + ab - ax - bx = x^2 - ax - bx + ab = x(x - a) - b(x - a) = (x - a)(x - b)$
51. Factoring by grouping: $ax + ay + bx + by + cx + cy = a(x + y) + b(x + y) + c(x + y) = (x + y)(a + b + c)$

53. Factoring by grouping: $6x^2 + 9x + 4x + 6 = 3x(2x+3) + 2(2x+3) = (2x+3)(3x+2)$

55. Factoring by grouping: $20x^2 - 2x + 50x - 5 = 2x(10x-1) + 5(10x-1) = (10x-1)(2x+5)$

57. Factoring by grouping: $20x^2 + 4x + 25x + 5 = 4x(5x+1) + 5(5x+1) = (5x+1)(4x+5)$

59. Factoring by grouping: $x^3 + 2x^2 + 3x + 6 = x^2(x+2) + 3(x+2) = (x+2)(x^2+3)$

61. Factoring by grouping: $6x^3 - 4x^2 + 15x - 10 = 2x^2(3x-2) + 5(3x-2) = (3x-2)(2x^2+5)$

63. Its greatest common factor is $3 \cdot 2 = 6$.

65. The correct factoring is: $12x^2 + 6x + 3 = 3(4x^2 + 2x + 1)$

67. Factoring: $1{,}000 + 1{,}000r = 1{,}000(1+r)$

    Evaluating when $r = 0.12$: $1{,}000(1+0.12) = 1{,}000(1.12) = \$1{,}120$

69. a. Factoring: $A = 1{,}000{,}000 + 1{,}000{,}000r = 1{,}000{,}000(1+r)$

    b. Evaluating when $r = 0.30$: $A = 1{,}000{,}000(1+0.30) = 1{,}300{,}000$

71. Multiplying using the FOIL method: $(x-7)(x+2) = x^2 + 2x - 7x - 14 = x^2 - 5x - 14$

73. Multiplying using the FOIL method: $(x-3)(x+2) = x^2 + 2x - 3x - 6 = x^2 - x - 6$

75. Multiplying using the column method:
$$\begin{array}{r} x^2 - 3x + 9 \\ x + 3 \\ \hline x^3 - 3x^2 + 9x \\ 3x^2 - 9x + 27 \\ \hline x^3 + 27 \end{array}$$

77. Multiplying using the column method:
$$\begin{array}{r} x^2 + 4x - 3 \\ 2x + 1 \\ \hline 2x^3 + 8x^2 - 6x \\ x^2 + 4x - 3 \\ \hline 2x^3 + 9x^2 - 2x - 3 \end{array}$$

79. Multiplying: $3x^4(6x^3 - 4x^2 + 2x) = 3x^4 \cdot 6x^3 - 3x^4 \cdot 4x^2 + 3x^4 \cdot 2x = 18x^7 - 12x^6 + 6x^5$

81. Multiplying: $\left(x + \frac{1}{3}\right)\left(x + \frac{2}{3}\right) = x^2 + \frac{2}{3}x + \frac{1}{3}x + \frac{2}{9} = x^2 + x + \frac{2}{9}$

83. Multiplying: $(6x+4y)(2x-3y) = 12x^2 - 18xy + 8xy - 12y^2 = 12x^2 - 10xy - 12y^2$

85. Multiplying: $(9a+1)(9a-1) = 81a^2 - 9a + 9a - 1 = 81a^2 - 1$

87. Multiplying: $(x-9)(x-9) = x^2 - 9x - 9x + 81 = x^2 - 18x + 81$

89. Multiplying: $(x+2)(x^2 - 2x + 4) = x(x^2 - 2x + 4) + 2(x^2 - 2x + 4) = x^3 - 2x^2 + 4x + 2x^2 - 4x + 8 = x^3 + 8$

5.2 Factoring Trinomials

1. Factoring the trinomial: $x^2 + 7x + 12 = (x+3)(x+4)$
3. Factoring the trinomial: $x^2 + 3x + 2 = (x+2)(x+1)$
5. Factoring the trinomial: $a^2 + 10a + 21 = (a+7)(a+3)$
7. Factoring the trinomial: $x^2 - 7x + 10 = (x-5)(x-2)$
9. Factoring the trinomial: $y^2 - 10y + 21 = (y-7)(y-3)$
11. Factoring the trinomial: $x^2 - x - 12 = (x-4)(x+3)$
13. Factoring the trinomial: $y^2 + y - 12 = (y+4)(y-3)$
15. Factoring the trinomial: $x^2 + 5x - 14 = (x+7)(x-2)$
17. Factoring the trinomial: $r^2 - 8r - 9 = (r-9)(r+1)$
19. Factoring the trinomial: $x^2 - x - 30 = (x-6)(x+5)$
21. Factoring the trinomial: $a^2 + 15a + 56 = (a+7)(a+8)$
23. Factoring the trinomial: $y^2 - y - 42 = (y-7)(y+6)$
25. Factoring the trinomial: $x^2 + 13x + 42 = (x+7)(x+6)$
27. Factoring the trinomial: $2x^2 + 6x + 4 = 2(x^2 + 3x + 2) = 2(x+2)(x+1)$
29. Factoring the trinomial: $3a^2 - 3a - 60 = 3(a^2 - a - 20) = 3(a-5)(a+4)$

31. Factoring the trinomial: $100x^2 - 500x + 600 = 100(x^2 - 5x + 6) = 100(x-3)(x-2)$

33. Factoring the trinomial: $100p^2 - 1,300p + 4,000 = 100(p^2 - 13p + 40) = 100(p-8)(p-5)$

35. Factoring the trinomial: $x^4 - x^3 - 12x^2 = x^2(x^2 - x - 12) = x^2(x-4)(x+3)$

37. Factoring the trinomial: $2r^3 + 4r^2 - 30r = 2r(r^2 + 2r - 15) = 2r(r+5)(r-3)$

39. Factoring the trinomial: $2y^4 - 6y^3 - 8y^2 = 2y^2(y^2 - 3y - 4) = 2y^2(y-4)(y+1)$

41. Factoring the trinomial: $x^5 + 4x^4 + 4x^3 = x^3(x^2 + 4x + 4) = x^3(x+2)(x+2) = x^3(x+2)^2$

43. Factoring the trinomial: $3y^4 - 12y^3 - 15y^2 = 3y^2(y^2 - 4y - 5) = 3y^2(y-5)(y+1)$

45. Factoring the trinomial: $4x^4 - 52x^3 + 144x^2 = 4x^2(x^2 - 13x + 36) = 4x^2(x-9)(x-4)$

47. Factoring the trinomial: $x^2 + 5xy + 6y^2 = (x+2y)(x+3y)$

49. Factoring the trinomial: $x^2 - 9xy + 20y^2 = (x-4y)(x-5y)$

51. Factoring the trinomial: $a^2 + 2ab - 8b^2 = (a+4b)(a-2b)$

53. Factoring the trinomial: $a^2 - 10ab + 25b^2 = (a-5b)(a-5b) = (a-5b)^2$

55. Factoring the trinomial: $a^2 + 10ab + 25b^2 = (a+5b)(a+5b) = (a+5b)^2$

57. Factoring the trinomial: $x^2 + 2xa - 48a^2 = (x+8a)(x-6a)$

59. Factoring the trinomial: $x^2 - 5xb - 36b^2 = (x-9b)(x+4b)$

61. Factoring the trinomial: $x^4 - 5x^2 + 6 = (x^2 - 2)(x^2 - 3)$

63. Factoring the trinomial: $x^2 - 80x - 2,000 = (x-100)(x+20)$

65. Factoring the trinomial: $x^2 - x + \dfrac{1}{4} = \left(x - \dfrac{1}{2}\right)\left(x - \dfrac{1}{2}\right) = \left(x - \dfrac{1}{2}\right)^2$

67. Factoring the trinomial: $x^2 + 0.6x + 0.08 = (x+0.4)(x+0.2)$

69. We can use long division to find the other factor:

$$\begin{array}{r} x+16 \\ x+8{\overline{\smash{\big)}\,x^2+24x+128}} \\ \underline{x^2+8x} \\ 16x+128 \\ \underline{16x+128} \\ 0 \end{array}$$

The other factor is $x+16$.

71. Using FOIL to multiply out the factors: $(4x+3)(x-1) = 4x^2 - 4x + 3x - 3 = 4x^2 - x - 3$

73. Multiplying using the FOIL method: $(6a+1)(a+2) = 6a^2 + 12a + a + 2 = 6a^2 + 13a + 2$

75. Multiplying using the FOIL method: $(3a+2)(2a+1) = 6a^2 + 3a + 4a + 2 = 6a^2 + 7a + 2$

77. Multiplying using the FOIL method: $(6a+2)(a+1) = 6a^2 + 6a + 2a + 2 = 6a^2 + 8a + 2$

5.3 More Trinomials to Factor

1. Factoring the trinomial: $2x^2 + 7x + 3 = (2x+1)(x+3)$
3. Factoring the trinomial: $2a^2 - a - 3 = (2a-3)(a+1)$
5. Factoring the trinomial: $3x^2 + 2x - 5 = (3x+5)(x-1)$
7. Factoring the trinomial: $3y^2 - 14y - 5 = (3y+1)(y-5)$
9. Factoring the trinomial: $6x^2 + 13x + 6 = (3x+2)(2x+3)$
11. Factoring the trinomial: $4x^2 - 12xy + 9y^2 = (2x-3y)(2x-3y) = (2x-3y)^2$
13. Factoring the trinomial: $4y^2 - 11y - 3 = (4y+1)(y-3)$
15. Factoring the trinomial: $20x^2 - 41x + 20 = (4x-5)(5x-4)$
17. Factoring the trinomial: $20a^2 + 48ab - 5b^2 = (10a-b)(2a+5b)$
19. Factoring the trinomial: $20x^2 - 21x - 5 = (4x-5)(5x+1)$
21. Factoring the trinomial: $12m^2 + 16m - 3 = (6m-1)(2m+3)$
23. Factoring the trinomial: $20x^2 + 37x + 15 = (4x+5)(5x+3)$
25. Factoring the trinomial: $12a^2 - 25ab + 12b^2 = (3a-4b)(4a-3b)$
27. Factoring the trinomial: $3x^2 - xy - 14y^2 = (3x-7y)(x+2y)$
29. Factoring the trinomial: $14x^2 + 29x - 15 = (2x+5)(7x-3)$
31. Factoring the trinomial: $6x^2 - 43x + 55 = (3x-5)(2x-11)$
33. Factoring the trinomial: $15t^2 - 67t + 38 = (5t-19)(3t-2)$
35. Factoring the trinomial: $4x^2 + 2x - 6 = 2(2x^2 + x - 3) = 2(2x+3)(x-1)$
37. Factoring the trinomial: $24a^2 - 50a + 24 = 2(12a^2 - 25a + 12) = 2(4a-3)(3a-4)$
39. Factoring the trinomial: $10x^3 - 23x^2 + 12x = x(10x^2 - 23x + 12) = x(5x-4)(2x-3)$
41. Factoring the trinomial: $6x^4 - 11x^3 - 10x^2 = x^2(6x^2 - 11x - 10) = x^2(3x+2)(2x-5)$
43. Factoring the trinomial: $10a^3 - 6a^2 - 4a = 2a(5a^2 - 3a - 2) = 2a(5a+2)(a-1)$
45. Factoring the trinomial: $15x^3 - 102x^2 - 21x = 3x(5x^2 - 34x - 7) = 3x(5x+1)(x-7)$
47. Factoring the trinomial: $35y^3 - 60y^2 - 20y = 5y(7y^2 - 12y - 4) = 5y(7y+2)(y-2)$
49. Factoring the trinomial: $15a^4 - 2a^3 - a^2 = a^2(15a^2 - 2a - 1) = a^2(5a+1)(3a-1)$
51. Factoring the trinomial: $24x^2y - 6xy - 45y = 3y(8x^2 - 2x - 15) = 3y(4x+5)(2x-3)$
53. Factoring the trinomial: $12x^2y - 34xy^2 + 14y^3 = 2y(6x^2 - 17xy + 7y^2) = 2y(2x-y)(3x-7y)$
55. Evaluating each expression when $x = 2$:
    $2x^2 + 7x + 3 = 2(2)^2 + 7(2) + 3 = 8 + 14 + 3 = 25$
    $(2x+1)(x+3) = (2 \cdot 2 + 1)(2+3) = (5)(5) = 25$
    Both expressions equal 25.
57. Multiplying using the difference of squares formula: $(2x+3)(2x-3) = (2x)^2 - (3)^2 = 4x^2 - 9$
59. Multiplying using the difference of squares formula: $(x+3)(x-3)(x^2+9) = (x^2-9)(x^2+9) = (x^2)^2 - (9)^2 = x^4 - 81$

61. a. Factoring: $h = 8 + 62t - 16t^2 = -2(8t^2 - 31t - 4) = -2(t-4)(8t+1)$

    b. Completing the table:

Time t (seconds)	0	1	2	3	4
Height h (feet)	8	54	68	50	0

63. a. Factoring: $V = x(99 - 40x + 4x^2) = x(11-2x)(9-2x)$

    b. Since $2x$ is cut from each side, the original box had dimensions of 11 inches by 9 inches.

65. Multiplying: $(x+3)(x-3) = x^2 - (3)^2 = x^2 - 9$ 67. Multiplying: $(x+5)(x-5) = x^2 - (5)^2 = x^2 - 25$

69. Multiplying: $(x+7)(x-7) = x^2 - (7)^2 = x^2 - 49$ 71. Multiplying: $(x+9)(x-9) = x^2 - (9)^2 = x^2 - 81$

73. Multiplying: $(2x-3y)(2x+3y) = (2x)^2 - (3y)^2 = 4x^2 - 9y^2$

75. Multiplying: $(x^2+4)(x+2)(x-2) = (x^2+4)(x^2-4) = (x^2)^2 - (4)^2 = x^4 - 16$

77. Multiplying: $(x+3)^2 = x^2 + 2(x)(3) + (3)^2 = x^2 + 6x + 9$

79. Multiplying: $(x+5)^2 = x^2 + 2(x)(5) + (5)^2 = x^2 + 10x + 25$

81. Multiplying: $(x+7)^2 = x^2 + 2(x)(7) + (7)^2 = x^2 + 14x + 49$

83. Multiplying: $(x+9)^2 = x^2 + 2(x)(9) + (9)^2 = x^2 + 18x + 81$

85. Multiplying: $(2x+3)^2 = (2x)^2 + 2(2x)(3) + (3)^2 = 4x^2 + 12x + 9$

87. Multiplying: $(4x-2y)^2 = (4x)^2 - 2(4x)(2y) + (2y)^2 = 16x^2 - 16xy + 4y^2$

5.4 The Difference of Two Squares

1. Factoring the binomial: $x^2 - 9 = (x+3)(x-3)$

3. Factoring the binomial: $a^2 - 36 = (a+6)(a-6)$

5. Factoring the binomial: $x^2 - 49 = (x+7)(x-7)$

7. Factoring the binomial: $4a^2 - 16 = 4(a^2 - 4) = 4(a+2)(a-2)$

9. The expression $9x^2 + 25$ cannot be factored.

11. Factoring the binomial: $25x^2 - 169 = (5x+13)(5x-13)$

13. Factoring the binomial: $9a^2 - 16b^2 = (3a+4b)(3a-4b)$

15. Factoring the binomial: $9 - m^2 = (3+m)(3-m)$

17. Factoring the binomial: $25 - 4x^2 = (5+2x)(5-2x)$

19. Factoring the binomial: $2x^2 - 18 = 2(x^2-9) = 2(x+3)(x-3)$

21. Factoring the binomial: $32a^2 - 128 = 32(a^2-4) = 32(a+2)(a-2)$

23. Factoring the binomial: $8x^2y - 18y = 2y(4x^2-9) = 2y(2x+3)(2x-3)$

25. Factoring the binomial: $a^4 - b^4 = (a^2+b^2)(a^2-b^2) = (a^2+b^2)(a+b)(a-b)$

27. Factoring the binomial: $16m^4 - 81 = (4m^2+9)(4m^2-9) = (4m^2+9)(2m+3)(2m-3)$

29. Factoring the binomial: $3x^3y - 75xy^3 = 3xy(x^2-25y^2) = 3xy(x+5y)(x-5y)$

31. Factoring the trinomial: $x^2 - 2x + 1 = (x-1)(x-1) = (x-1)^2$

33. Factoring the trinomial: $x^2 + 2x + 1 = (x+1)(x+1) = (x+1)^2$

35. Factoring the trinomial: $a^2 - 10a + 25 = (a-5)(a-5) = (a-5)^2$

37. Factoring the trinomial: $y^2 + 4y + 4 = (y+2)(y+2) = (y+2)^2$

39. Factoring the trinomial: $x^2 - 4x + 4 = (x-2)(x-2) = (x-2)^2$

41. Factoring the trinomial: $m^2 - 12m + 36 = (m-6)(m-6) = (m-6)^2$

43. Factoring the trinomial: $4a^2 + 12a + 9 = (2a+3)(2a+3) = (2a+3)^2$

45. Factoring the trinomial: $49x^2 - 14x + 1 = (7x-1)(7x-1) = (7x-1)^2$

47. Factoring the trinomial: $9y^2 - 30y + 25 = (3y-5)(3y-5) = (3y-5)^2$

49. Factoring the trinomial: $x^2 + 10xy + 25y^2 = (x+5y)(x+5y) = (x+5y)^2$

51. Factoring the trinomial: $9a^2 + 6ab + b^2 = (3a+b)(3a+b) = (3a+b)^2$

53. Factoring the trinomial: $3a^2 + 18a + 27 = 3(a^2 + 6a + 9) = 3(a+3)(a+3) = 3(a+3)^2$

55. Factoring the trinomial: $2x^2 + 20xy + 50y^2 = 2(x^2 + 10xy + 25y^2) = 2(x+5y)(x+5y) = 2(x+5y)^2$

57. Factoring the trinomial: $5x^3 + 30x^2y + 45xy^2 = 5x(x^2 + 6xy + 9y^2) = 5x(x+3y)(x+3y) = 5x(x+3y)^2$

59. Factoring by grouping: $x^2 + 6x + 9 - y^2 = (x+3)^2 - y^2 = (x+3+y)(x+3-y)$

61. Factoring by grouping: $x^2 + 2xy + y^2 - 9 = (x+y)^2 - 9 = (x+y+3)(x+y-3)$

63. Since $(x+7)^2 = x^2 + 14x + 49$, the value is $b = 14$.

65. Since $(x+5)^2 = x^2 + 10x + 25$, the value is $c = 25$.

67. a. Subtracting the area of the missing square, the area is $A = x^2 - 16$.
    b. Factoring: $A = x^2 - 16 = (x+4)(x-4)$
    c. Make a single rectangle of dimensions $x - 4$ by $x + 4$.

69. Subtracting the area of the missing square: $A = a^2 - b^2 = (a+b)(a-b)$

71. a. Multiplying: $1^3 = 1$ b. Multiplying: $2^3 = 8$
    c. Multiplying: $3^3 = 27$ d. Multiplying: $4^3 = 64$
    e. Multiplying: $5^3 = 125$

73. a. Multiplying: $x(x^2 - x + 1) = x^3 - x^2 + x$ b. Multiplying: $1(x^2 - x + 1) = x^2 - x + 1$
    c. Multiplying: $(x+1)(x^2 - x + 1) = x(x^2 - x + 1) + 1(x^2 - x + 1) = x^3 - x^2 + x + x^2 - x + 1 = x^3 + 1$

75. a. Multiplying: $x(x^2 - 2x + 4) = x^3 - 2x^2 + 4x$ b. Multiplying: $2(x^2 - 2x + 4) = 2x^2 - 4x + 8$
    c. Multiplying: $(x+2)(x^2 - 2x + 4) = x(x^2 - 2x + 4) + 2(x^2 - 2x + 4) = x^3 - 2x^2 + 4x + 2x^2 - 4x + 8 = x^3 + 8$

77. a. Multiplying: $x(x^2 - 3x + 9) = x^3 - 3x^2 + 9x$ b. Multiplying: $3(x^2 - 3x + 9) = 3x^2 - 9x + 27$
    c. Multiplying: $(x+3)(x^2 - 3x + 9) = x(x^2 - 3x + 9) + 3(x^2 - 3x + 9) = x^3 - 3x^2 + 9x + 3x^2 - 9x + 27 = x^3 + 27$

5.5 The Sum and Difference of Two Cubes

1. Factoring: $x^3 - y^3 = (x-y)(x^2 + xy + y^2)$
3. Factoring: $a^3 + 8 = (a+2)(a^2 - 2a + 4)$
5. Factoring: $27 + x^3 = (3+x)(9 - 3x + x^2)$
7. Factoring: $y^3 - 1 = (y-1)(y^2 + y + 1)$
9. Factoring: $y^3 - 64 = (y-4)(y^2 + 4y + 16)$
11. Factoring: $125h^3 - t^3 = (5h - t)(25h^2 + 5ht + t^2)$
13. Factoring: $x^3 - 216 = (x-6)(x^2 + 6x + 36)$
15. Factoring: $2y^3 - 54 = 2(y^3 - 27) = 2(y-3)(y^2 + 3y + 9)$
17. Factoring: $2a^3 - 128b^3 = 2(a^3 - 64b^3) = 2(a - 4b)(a^2 + 4ab + 16b^2)$
19. Factoring: $2x^3 + 432y^3 = 2(x^3 + 216y^3) = 2(x + 6y)(x^2 - 6xy + 36y^2)$
21. Factoring: $10a^3 - 640b^3 = 10(a^3 - 64b^3) = 10(a - 4b)(a^2 + 4ab + 16b^2)$
23. Factoring: $10r^3 - 1,250 = 10(r^3 - 125) = 10(r - 5)(r^2 + 5r + 25)$
25. Factoring: $64 + 27a^3 = (4 + 3a)(16 - 12a + 9a^2)$
27. Factoring: $8x^3 - 27y^3 = (2x - 3y)(4x^2 + 6xy + 9y^2)$
29. Factoring: $t^3 + \dfrac{1}{27} = \left(t + \dfrac{1}{3}\right)\left(t^2 - \dfrac{1}{3}t + \dfrac{1}{9}\right)$
31. Factoring: $27x^3 - \dfrac{1}{27} = \left(3x - \dfrac{1}{3}\right)\left(9x^2 + x + \dfrac{1}{9}\right)$
33. Factoring: $64a^3 + 125b^3 = (4a + 5b)(16a^2 - 20ab + 25b^2)$
35. Factoring: $\dfrac{1}{8}x^3 - \dfrac{1}{27}y^3 = \left(\dfrac{1}{2}x - \dfrac{1}{3}y\right)\left(\dfrac{1}{4}x^2 + \dfrac{1}{6}xy + \dfrac{1}{9}y^2\right)$
37. Factoring: $a^6 - b^6 = (a^3 + b^3)(a^3 - b^3) = (a + b)(a^2 - ab + b^2)(a - b)(a^2 + ab + b^2)$
39. Factoring: $64x^6 - y^6 = (8x^3 + y^3)(8x^3 - y^3) = (2x + y)(4x^2 - 2xy + y^2)(2x - y)(4x^2 + 2xy + y^2)$
41. Factoring: $x^6 - (5y)^6 = (x^3 + (5y)^3)(x^3 - (5y)^3) = (x + 5y)(x^2 - 5xy + 25y^2)(x - 5y)(x^2 + 5xy + 25y^2)$
43. Multiplying: $2x^3(x+2)(x-2) = 2x^3(x^2 - 4) = 2x^5 - 8x^3$
45. Multiplying: $3x^2(x-3)^2 = 3x^2(x^2 - 6x + 9) = 3x^4 - 18x^3 + 27x^2$
47. Multiplying: $y(y^2 + 25) = y^3 + 25y$
49. Multiplying: $(5a - 2)(3a + 1) = 15a^2 + 5a - 6a - 2 = 15a^2 - a - 2$
51. Multiplying: $4x^2(x - 5)(x + 2) = 4x^2(x^2 - 3x - 10) = 4x^4 - 12x^3 - 40x^2$
53. Multiplying: $2ab^3(b^2 - 4b + 1) = 2ab^5 - 8ab^4 + 2ab^3$

5.6 Factoring: A General Review

1. Factoring the polynomial: $x^2 - 81 = (x+9)(x-9)$
3. Factoring the polynomial: $x^2 + 2x - 15 = (x+5)(x-3)$
5. Factoring the polynomial: $x^2 + 6x + 9 = (x+3)(x+3) = (x+3)^2$
7. Factoring the polynomial: $y^2 - 10y + 25 = (y-5)(y-5) = (y-5)^2$
9. Factoring the polynomial: $2a^3b + 6a^2b + 2ab = 2ab(a^2 + 3a + 1)$
11. The polynomial $x^2 + x + 1$ cannot be factored.
13. Factoring the polynomial: $12a^2 - 75 = 3(4a^2 - 25) = 3(2a+5)(2a-5)$
15. Factoring the polynomial: $9x^2 - 12xy + 4y^2 = (3x-2y)(3x-2y) = (3x-2y)^2$
17. Factoring the polynomial: $4x^3 + 16xy^2 = 4x(x^2 + 4y^2)$
19. Factoring the polynomial: $2y^3 + 20y^2 + 50y = 2y(y^2 + 10y + 25) = 2y(y+5)(y+5) = 2y(y+5)^2$
21. Factoring the polynomial: $a^6 + 4a^4b^2 = a^4(a^2 + 4b^2)$
23. Factoring the polynomial: $xy + 3x + 4y + 12 = x(y+3) + 4(y+3) = (y+3)(x+4)$
25. Factoring the polynomial: $x^4 - 16 = (x^2+4)(x^2-4) = (x^2+4)(x+2)(x-2)$
27. Factoring the polynomial: $xy - 5x + 2y - 10 = x(y-5) + 2(y-5) = (y-5)(x+2)$
29. Factoring the polynomial: $5a^2 + 10ab + 5b^2 = 5(a^2 + 2ab + b^2) = 5(a+b)(a+b) = 5(a+b)^2$
31. The polynomial $x^2 + 49$ cannot be factored.
33. Factoring the polynomial: $3x^2 + 15xy + 18y^2 = 3(x^2 + 5xy + 6y^2) = 3(x+2y)(x+3y)$
35. Factoring the polynomial: $2x^2 + 15x - 38 = (2x+19)(x-2)$
37. Factoring the polynomial: $100x^2 - 300x + 200 = 100(x^2 - 3x + 2) = 100(x-2)(x-1)$
39. Factoring the polynomial: $x^2 - 64 = (x+8)(x-8)$
41. Factoring the polynomial: $x^2 + 3x + ax + 3a = x(x+3) + a(x+3) = (x+3)(x+a)$
43. Factoring the polynomial: $49a^7 - 9a^5 = a^5(49a^2 - 9) = a^5(7a+3)(7a-3)$
45. The polynomial $49x^2 + 9y^2$ cannot be factored.
47. Factoring the polynomial: $25a^3 + 20a^2 + 3a = a(25a^2 + 20a + 3) = a(5a+3)(5a+1)$
49. Factoring the polynomial: $xa - xb + ay - by = x(a-b) + y(a-b) = (a-b)(x+y)$
51. Factoring the polynomial: $48a^4b - 3a^2b = 3a^2b(16a^2 - 1) = 3a^2b(4a+1)(4a-1)$
53. Factoring the polynomial: $20x^4 - 45x^2 = 5x^2(4x^2 - 9) = 5x^2(2x+3)(2x-3)$
55. Factoring the polynomial: $3x^2 + 35xy - 82y^2 = (3x+41y)(x-2y)$
57. Factoring the polynomial: $16x^5 - 44x^4 + 30x^3 = 2x^3(8x^2 - 22x + 15) = 2x^3(2x-3)(4x-5)$
59. Factoring the polynomial: $2x^2 + 2ax + 3x + 3a = 2x(x+a) + 3(x+a) = (x+a)(2x+3)$
61. Factoring the polynomial: $y^4 - 1 = (y^2+1)(y^2-1) = (y^2+1)(y+1)(y-1)$
63. Factoring the polynomial:
$$12x^4y^2 + 36x^3y^3 + 27x^2y^4 = 3x^2y^2(4x^2 + 12xy + 9y^2) = 3x^2y^2(2x+3y)(2x+3y) = 3x^2y^2(2x+3y)^2$$

65. Solving the equation:
$$3x - 6 = 9$$
$$3x = 15$$
$$x = 5$$

67. Solving the equation:
$$2x + 3 = 0$$
$$2x = -3$$
$$x = -\frac{3}{2}$$

69. Solving the equation:
$$4x + 3 = 0$$
$$4x = -3$$
$$x = -\frac{3}{4}$$

5.7 Solving Equations by Factoring

1. Setting each factor equal to 0:
$$x + 2 = 0 \qquad x - 1 = 0$$
$$x = -2 \qquad x = 1$$
The solutions are -2 and 1.

3. Setting each factor equal to 0:
$$a - 4 = 0 \qquad a - 5 = 0$$
$$a = 4 \qquad a = 5$$
The solutions are 4 and 5.

5. Setting each factor equal to 0:
$$x = 0 \qquad x + 1 = 0 \qquad x - 3 = 0$$
$$x = -1 \qquad x = 3$$
The solutions are $0, -1$ and 3.

7. Setting each factor equal to 0:
$$3x + 2 = 0 \qquad 2x + 3 = 0$$
$$3x = -2 \qquad 2x = -3$$
$$x = -\frac{2}{3} \qquad x = -\frac{3}{2}$$
The solutions are $-\frac{2}{3}$ and $-\frac{3}{2}$.

9. Setting each factor equal to 0:
$$m = 0 \qquad 3m + 4 = 0 \qquad 3m - 4 = 0$$
$$3m = -4 \qquad 3m = 4$$
$$m = -\frac{4}{3} \qquad m = \frac{4}{3}$$
The solutions are $0, -\frac{4}{3}$ and $\frac{4}{3}$.

11. Setting each factor equal to 0:
$$2y = 0 \qquad 3y + 1 = 0 \qquad 5y + 3 = 0$$
$$y = 0 \qquad 3y = -1 \qquad 5y = -3$$
$$y = -\frac{1}{3} \qquad y = -\frac{3}{5}$$
The solutions are $0, -\frac{1}{3}$ and $-\frac{3}{5}$.

13. Solving by factoring:
$$x^2 + 3x + 2 = 0$$
$$(x+2)(x+1) = 0$$
$$x = -2, -1$$

15. Solving by factoring:
$$x^2 - 9x + 20 = 0$$
$$(x-4)(x-5) = 0$$
$$x = 4, 5$$

17. Solving by factoring:
$$a^2 - 2a - 24 = 0$$
$$(a-6)(a+4) = 0$$
$$a = 6, -4$$

19. Solving by factoring:
$$100x^2 - 500x + 600 = 0$$
$$100(x^2 - 5x + 6) = 0$$
$$100(x-2)(x-3) = 0$$
$$x = 2, 3$$

21. Solving by factoring:
$$x^2 = -6x - 9$$
$$x^2 + 6x + 9 = 0$$
$$(x+3)^2 = 0$$
$$x + 3 = 0$$
$$x = -3$$

23. Solving by factoring:
$$a^2 - 16 = 0$$
$$(a+4)(a-4) = 0$$
$$a = -4, 4$$

25. Solving by factoring:
$$2x^2 + 5x - 12 = 0$$
$$(2x-3)(x+4) = 0$$
$$x = \frac{3}{2}, -4$$

27. Solving by factoring:
$$9x^2 + 12x + 4 = 0$$
$$(3x+2)^2 = 0$$
$$3x + 2 = 0$$
$$x = -\frac{2}{3}$$

29. Solving by factoring:
$$a^2 + 25 = 10a$$
$$a^2 - 10a + 25 = 0$$
$$(a-5)^2 = 0$$
$$a - 5 = 0$$
$$a = 5$$

31. Solving by factoring:
$$0 = 20 + 3x - 2x^2$$
$$2x^2 - 3x - 20 = 0$$
$$(2x+5)(x-4) = 0$$
$$x = -\frac{5}{2}, 4$$

33. Solving by factoring:
$$3m^2 = 20 - 7m$$
$$3m^2 + 7m - 20 = 0$$
$$(3m-5)(m+4) = 0$$
$$m = \frac{5}{3}, -4$$

35. Solving by factoring:
$$4x^2 - 49 = 0$$
$$(2x+7)(2x-7) = 0$$
$$x = -\frac{7}{2}, \frac{7}{2}$$

37. Solving by factoring:
$$x^2 + 6x = 0$$
$$x(x+6) = 0$$
$$x = 0, -6$$

39. Solving by factoring:
$$x^2 - 3x = 0$$
$$x(x-3) = 0$$
$$x = 0, 3$$

41. Solving by factoring:
$$2x^2 = 8x$$
$$2x^2 - 8x = 0$$
$$2x(x-4) = 0$$
$$x = 0, 4$$

43. Solving by factoring:
$$3x^2 = 15x$$
$$3x^2 - 15x = 0$$
$$3x(x-5) = 0$$
$$x = 0, 5$$

45. Solving by factoring:
$$1{,}400 = 400 + 700x - 100x^2$$
$$100x^2 - 700x + 1{,}000 = 0$$
$$100(x^2 - 7x + 10) = 0$$
$$100(x-5)(x-2) = 0$$
$$x = 2, 5$$

47. Solving by factoring:
$$6x^2 = -5x + 4$$
$$6x^2 + 5x - 4 = 0$$
$$(3x+4)(2x-1) = 0$$
$$x = -\frac{4}{3}, \frac{1}{2}$$

49. Solving by factoring:
$$x(2x-3) = 20$$
$$2x^2 - 3x = 20$$
$$2x^2 - 3x - 20 = 0$$
$$(2x+5)(x-4) = 0$$
$$x = -\frac{5}{2}, 4$$

51. Solving by factoring:
$$t(t+2) = 80$$
$$t^2 + 2t = 80$$
$$t^2 + 2t - 80 = 0$$
$$(t+10)(t-8) = 0$$
$$t = -10, 8$$

53. Solving by factoring:
$$4{,}000 = (1{,}300 - 100p)p$$
$$4{,}000 = 1{,}300p - 100p^2$$
$$100p^2 - 1{,}300p + 4{,}000 = 0$$
$$100(p^2 - 13p + 40) = 0$$
$$100(p-8)(p-5) = 0$$
$$p = 5, 8$$

55. Solving by factoring:
$$x(14-x) = 48$$
$$14x - x^2 = 48$$
$$-x^2 + 14x - 48 = 0$$
$$x^2 - 14x + 48 = 0$$
$$(x-6)(x-8) = 0$$
$$x = 6, 8$$

57. Solving by factoring:
$$(x+5)^2 = 2x + 9$$
$$x^2 + 10x + 25 = 2x + 9$$
$$x^2 + 8x + 16 = 0$$
$$(x+4)^2 = 0$$
$$x + 4 = 0$$
$$x = -4$$

59. Solving by factoring:
$$(y-6)^2 = y - 4$$
$$y^2 - 12y + 36 = y - 4$$
$$y^2 - 13y + 40 = 0$$
$$(y-5)(y-8) = 0$$
$$y = 5, 8$$

61. Solving by factoring:
$$10^2 = (x+2)^2 + x^2$$
$$100 = x^2 + 4x + 4 + x^2$$
$$100 = 2x^2 + 4x + 4$$
$$0 = 2x^2 + 4x - 96$$
$$0 = 2(x^2 + 2x - 48)$$
$$0 = 2(x+8)(x-6)$$
$$x = -8, 6$$

63. Solving by factoring:
$$2x^3 + 11x^2 + 12x = 0$$
$$x(2x^2 + 11x + 12) = 0$$
$$x(2x+3)(x+4) = 0$$
$$x = 0, -\frac{3}{2}, -4$$

65. Solving by factoring:
$$4y^3 - 2y^2 - 30y = 0$$
$$2y(2y^2 - y - 15) = 0$$
$$2y(2y+5)(y-3) = 0$$
$$y = 0, -\frac{5}{2}, 3$$

67. Solving by factoring:
$$8x^3 + 16x^2 = 10x$$
$$8x^3 + 16x^2 - 10x = 0$$
$$2x(4x^2 + 8x - 5) = 0$$
$$2x(2x-1)(2x+5) = 0$$
$$x = 0, \frac{1}{2}, -\frac{5}{2}$$

69. Solving by factoring:
$$20a^3 = -18a^2 + 18a$$
$$20a^3 + 18a^2 - 18a = 0$$
$$2a(10a^2 + 9a - 9) = 0$$
$$2a(5a-3)(2a+3) = 0$$
$$a = 0, \frac{3}{5}, -\frac{3}{2}$$

71. Solving by factoring:
$$16t^2 - 32t + 12 = 0$$
$$4(4t^2 - 8t + 3) = 0$$
$$4(2t-1)(2t-3) = 0$$
$$t = \frac{1}{2}, \frac{3}{2}$$

73. Solving the equation:
$$(a-5)(a+4) = -2a$$
$$a^2 - a - 20 = -2a$$
$$a^2 + a - 20 = 0$$
$$(a+5)(a-4) = 0$$
$$a = -5, 4$$

75. Solving the equation:
$$3x(x+1) - 2x(x-5) = -42$$
$$3x^2 + 3x - 2x^2 + 10x = -42$$
$$x^2 + 13x + 42 = 0$$
$$(x+7)(x+6) = 0$$
$$x = -7, -6$$

77. Solving the equation:
$$2x(x+3) = x(x+2) - 3$$
$$2x^2 + 6x = x^2 + 2x - 3$$
$$x^2 + 4x + 3 = 0$$
$$(x+3)(x+1) = 0$$
$$x = -3, -1$$

79. Solving the equation:
$$a(a-3) + 6 = 2a$$
$$a^2 - 3a + 6 = 2a$$
$$a^2 - 5a + 6 = 0$$
$$(a-2)(a-3) = 0$$
$$a = 2, 3$$

81. Solving the equation:
$$15(x+20) + 15x = 2x(x+20)$$
$$15x + 300 + 15x = 2x^2 + 40x$$
$$30x + 300 = 2x^2 + 40x$$
$$0 = 2x^2 + 10x - 300$$
$$0 = 2(x^2 + 5x - 150)$$
$$0 = 2(x+15)(x-10)$$
$$x = -15, 10$$

83. Solving the equation:
$$15 = a(a+2)$$
$$15 = a^2 + 2a$$
$$0 = a^2 + 2a - 15$$
$$0 = (a+5)(a-3)$$
$$a = -5, 3$$

85. Solving by factoring:
$$x^3 + 3x^2 - 4x - 12 = 0$$
$$x^2(x+3) - 4(x+3) = 0$$
$$(x+3)(x^2 - 4) = 0$$
$$(x+3)(x+2)(x-2) = 0$$
$$x = -3, -2, 2$$

87. Solving by factoring:
$$x^3 + x^2 - 16x - 16 = 0$$
$$x^2(x+1) - 16(x+1) = 0$$
$$(x+1)(x^2 - 16) = 0$$
$$(x+1)(x+4)(x-4) = 0$$
$$x = -4, -1, 4$$

89. **a.** Factoring: $x^2 - x - 2 = (x-2)(x+1)$

c. Solving by factoring:
$$x^2 - x - 2 = 0$$
$$(x-2)(x+1) = 0$$
$$x = -1, 2$$

d. Solving the equation:
$$x^2 - x - 2 = 4$$
$$x^2 - x - 6 = 0$$
$$(x-3)(x+2) = 0$$
$$x = -2, 3$$

91. Let x and $x + 1$ represent the two consecutive integers. The equation is: $x(x+1) = 72$

93. Let x and $x + 2$ represent the two consecutive odd integers. The equation is: $x(x+2) = 99$

95. Let x and $x + 2$ represent the two consecutive even integers. The equation is: $x(x+2) = 5[x+(x+2)] - 10$

97. Let x represent the cost of the suit and $5x$ represent the cost of the bicycle. The equation is:
$$x + 5x = 90$$
$$6x = 90$$
$$x = 15$$
$$5x = 75$$
The suit costs $15 and the bicycle costs $75.

99. Let x represent the cost of the lot and $4x$ represent the cost of the house. The equation is:
$$x + 4x = 3000$$
$$5x = 3000$$
$$x = 600$$
$$4x = 2400$$
The lot costs $600 and the house costs $2,400.

5.8 Applications

1. Let x and $x + 2$ represent the two integers. The equation is:
$$x(x+2) = 80$$
$$x^2 + 2x = 80$$
$$x^2 + 2x - 80 = 0$$
$$(x+10)(x-8) = 0$$
$$x = -10, 8$$
$$x + 2 = -8, 10$$
The two numbers are either −10 and −8, or 8 and 10.

3. Let x and $x + 2$ represent the two integers. The equation is:
$$x(x+2) = 99$$
$$x^2 + 2x = 99$$
$$x^2 + 2x - 99 = 0$$
$$(x+11)(x-9) = 0$$
$$x = -11, 9$$
$$x + 2 = -9, 11$$
The two numbers are either −11 and −9, or 9 and 11.

5. Let x and $x + 2$ represent the two integers. The equation is:
$$x(x+2) = 5(x+x+2) - 10$$
$$x^2 + 2x = 5(2x+2) - 10$$
$$x^2 + 2x = 10x + 10 - 10$$
$$x^2 + 2x = 10x$$
$$x^2 - 8x = 0$$
$$x(x-8) = 0$$
$$x = 0, 8$$
$$x + 2 = 2, 10$$
The two numbers are either 0 and 2, or 8 and 10.

7. Let x and $14 - x$ represent the two numbers. The equation is:
$$x(14-x) = 48$$
$$14x - x^2 = 48$$
$$0 = x^2 - 14x + 48$$
$$0 = (x-8)(x-6)$$
$$x = 8, 6$$
$$14 - x = 6, 8$$
The two numbers are 6 and 8.

9. Let x and $5x + 2$ represent the two numbers. The equation is:
$$x(5x+2) = 24$$
$$5x^2 + 2x = 24$$
$$5x^2 + 2x - 24 = 0$$
$$(5x+12)(x-2) = 0$$
$$x = -\frac{12}{5}, 2$$
$$5x + 2 = -10, 12$$
The two numbers are either $-\frac{12}{5}$ and -10, or 2 and 12.

11. Let x and $4x$ represent the two numbers. The equation is:
$$x(4x) = 4(x+4x)$$
$$4x^2 = 4(5x)$$
$$4x^2 = 20x$$
$$4x^2 - 20x = 0$$
$$4x(x-5) = 0$$
$$x = 0, 5$$
$$4x = 0, 20$$
The two numbers are either 0 and 0, or 5 and 20.

13. Let w represent the width and $w + 1$ represent the length. The equation is:
$$w(w+1) = 12$$
$$w^2 + w = 12$$
$$w^2 + w - 12 = 0$$
$$(w+4)(w-3) = 0$$
$$w = 3 \quad (w = -4 \text{ is impossible})$$
$$w + 1 = 4$$
The width is 3 inches and the length is 4 inches.

15. Let b represent the base and $2b$ represent the height. The equation is:
$$\frac{1}{2}b(2b) = 9$$
$$b^2 = 9$$
$$b^2 - 9 = 0$$
$$(b+3)(b-3) = 0$$
$$b = 3 \quad (b = -3 \text{ is impossible})$$
The base is 3 inches.

17. Let x and $x + 2$ represent the two legs. The equation is:
$$x^2 + (x+2)^2 = 10^2$$
$$x^2 + x^2 + 4x + 4 = 100$$
$$2x^2 + 4x + 4 = 100$$
$$2x^2 + 4x - 96 = 0$$
$$2(x^2 + 2x - 48) = 0$$
$$2(x+8)(x-6) = 0$$
$$x = 6 \quad (x = -8 \text{ is impossible})$$
$$x + 2 = 8$$
The legs are 6 inches and 8 inches.

19. Let x represent the longer leg and $x + 1$ represent the hypotenuse. The equation is:
$$5^2 + x^2 = (x+1)^2$$
$$25 + x^2 = x^2 + 2x + 1$$
$$25 = 2x + 1$$
$$24 = 2x$$
$$x = 12$$
The longer leg is 12 meters.

21. Setting $C = \$1,400$:
$$1400 = 400 + 700x - 100x^2$$
$$100x^2 - 700x + 1000 = 0$$
$$100(x^2 - 7x + 10) = 0$$
$$100(x-5)(x-2) = 0$$
$$x = 2, 5$$
The company can manufacture either 200 items or 500 items.

23. The revenue is given by: $R = xp = (1,700 - 100p)p$. Setting $R = \$7,000$:
$$7,000 = (1,700 - 100p)p$$
$$7,000 = 1,700p - 100p^2$$
$$100p^2 - 1,700p + 7,000 = 0$$
$$100(p^2 - 17p + 70) = 0$$
$$100(p - 7)(p - 10) = 0$$
$$p = 7, 10$$
The calculators should be sold for either $7 or $10.

25. a. Let x represent the distance from the base to the wall, and $2x + 2$ represent the height on the wall. Using the Pythagorean theorem:
$$x^2 + (2x + 2)^2 = 13^2$$
$$x^2 + 4x^2 + 8x + 4 = 169$$
$$5x^2 + 8x - 165 = 0$$
$$(5x + 33)(x - 5) = 0$$
$$x = 5 \quad \left(x = -\frac{33}{5} \text{ is impossible}\right)$$
The base of the ladder is 5 feet from the wall.

b. Since $2x + 2 = 2 \cdot 5 + 2 = 12$, the ladder reaches a height of 12 feet.

27. a. Finding when $h = 0$:
$$0 = -16t^2 + 396t + 100$$
$$16t^2 - 396t - 100 = 0$$
$$4t^2 - 99t - 25 = 0$$
$$(4t + 1)(t - 25) = 0$$
$$t = 25 \quad \left(t = -\frac{1}{4} \text{ is impossible}\right)$$
The bullet will land on the ground after 25 seconds.

b. Completing the table:

t (seconds)	h (feet)
0	100
5	1,680
10	2,460
15	2,440
20	1,620
25	0

29. Simplifying the expression: $(5x^3)^2 (2x^6)^3 = 25x^6 \cdot 8x^{18} = 200x^{24}$

31. Simplifying the expression: $\dfrac{x^4}{x^{-3}} = x^{4-(-3)} = x^{4+3} = x^7$

33. Simplifying the expression: $(2 \times 10^{-4})(4 \times 10^5) = 8 \times 10^1 = 80$

35. Simplifying the expression: $20ab^2 - 16ab^2 + 6ab^2 = 10ab^2$

Chapter 6
Functions and Function Notation

6.1 Introduction to Functions

1. The domain is $\{1,3,5,7\}$ and the range is $\{2,4,6,8\}$. This is a function.
3. The domain is $\{0,1,2,3\}$ and the range is $\{4,5,6\}$. This is a function.
5. The domain is $\{a,b,c,d\}$ and the range is $\{3,4,5\}$. This is a function.
7. The domain is $\{a\}$ and the range is $\{1,2,3,4\}$. This is not a function.
9. Yes, since it passes the vertical line test.
11. No, since it fails the vertical line test.
13. No, since it fails the vertical line test.
15. Yes, since it passes the vertical line test.
17. Yes, since it passes the vertical line test.
19. The domain is $\{x \mid -5 \leq x \leq 5\} = [-5,5]$ and the range is $\{y \mid 0 \leq y \leq 5\} = [0,5]$.
21. The domain is $\{x \mid -5 \leq x \leq 3\} = [-5,3]$ and the range is $\{y \mid y = 3\}$.
23. The domain is all real numbers and the range is $\{y \mid y \geq -1\} = [-1,\infty)$. This is a function.

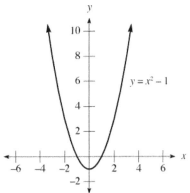

25. The domain is all real numbers and the range is $\{y \mid y \geq 4\} = [4,\infty)$. This is a function.

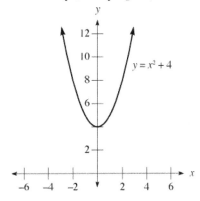

27. The domain is $\{x \mid x \geq -1\} = [-1, \infty)$ and the range is all real numbers. This is not a function.

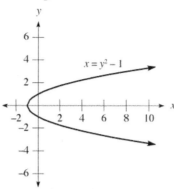

29. The domain is all real numbers and the range is $\{y \mid y \geq 0\} = [0, \infty)$. This is a function.

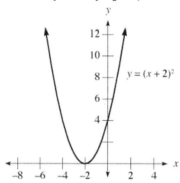

31. The domain is $\{x \mid x \geq 0\} = [0, \infty)$ and the range is all real numbers. This is not a function.

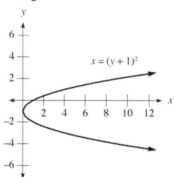

33.   **a.**   The equation is $y = 8.5x$ for $10 \leq x \leq 40$.

     b.   Completing the table:

Hours Worked x	Function Rule $y = 8.5x$	Gross Pay (\$) y
10	$y = 8.5(10) = 85$	85
20	$y = 8.5(20) = 170$	170
30	$y = 8.5(30) = 255$	255
40	$y = 8.5(40) = 340$	340

c. Constructing a line graph:

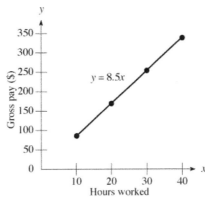

d. The domain is $\{x \mid 10 \leq x \leq 40\} = [10,40]$ and the range is $\{y \mid 85 \leq y \leq 340\} = [85,340]$.

e. The minimum is $85 and the maximum is $340.

35. The domain is {2004,2005,2006,2007,2008,2009,2010} and the range is {680,730,800,900,920,990,1030}.

37. a. Figure III   b. Figure I
     c. Figure II   d. Figure IV

39. Simplifying: $4(3.14)(9) = 113.04 \approx 113$ 41. Simplifying: $4(-2) - 1 = -8 - 1 = -9$

43. a. Substituting $t = 10$: $s = \dfrac{60}{10} = 6$ b. Substituting $t = 8$: $s = \dfrac{60}{8} = 7.5$

45. a. Substituting $x = 5$: $(5)^2 + 2 = 25 + 2 = 27$ b. Substituting $x = -2$: $(-2)^2 + 2 = 4 + 2 = 6$

47. Substituting $x = 2$: $y = (2)^2 - 3 = 4 - 3 = 1$ 49. Substituting $x = 0$: $y = (0)^2 - 3 = 0 - 3 = -3$

51. Solving for y:
$$\frac{8}{5} - 2y = 4$$
$$-2y = \frac{12}{5}$$
$$y = -\frac{6}{5}$$

53. Substituting $x = 0$ and $y = 0$:
$$0 = a(0-8)^2 + 70$$
$$0 = 64a + 70$$
$$64a = -70$$
$$a = -\frac{70}{64} = -\frac{35}{32}$$

6.2 Function Notation

1. Evaluating the function: $f(2) = 2(2) - 5 = 4 - 5 = -1$

3. Evaluating the function: $f(-3) = 2(-3) - 5 = -6 - 5 = -11$

5. Evaluating the function: $g(-1) = (-1)^2 + 3(-1) + 4 = 1 - 3 + 4 = 2$

7. Evaluating the function: $g(-3) = (-3)^2 + 3(-3) + 4 = 9 - 9 + 4 = 4$

9. Evaluating the function: $g(a) = a^2 + 3a + 4$

11. Evaluating the function: $f(a+6) = 2(x+6) - 5 = 2a + 12 - 5 = 2a + 7$

13. Evaluating the function: $f(0) = 3(0)^2 - 4(0) + 1 = 0 - 0 + 1 = 1$

15. Evaluating the function: $g(-4) = 2(-4) - 1 = -8 - 1 = -9$

17. Evaluating the function: $f(-1) = 3(-1)^2 - 4(-1) + 1 = 3 + 4 + 1 = 8$

19. Evaluating the function: $g\left(\dfrac{1}{2}\right) = 2\left(\dfrac{1}{2}\right) - 1 = 1 - 1 = 0$

21. Evaluating the function: $f(a) = 3a^2 - 4a + 1$

23. Evaluating the function: $f(a+2) = 3(a+2)^2 - 4(a+2) + 1 = 3a^2 + 12a + 12 - 4a - 8 + 1 = 3a^2 + 8a + 5$

25. Evaluating: $f(1) = 4$

27. Evaluating: $g\left(\dfrac{1}{2}\right) = 0$

29. Evaluating: $g(-2) = 2$

31. Evaluating the function: $f(-4) = (-4)^2 - 2(-4) = 16 + 8 = 24$

33. First evaluate each function:
$$f(-2) = (-2)^2 - 2(-2) = 4 + 4 = 8 \qquad g(-1) = 5(-1) - 4 = -5 - 4 = -9$$
Now evaluating: $f(-2) + g(-1) = 8 - 9 = -1$

35. Evaluating the function: $2f(x) - 3g(x) = 2(x^2 - 2x) - 3(5x - 4) = 2x^2 - 4x - 15x + 12 = 2x^2 - 19x + 12$

37. Evaluating the function: $f[g(3)] = f[5(3) - 4] = f(11) = (11)^2 - 2(11) = 121 - 22 = 99$

39. Evaluating the function: $f\left(\dfrac{1}{3}\right) = \dfrac{1}{\frac{1}{3}+3} \cdot \dfrac{3}{3} = \dfrac{3}{1+9} = \dfrac{3}{10}$

41. Evaluating the function: $f\left(-\dfrac{1}{2}\right) = \dfrac{1}{-\frac{1}{2}+3} \cdot \dfrac{2}{2} = \dfrac{2}{-1+6} = \dfrac{2}{5}$

43. Evaluating the function: $f(-3) = \dfrac{1}{-3+3} = \dfrac{1}{0}$, which is undefined

45. a. Evaluating the function: $f(a) - 3 = (a^2 - 4) - 3 = a^2 - 7$

    b. Evaluating the function: $f(a - 3) = (a - 3)^2 - 4 = a^2 - 6a + 9 - 4 = a^2 - 6a + 5$

    c. Evaluating the function: $f(x) + 2 = (x^2 - 4) + 2 = x^2 - 2$

    d. Evaluating the function: $f(x + 2) = (x + 2)^2 - 4 = x^2 + 4x + 4 - 4 = x^2 + 4x$

    e. Evaluating the function: $f(a + b) = (a + b)^2 - 4 = a^2 + 2ab + b^2 - 4$

    f. Evaluating the function: $f(x + h) = (x + h)^2 - 4 = x^2 + 2xh + h^2 - 4$

47. Graphing the function:

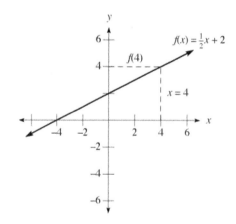

Note that $f(4) = 4$.

49. Finding where $f(x) = x$:
$$\frac{1}{2}x + 2 = x$$
$$2 = \frac{1}{2}x$$
$$x = 4$$

51. Graphing the function:

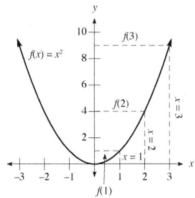

Note that $f(1) = 1$, $f(2) = 4$, and $f(3) = 9$.

53. Evaluating: $V(3) = 150 \cdot 2^{3/3} = 150 \cdot 2 = 300$; The painting is worth $300 in 3 years.

Evaluating: $V(6) = 150 \cdot 2^{6/3} = 150 \cdot 4 = 600$; The painting is worth $600 in 6 years.

55.
a. This statement is true.
b. This statement is false.
c. This statement is true.
d. This statement is false.
e. This statement is true.

57.
a. Evaluating: $V(3.75) = -3300(3.75) + 18000 = \$5,625$

b. Evaluating: $V(5) = -3300(5) + 18000 = \$1,500$

c. The domain of this function is $\{t \mid 0 \le t \le 5\} = [0, 5]$.

d. Sketching the graph:

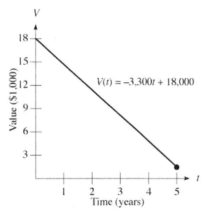

e. The range of this function is $\{V(t) \mid 1,500 \le V(t) \le 18,000\} = [1500, 18000]$.

f. Solving $V(t) = 10000$:
$$-3300t + 18000 = 10000$$
$$-3300t = -8000$$
$$t \approx 2.42$$
The copier will be worth $10,000 after approximately 2.42 years.

59. Simplifying: $16(3.5)^2 = 16(12.25) = 196$

61. Simplifying: $\dfrac{180}{45} = 4$

63. Simplifying: $\dfrac{0.0005(200)}{(0.25)^2} = \dfrac{0.1}{0.0625} = 1.6$

65. Solving for K:
$$15 = K(5)$$
$$K = 3$$

67. Solving for K:
$$50 = \dfrac{K}{48}$$
$$K = 50 \cdot 48 = 2{,}400$$

6.3 Variation

1. The variation equation is $y = Kx$. Substituting $x = 2$ and $y = 10$:
$$10 = K \cdot 2$$
$$K = 5$$
So $y = 5x$. Substituting $x = 6$: $y = 5 \cdot 6 = 30$

3. The variation equation is $r = \dfrac{K}{s}$. Substituting $s = 4$ and $r = -3$:
$$-3 = \dfrac{K}{4}$$
$$K = -12$$
So $r = \dfrac{-12}{s}$. Substituting $s = 2$: $r = \dfrac{-12}{2} = -6$

5. The variation equation is $d = Kr^2$. Substituting $r = 5$ and $d = 10$:
$$10 = K \cdot 5^2$$
$$10 = 25K$$
$$K = \dfrac{2}{5}$$
So $d = \dfrac{2}{5}r^2$. Substituting $r = 10$: $d = \dfrac{2}{5}(10)^2 = \dfrac{2}{5} \cdot 100 = 40$

7. The variation equation is $y = \dfrac{K}{x^2}$. Substituting $x = 3$ and $y = 45$:
$$45 = \dfrac{K}{3^2}$$
$$45 = \dfrac{K}{9}$$
$$K = 405$$
So $y = \dfrac{405}{x^2}$. Substituting $x = 5$: $y = \dfrac{405}{5^2} = \dfrac{405}{25} = \dfrac{81}{5}$

9. The variation equation is $z = Kxy^2$. Substituting $x = 3$, $y = 3$, and $z = 54$:
$$54 = K(3)(3)^2$$
$$54 = 27K$$
$$K = 2$$
So $z = 2xy^2$. Substituting $x = 2$ and $y = 4$: $z = 2(2)(4)^2 = 64$

11. The variation equation is $I = \frac{K}{w^3}$. Substituting $w = \frac{1}{2}$ and $I = 32$:

$$32 = \frac{K}{\left(\frac{1}{2}\right)^3}$$

$$32 = \frac{K}{1/8}$$

$$K = 4$$

So $I = \frac{4}{w^3}$. Substituting $w = \frac{1}{3}$: $I = \frac{4}{\left(\frac{1}{3}\right)^3} = \frac{4}{1/27} = 108$

13. The variation equation is $z = Kyx^2$. Substituting $x = 3, y = 2$, and $z = 72$:

$$72 = K(2)(3)^2$$

$$72 = 18K$$

$$K = 4$$

So $z = 4yx^2$. Substituting $x = 5$ and $y = 3$: $z = 4(3)(5)^2 = 300$

15. The variation equation is $z = Kyx^2$. Substituting $x = 1, y = 5$, and $z = 25$:

$$25 = K(5)(1)^2$$

$$25 = 5K$$

$$K = 5$$

So $z = 5yx^2$. Substituting $z = 160$ and $y = 8$:

$$160 = 5(8)x^2$$

$$160 = 40x^2$$

$$x^2 = 4$$

$$x = \pm 2$$

17. The variation equation is $F = \frac{Km}{d^2}$. Substituting $F = 150, m = 240$, and $d = 8$:

$$150 = \frac{K(240)}{8^2}$$

$$150 = \frac{240K}{64}$$

$$240K = 9{,}600$$

$$K = 40$$

So $F = \frac{40m}{d^2}$. Substituting $m = 360$ and $d = 3$: $F = \frac{40(360)}{3^2} = \frac{14{,}400}{9} = 1{,}600$

19. The variation equation is $F = \dfrac{Km}{d^2}$. Substituting $F = 24$, $m = 20$, and $d = 5$:

$$24 = \dfrac{K(20)}{5^2}$$

$$24 = \dfrac{20K}{25}$$

$$20K = 600$$

$$K = 30$$

So $F = \dfrac{30m}{d^2}$. Substituting $F = 18.75$ and $m = 40$:

$$18.75 = \dfrac{30(40)}{d^2}$$

$$18.75 = \dfrac{1{,}200}{d^2}$$

$$18.75 d^2 = 1{,}200$$

$$d^2 = 64$$

$$d = \pm 8$$

21. Let l represent the length and f represent the force. The variation equation is $l = Kf$. Substituting $f = 5$ and $l = 7$:

$$7 = K \cdot 5$$

$$K = \dfrac{7}{5}$$

So $l = \dfrac{7}{5}f$. Substituting $l = 10$:

$$10 = \dfrac{7}{5}f$$

$$50 = 7f$$

$$f = \dfrac{50}{7}$$

The force required is $\dfrac{50}{7}$ pounds.

23. a. The variation equation is $T = 4P$.
    b. Graphing the equation:

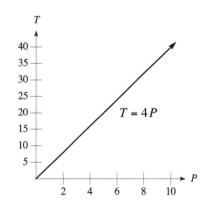

   c. Substituting $T = 280$:

   $$280 = 4P$$

   $$P = 70$$

   The pressure is 70 pounds per square inch.

25. Let v represent the volume and p represent the pressure. The variation equation is $v = \dfrac{K}{p}$.

   Substituting $p = 36$ and $v = 25$:
   $$25 = \dfrac{K}{36}$$
   $$K = 900$$
   The equation is $v = \dfrac{900}{p}$. Substituting $v = 75$:
   $$75 = \dfrac{900}{p}$$
   $$75p = 900$$
   $$p = 12$$
   The pressure is 12 pounds per square inch.

27. **a.** The variation equation is $f = \dfrac{80}{d}$.

   b. Graphing the equation:

   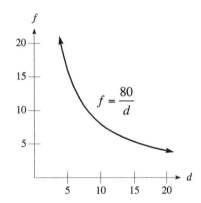

   c. Substituting $d = 10$:
   $$f = \dfrac{80}{10}$$
   $$f = 8$$
   The f-stop is 8.

29. Let A represent the surface area, h represent the height, and r represent the radius. The variation equation is $A = Khr$.
   Substituting $A = 94$, $r = 3$, and $h = 5$:
   $$94 = K(3)(5)$$
   $$94 = 15K$$
   $$K = \dfrac{94}{15}$$
   The equation is $A = \dfrac{94}{15}hr$. Substituting $r = 2$ and $h = 8$: $A = \dfrac{94}{15}(8)(2) = \dfrac{1{,}504}{15}$.

   The surface area is $\dfrac{1{,}504}{15} \approx 100.3$ square inches.

31. Let R represent the resistance, l represent the length, and d represent the diameter. The variation equation is $R = \dfrac{Kl}{d^2}$.

 Substituting $R = 10$, $l = 100$, and $d = 0.01$:
 $$10 = \dfrac{K(100)}{(0.01)^2}$$
 $$0.001 = 100K$$
 $$K = 0.00001$$

 The equation is $R = \dfrac{0.00001l}{d^2}$. Substituting $l = 60$ and $d = 0.02$: $R = \dfrac{0.00001(60)}{(0.02)^2} = 1.5$. The resistance is 1.5 ohms.

33. a. The variation equation is $P = 0.21\sqrt{L}$.
    b. Graphing the equation:

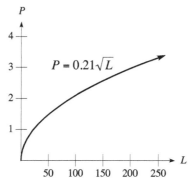

   c. Substituting $L = 225$: $P = 0.21\sqrt{225} = 3.15$
      The period is 3.15 seconds.

35. Multiplying: $0.6(M - 70) = 0.6M - 42$

37. Multiplying: $(4x - 3)(4x^2 - 7x + 3) = 16x^3 - 28x^2 + 12x - 12x^2 + 21x - 9 = 16x^3 - 40x^2 + 33x - 9$

39. Simplifying: $(4x - 3) + (4x^2 - 7x + 3) = 4x - 3 + 4x^2 - 7x + 3 = 4x^2 - 3x$

41. Simplifying: $(4x^2 + 3x + 2) + (2x^2 - 5x - 6) = 4x^2 + 3x + 2 + 2x^2 - 5x - 6 = 6x^2 - 2x - 4$

43. Simplifying: $4(-1)^2 - 7(-1) = 4 + 7 = 11$

6.4 Algebra and Composition with Functions

1. Writing the formula: $(f + g)(x) = f(x) + g(x) = (4x - 3) + (2x + 5) = 6x + 2$
3. Writing the formula: $(g - f)(x) = g(x) - f(x) = (2x + 5) - (4x - 3) = -2x + 8$
5. Writing the formula: $(f \cdot g)(x) = f(x) \cdot g(x) = (4x - 3)(2x + 5) = 8x^2 + 14x - 15$
7. Writing the formula: $\left(\dfrac{g}{f}\right)(x) = \dfrac{g(x)}{f(x)} = \dfrac{2x + 5}{4x - 3}$
9. Writing the formula: $(g + f)(x) = g(x) + f(x) = (x - 2) + (3x - 5) = 4x - 7$
11. Writing the formula: $(g + h)(x) = g(x) + h(x) = (x - 2) + (3x^2 - 11x + 10) = 3x^2 - 10x + 8$
13. Writing the formula: $(g - f)(x) = g(x) - f(x) = (x - 2) - (3x - 5) = -2x + 3$
15. Writing the formula: $(f \cdot g)(x) = f(x) \cdot g(x) = (3x - 5)(x - 2) = 3x^2 - 11x + 10$
17. Writing the formula:
 $(f \cdot h)(x) = f(x) \cdot h(x) = (3x - 5)(3x^2 - 11x + 10) = 9x^3 - 33x^2 + 30x - 15x^2 + 55x - 50 = 9x^3 - 48x^2 + 85x - 50$

19. Writing the formula: $\left(\dfrac{h}{f}\right)(x) = \dfrac{h(x)}{f(x)} = \dfrac{3x^2 - 11x + 10}{3x - 5} = \dfrac{(3x-5)(x-2)}{3x-5} = x - 2$

21. Writing the formula: $\left(\dfrac{f}{h}\right)(x) = \dfrac{f(x)}{h(x)} = \dfrac{3x-5}{3x^2 - 11x + 10} = \dfrac{3x-5}{(3x-5)(x-2)} = \dfrac{1}{x-2}$

23. Writing the formula: $(f + g + h)(x) = f(x) + g(x) + h(x) = (3x - 5) + (x - 2) + (3x^2 - 11x + 10) = 3x^2 - 7x + 3$

25. Writing the formula:
$$(h + f \cdot g)(x) = h(x) + f(x) \cdot g(x)$$
$$= (3x^2 - 11x + 10) + (3x - 5)(x - 2)$$
$$= 3x^2 - 11x + 10 + 3x^2 - 11x + 10$$
$$= 6x^2 - 22x + 20$$

27. Evaluating: $(f + g)(2) = f(2) + g(2) = (2 \cdot 2 + 1) + (4 \cdot 2 + 2) = 5 + 10 = 15$

29. Evaluating: $(f \cdot g)(3) = f(3) \cdot g(3) = (2 \cdot 3 + 1)(4 \cdot 3 + 2) = 7 \cdot 14 = 98$

31. Evaluating: $\left(\dfrac{h}{g}\right)(1) = \dfrac{h(1)}{g(1)} = \dfrac{4(1)^2 + 4(1) + 1}{4(1) + 2} = \dfrac{9}{6} = \dfrac{3}{2}$

33. Evaluating: $(f \cdot h)(0) = f(0) \cdot h(0) = (2(0) + 1)(4(0)^2 + 4(0) + 1) = (1)(1) = 1$

35. Evaluating: $(f + g + h)(2) = f(2) + g(2) + h(2) = (2(2) + 1) + (4(2) + 2) + (4(2)^2 + 4(2) + 1) = 5 + 10 + 25 = 40$

37. Evaluating:
$$(h + f \cdot g)(3) = h(3) + f(3) \cdot g(3) = (4(3)^2 + 4(3) + 1) + (2(3) + 1) \cdot (4(3) + 2) = 49 + 7 \cdot 14 = 49 + 98 = 147$$

39.
  a. Evaluating: $(f \circ g)(5) = f(g(5)) = f(5 + 4) = f(9) = 9^2 = 81$
  b. Evaluating: $(g \circ f)(5) = g(f(5)) = g(5^2) = g(25) = 25 + 4 = 29$
  c. Evaluating: $(f \circ g)(x) = f(g(x)) = f(x + 4) = (x + 4)^2$
  d. Evaluating: $(g \circ f)(x) = g(f(x)) = g(x^2) = x^2 + 4$

41.
  a. Evaluating: $(f \circ g)(0) = f(g(0)) = f(4 \cdot 0 - 1) = f(-1) = (-1)^2 + 3(-1) = 1 - 3 = -2$
  b. Evaluating: $(g \circ f)(0) = g(f(0)) = g(0^2 + 3 \cdot 0) = g(0) = 4(0) - 1 = -1$
  c. Evaluating: $(f \circ g)(x) = f(g(x)) = f(4x - 1) = (4x - 1)^2 + 3(4x - 1) = 16x^2 - 8x + 1 + 12x - 3 = 16x^2 + 4x - 2$
  d. Evaluating: $(g \circ f)(x) = g(f(x)) = g(x^2 + 3x) = 4(x^2 + 3x) - 1 = 4x^2 + 12x - 1$

43. Evaluating each composition:
$$(f \circ g)(x) = f(g(x)) = f\left(\dfrac{x+4}{5}\right) = 5\left(\dfrac{x+4}{5}\right) - 4 = x + 4 - 4 = x$$
$$(g \circ f)(x) = g(f(x)) = g(5x - 4) = \dfrac{5x - 4 + 4}{5} = \dfrac{5x}{5} = x$$
Thus $(f \circ g)(x) = (g \circ f)(x) = x$.

45.
  a. Finding the revenue: $R(x) = x(11.5 - 0.05x) = 11.5x - 0.05x^2$
  b. Finding the cost: $C(x) = 2x + 200$
  c. Finding the profit: $P(x) = R(x) - C(x) = (11.5x - 0.05x^2) - (2x + 200) = -0.05x^2 + 9.5x - 200$
  d. Finding the average cost: $\overline{C}(x) = \dfrac{C(x)}{x} = \dfrac{2x + 200}{x} = 2 + \dfrac{200}{x}$

47. a. The function is $M(x) = 220 - x$.
  b. Evaluating: $M(24) = 220 - 24 = 196$ beats per minute
  c. The training heart rate function is: $T(M) = 62 + 0.6(M - 62) = 0.6M + 24.8$
     Finding the composition: $T(M(x)) = T(220 - x) = 0.6(220 - x) + 24.8 = 156.8 - 0.6x$
     Evaluating: $T(M(24)) = 156.8 - 0.6(24) \approx 142$ beats per minute
  d. Evaluating: $T(M(36)) = 156.8 - 0.6(36) \approx 135$ beats per minute
  e. Evaluating: $T(M(48)) = 156.8 - 0.6(48) \approx 128$ beats per minute

49. Solving for y:
$$y - 2.74 = 0.055(x - 2010)$$
$$y - 2.74 = 0.055x - 110.55$$
$$y = 0.055x - 107.81$$

51. Solving for y:
$$y - 180{,}000 = -3{,}400(x - 10)$$
$$y - 180{,}000 = -3{,}400x + 34{,}000$$
$$y = -3{,}400x + 214{,}000$$

53. Simplifying: $\dfrac{64.7 - 1.32}{4{,}860 - 221} = \dfrac{63.38}{4{,}639} \approx 0.0137$

6.5 Regression Analysis and the Coefficient of Correlation

Note: Answers in this section may vary slightly from the textbook. More exact regression lines were used in the calculations for this section.

1. The regression line is $y = 65x + 42$. Drawing the data and regression line:

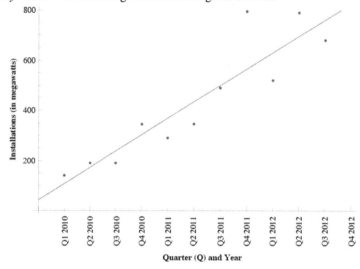

3. ab. The data is not linear. The regression line is $y = 1.1x + 19.9$. Drawing the data and regression line:

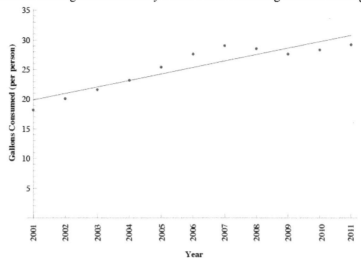

5. ab. The data is not linear. The regression line is $y = 0.017x + 0.160$. Drawing the data and regression line:

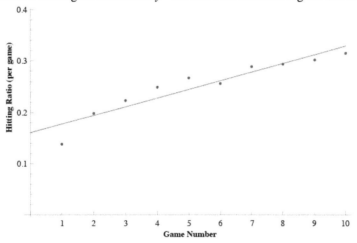

    c. The slope tells us the hitting ratio is increasing by approximately 1.7% per game.

7. a. The regression line is $y = 1,999x + 57,762$. Plotting the data and the regression line:

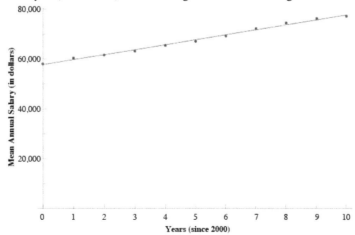

b. The points are $(2, 61760)$ and $(10, 77752)$.

c. The regression line is $y = 1,999x + 57,762$.

d. For 2015, substitute $x = 15$: $y = 1,999(15) + 57,762 = \$87,747$

For 2030, substitute $x = 30$: $y = 1,999(30) + 57,762 = \$117,732$

9. First complete the table of values needed for the computation of m and b:

x	y	xy	x^2
1	4	4	1
3	3	9	9
5	6	30	25
8	7	56	64
10	8	80	100
27	28	179	199

Now compute the values of m and b (for b, the exact value of m was used):

$$m = \frac{n\left(\sum xy\right) - \left(\sum x\right)\left(\sum y\right)}{n\left(\sum x^2\right) - \left(\sum x\right)^2} = \frac{5(179) - (27)(28)}{5(199) - (27)^2} = \frac{139}{266} \approx 0.5226$$

$$b = \frac{\sum y - m\left(\sum x\right)}{n} = \frac{28 - 0.5226(27)}{5} = \frac{13.8910}{5} \approx 2.7782$$

The regression line is $y = 0.5226x + 2.7782$.

11. First complete the table of values needed for the computation of m and b:

x	y	xy	x^2
0.5	1	0.5	0.25
1	3	3	1
1.75	3	5.25	3.0625
2	5	10	4
2.25	7	15.75	5.0625
2.5	7	17.5	6.25
3	8	24	9
13	34	76	28.625

Now compute the values of m and b (for b, the exact value of m was used):

$$m = \frac{n\left(\sum xy\right) - \left(\sum x\right)\left(\sum y\right)}{n\left(\sum x^2\right) - \left(\sum x\right)^2} = \frac{7(76) - (13)(34)}{7(28.625) - (13)^2} = \frac{90}{31.375} \approx 2.8685$$

$$b = \frac{\sum y - m\left(\sum x\right)}{n} = \frac{34 - 2.8685(13)}{7} = \frac{-3.2908}{7} \approx -0.4701$$

The regression line is $y = 2.8685x - 0.4701$.

13. a. The regression line is $y = 0.4963x + 17.7339$.
     b. Plotting the data and the regression line:

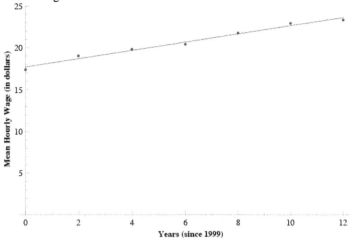

   c. For 2008, substitute $x = 9$: $y = 0.4963(9) + 17.7339 \approx \$22.20$
   d. For 2020, substitute $x = 21$: $y = 0.4963(21) + 17.7339 \approx \$28.16$
   e. The correlation coefficient is $r \approx 0.9914$.

15. Solving the equation:

$$x - 5 = 7$$
$$x = 12$$

17. Solving the equation:

$$5 - \frac{4}{7}a = -11$$
$$7\left(5 - \frac{4}{7}a\right) = 7(-11)$$
$$35 - 4a = -77$$
$$-4a = -112$$
$$a = 28$$

19. Solving the equation:

$$5(x-1) - 2(2x+3) = 5x - 4$$
$$5x - 5 - 4x - 6 = 5x - 4$$
$$x - 11 = 5x - 4$$
$$-4x = 7$$
$$x = -\frac{7}{4}$$

21. Solving for w:

$$P = 2l + 2w$$
$$P - 2l = 2w$$
$$w = \frac{P - 2l}{2}$$

23. Solving the inequality:

$$-5t \leq 30$$
$$t \geq -6$$

The solution set is $[-6, \infty)$. Graphing:

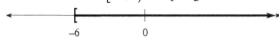

25. Solving the inequality:

$$1.6x - 2 < 0.8x + 2.8$$
$$0.8x - 2 < 2.8$$
$$0.8x < 4.8$$
$$x < 6$$

The solution set is $(-\infty, 6)$. Graphing:

27. Solving the equation:
$$\left|\frac{1}{4}x - 1\right| = \frac{1}{2}$$
$$\frac{1}{4}x - 1 = -\frac{1}{2}, \frac{1}{2}$$
$$\frac{1}{4}x = \frac{1}{2}, \frac{3}{2}$$
$$x = 2, 6$$

29. Solving the equation:
$$|3 - 2x| + 5 = 2$$
$$|3 - 2x| = -3$$
Since this statement is false, there is no solution, or $\varnothing$.

Chapter 6 Test

1. The domain is $\{-3, -2\}$ and the range is $\{0, 1\}$. This is not a function.
2. The domain is all real numbers and the range is $\{y \mid y \geq -9\} = [-9, \infty)$. This is a function.
3. Evaluating the function: $f(3) + g(2) = [3 - 2] + [3 \cdot 2 + 4] = 1 + 10 = 11$
4. Evaluating the function: $h(0) + g(0) = (-8) + 4 = -4$
5. Evaluating the function: $(f \circ g)(2) = f(g(2)) = f(6 + 4) = f(10) = 10 - 2 = 8$
6. Evaluating the function: $(g \circ f)(2) = g(f(2)) = g(2 - 2) = g(0) = 4$
7. The variation equation is $y = Kx^2$. Substituting $x = 5$ and $y = 50$:
$$50 = K(5)^2$$
$$50 = 25K$$
$$K = 2$$
The equation is $y = 2x^2$. Substituting $x = 3$: $y = 2(3)^2 = 2 \cdot 9 = 18$

8. The variation equation is $z = Kxy^3$. Substituting $x = 5, y = 2$, and $z = 15$:
$$15 = K(5)(2)^3$$
$$15 = 40K$$
$$K = \frac{3}{8}$$
The equation is $z = \frac{3}{8}xy^3$. Substituting $x = 2$ and $y = 3$: $z = \frac{3}{8}(2)(3)^3 = \frac{3}{8} \cdot 54 = \frac{81}{4}$

9. The variation equation is $L = \dfrac{Kwd^2}{l}$. Substituting $l = 10$, $w = 3$, $d = 4$, and $L = 800$:

$$800 = \dfrac{K(3)(4)^2}{10}$$
$$8000 = 48K$$
$$K = \dfrac{500}{3}$$

The equation is $L = \dfrac{500wd^2}{3l}$. Substituting $l = 12$, $w = 3$, and $d = 4$: $L = \dfrac{500(3)(4)^2}{3(12)} = \dfrac{2000}{3}$

The beam can safely hold $\dfrac{2000}{3}$ pounds.

10. a. Drawing the line connecting the first point and the last point:

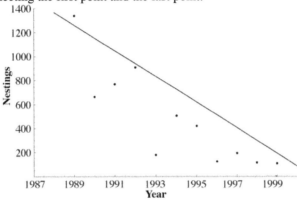

b. Finding the slope: $m = \dfrac{88 - 1{,}367}{2000 - 1988} = \dfrac{-1{,}279}{12} \approx -106.58$

c. Using the point (1,1367) in the point-slope formula:
$$y - 1367 = -106.58(x - 1)$$
$$y - 1367 = -106.58x + 106.58$$
$$y = 1473.58 - 106.58x$$

d. Since it only uses two points to form the equation, it is not as good of a fit as the regression line.

e. The regression line is $y = 1255.92 - 104.769x$. Sketching the graph of this data and the regression line:

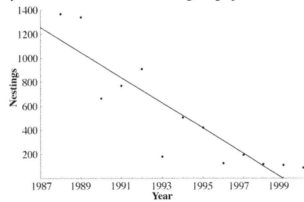

Chapter 7
Rational Expressions and Rational Functions

7.1 Basic Properties and Reducing to Lowest Terms

1. Finding each function value:
$$g(0) = \frac{0+3}{0-1} = \frac{3}{-1} = -3 \qquad g(-3) = \frac{-3+3}{-3-1} = \frac{0}{-4} = 0$$
$$g(3) = \frac{3+3}{3-1} = \frac{6}{2} = 3 \qquad g(-1) = \frac{-1+3}{-1-1} = \frac{2}{-2} = -1$$
$$g(1) = \frac{1+3}{1-1} = \frac{4}{0}, \text{ which is undefined}$$

3. Finding each function value:
$$h(0) = \frac{0-3}{0+1} = \frac{-3}{1} = -3 \qquad h(-3) = \frac{-3-3}{-3+1} = \frac{-6}{-2} = 3$$
$$h(3) = \frac{3-3}{3+1} = \frac{0}{4} = 0 \qquad h(-1) = \frac{-1-3}{-1+1} = \frac{-4}{0}, \text{ which is undefined}$$
$$h(1) = \frac{1-3}{1+1} = \frac{-2}{2} = -1$$

5. The domain is $\{x \mid x \neq 1\}$.

7. The domain is $\{x \mid x \neq 2\}$.

9. Setting the denominator equal to 0:
$$t^2 - 16 = 0$$
$$(t+4)(t-4) = 0$$
$$t = -4, 4$$
The domain is $\{t \mid t \neq -4, t \neq 4\}$.

11. Reducing to lowest terms: $\dfrac{x^2 - 16}{6x + 24} = \dfrac{(x+4)(x-4)}{6(x+4)} = \dfrac{x-4}{6}$

13. Reducing to lowest terms: $\dfrac{a^4 - 81}{a - 3} = \dfrac{(a^2+9)(a^2-9)}{a-3} = \dfrac{(a^2+9)(a+3)(a-3)}{a-3} = (a^2+9)(a+3)$

15. Reducing to lowest terms: $\dfrac{20y^2 - 45}{10y^2 - 5y - 15} = \dfrac{5(4y^2-9)}{5(2y^2-y-3)} = \dfrac{5(2y+3)(2y-3)}{5(2y-3)(y+1)} = \dfrac{2y+3}{y+1}$

17. Reducing to lowest terms: $\dfrac{12y - 2xy - 2x^2y}{6y - 4xy - 2x^2y} = \dfrac{2y(6-x-x^2)}{2y(3-2x-x^2)} = \dfrac{2y(3+x)(2-x)}{2y(3+x)(1-x)} = \dfrac{2-x}{1-x} = \dfrac{x-2}{x-1}$

19. Reducing to lowest terms: $\dfrac{(x-3)^2(x+2)}{(x+2)^2(x-3)} = \dfrac{x-3}{x+2}$

21. Reducing to lowest terms: $\dfrac{x^3 + 1}{x^2 - 1} = \dfrac{(x+1)(x^2-x+1)}{(x+1)(x-1)} = \dfrac{x^2-x+1}{x-1}$

23. Reducing to lowest terms: $\dfrac{4am-4an}{3n-3m} = \dfrac{4a(m-n)}{3(n-m)} = \dfrac{-4a}{3}$

25. Reducing to lowest terms: $\dfrac{ab-a+b-1}{ab+a+b+1} = \dfrac{a(b-1)+1(b-1)}{a(b+1)+1(b+1)} = \dfrac{(b-1)(a+1)}{(b+1)(a+1)} = \dfrac{b-1}{b+1}$

27. Reducing to lowest terms: $\dfrac{21x^2-23x+6}{21x^2+x-10} = \dfrac{(7x-3)(3x-2)}{(7x+5)(3x-2)} = \dfrac{7x-3}{7x+5}$

29. Reducing to lowest terms: $\dfrac{8x^2-6x-9}{8x^2-18x+9} = \dfrac{(4x+3)(2x-3)}{(4x-3)(2x-3)} = \dfrac{4x+3}{4x-3}$

31. Reducing to lowest terms: $\dfrac{4x^2+29x+45}{8x^2-10x-63} = \dfrac{(x+5)(4x+9)}{(2x-7)(4x+9)} = \dfrac{x+5}{2x-7}$

33. Reducing to lowest terms: $\dfrac{a^3+b^3}{a^2-b^2} = \dfrac{(a+b)(a^2-ab+b^2)}{(a+b)(a-b)} = \dfrac{a^2-ab+b^2}{a-b}$

35. Reducing to lowest terms: $\dfrac{8x^4-8x}{4x^4+4x^3+4x^2} = \dfrac{8x(x^3-1)}{4x^2(x^2+x+1)} = \dfrac{8x(x-1)(x^2+x+1)}{4x^2(x^2+x+1)} = \dfrac{2(x-1)}{x} = \dfrac{2x-2}{x}$

37. Reducing to lowest terms: $\dfrac{ax+2x+3a+6}{ay+2y-4a-8} = \dfrac{x(a+2)+3(a+2)}{y(a+2)-4(a+2)} = \dfrac{(a+2)(x+3)}{(a+2)(y-4)} = \dfrac{x+3}{y-4}$

39. Reducing to lowest terms: $\dfrac{x^3+3x^2-4x-12}{x^2+x-6} = \dfrac{x^2(x+3)-4(x+3)}{(x+3)(x-2)} = \dfrac{(x+3)(x^2-4)}{(x+3)(x-2)} = \dfrac{(x+3)(x+2)(x-2)}{(x+3)(x-2)} = x+2$

41. Reducing to lowest terms: $\dfrac{x^3-8}{x^2-4} = \dfrac{(x-2)(x^2+2x+4)}{(x-2)(x+2)} = \dfrac{x^2+2x+4}{x+2}$

43. Reducing to lowest terms: $\dfrac{8x^3-27}{4x^2-9} = \dfrac{(2x-3)(4x^2+6x+9)}{(2x-3)(2x+3)} = \dfrac{4x^2+6x+9}{2x+3}$

45. Reducing to lowest terms: $\dfrac{x-4}{4-x} = \dfrac{x-4}{-1(x-4)} = -1$

47. Reducing to lowest terms: $\dfrac{y^2-36}{6-y} = \dfrac{(y+6)(y-6)}{-1(y-6)} = -(y+6) = -y-6$

49. Reducing to lowest terms: $\dfrac{1-9a^2}{9a^2-6a+1} = \dfrac{-1(9a^2-1)}{(3a-1)^2} = \dfrac{-1(3a+1)(3a-1)}{(3a-1)^2} = -\dfrac{3a+1}{3a-1}$

51. Reducing to lowest terms: $\dfrac{(3x-5)-(3a-5)}{x-a} = \dfrac{3x-3a}{x-a} = \dfrac{3(x-a)}{x-a} = 3$

53. Reducing to lowest terms: $\dfrac{(x^2-4)-(a^2-4)}{x-a} = \dfrac{x^2-a^2}{x-a} = \dfrac{(x+a)(x-a)}{x-a} = x+a$

55. a. Evaluating the difference quotient: $\dfrac{f(x)-f(a)}{x-a} = \dfrac{4x-4a}{x-a} = \dfrac{4(x-a)}{x-a} = 4$

    b. Evaluating the difference quotient: $\dfrac{f(x+h)-f(x)}{h} = \dfrac{4x+4h-4x}{h} = \dfrac{4h}{h} = 4$

57. a. Evaluating the difference quotient: $\dfrac{f(x)-f(a)}{x-a} = \dfrac{(5x+3)-(5a+3)}{x-a} = \dfrac{5x-5a}{x-a} = \dfrac{5(x-a)}{x-a} = 5$

    b. Evaluating the difference quotient: $\dfrac{f(x+h)-f(x)}{h} = \dfrac{5(x+h)+3-(5x+3)}{h} = \dfrac{5x+5h+3-5x-3}{h} = \dfrac{5h}{h} = 5$

59. **a.** Evaluating the difference quotient: $\dfrac{f(x)-f(a)}{x-a} = \dfrac{x^2-a^2}{x-a} = \dfrac{(x+a)(x-a)}{x-a} = x+a$

b. Evaluating the difference quotient:
$$\dfrac{f(x+h)-f(x)}{h} = \dfrac{(x+h)^2 - x^2}{h} = \dfrac{x^2+2xh+h^2-x^2}{h} = \dfrac{2xh+h^2}{h} = \dfrac{h(2x+h)}{h} = 2x+h$$

61. **a.** Evaluating the difference quotient: $\dfrac{f(x)-f(a)}{x-a} = \dfrac{(x^2+1)-(a^2+1)}{x-a} = \dfrac{x^2-a^2}{x-a} = \dfrac{(x+a)(x-a)}{x-a} = x+a$

b. Evaluating the difference quotient:
$$\dfrac{f(x+h)-f(x)}{h} = \dfrac{(x+h)^2+1-(x^2+1)}{h} = \dfrac{x^2+2xh+h^2+1-x^2-1}{h} = \dfrac{2xh+h^2}{h} = \dfrac{h(2x+h)}{h} = 2x+h$$

63. **a.** Evaluating the difference quotient:
$$\dfrac{f(x)-f(a)}{x-a} = \dfrac{(x^2-3x+4)-(a^2-3a+4)}{x-a}$$
$$= \dfrac{x^2-a^2-3x+3a}{x-a}$$
$$= \dfrac{(x+a)(x-a)-3(x-a)}{x-a}$$
$$= \dfrac{(x-a)(x+a-3)}{x-a}$$
$$= x+a-3$$

b. Evaluating the difference quotient:
$$\dfrac{f(x+h)-f(x)}{h} = \dfrac{(x+h)^2 - 3(x+h) + 4 - (x^2 - 3x + 4)}{h}$$
$$= \dfrac{x^2 + 2xh + h^2 - 3x - 3h + 4 - x^2 + 3x - 4}{h}$$
$$= \dfrac{2xh + h^2 - 3h}{h}$$
$$= \dfrac{h(2x+h-3)}{h}$$
$$= 2x+h-3$$

65. **a.** From the graph: $f(2) = 2$ **b.** From the graph: $f(-1) = -4$

c. From the graph: $f(0)$ is undefined **d.** From the graph: $g(3) = 2$

67. Completing the table:

Weeks x	Weight (lb) $W(x)$
0	200
1	194
4	184
12	173
24	168

69. Multiplying: $\dfrac{6}{7} \cdot \dfrac{14}{18} = \dfrac{6}{7} \cdot \dfrac{2 \cdot 7}{3 \cdot 6} = \dfrac{2}{3}$

71. Multiplying: $5y^2 \cdot 4x^2 = 20x^2y^2$

73. Multiplying: $9x^4 \cdot 8y^5 = 72x^4y^5$

75. Factoring: $x^2 - 4 = (x+2)(x-2)$

77. Factoring: $x^3 - x^2y = x^2(x-y)$

79. Factoring: $2y^2 - 2 = 2(y^2-1) = 2(y+1)(y-1)$

7.2 Multiplication and Division of Rational Expressions

1. Performing the operations: $\dfrac{2}{9} \cdot \dfrac{3}{4} = \dfrac{2}{3 \cdot 3} \cdot \dfrac{3}{2 \cdot 2} = \dfrac{1}{2 \cdot 3} = \dfrac{1}{6}$

3. Performing the operations: $\dfrac{3}{4} \div \dfrac{1}{3} = \dfrac{3}{4} \cdot \dfrac{3}{1} = \dfrac{9}{4}$

5. Performing the operations: $\dfrac{3}{7} \cdot \dfrac{14}{24} \div \dfrac{1}{2} = \dfrac{1}{4} \div \dfrac{1}{2} = \dfrac{1}{4} \cdot \dfrac{2}{1} = \dfrac{2}{4} = \dfrac{1}{2}$

7. Performing the operations: $\dfrac{10x^2}{5y^2} \cdot \dfrac{15y^3}{2x^4} = \dfrac{150x^2y^3}{10x^4y^2} = \dfrac{15y}{x^2}$

9. Performing the operations: $\dfrac{11a^2b}{5ab^2} \div \dfrac{22a^3b^2}{10ab^4} = \dfrac{11a^2b}{5ab^2} \cdot \dfrac{10ab^4}{22a^3b^2} = \dfrac{110a^3b^5}{110a^4b^4} = \dfrac{b}{a}$

11. Performing the operations: $\dfrac{6x^2}{5y^3} \cdot \dfrac{11z^2}{2x^2} \div \dfrac{33z^5}{10y^8} = \dfrac{33z^2}{5y^3} \cdot \dfrac{10y^8}{33z^5} = \dfrac{2y^8z^2}{y^3z^5} = \dfrac{2y^5}{z^3}$

13. Performing the operations: $\dfrac{x^2-9}{x^2-4} \cdot \dfrac{x-2}{x-3} = \dfrac{(x+3)(x-3)}{(x+2)(x-2)} \cdot \dfrac{x-2}{x-3} = \dfrac{x+3}{x+2}$

15. Performing the operations: $\dfrac{y^2-1}{y+2} \cdot \dfrac{y^2+5y+6}{y^2+2y-3} = \dfrac{(y+1)(y-1)}{y+2} \cdot \dfrac{(y+2)(y+3)}{(y+3)(y-1)} = y+1$

17. Performing the operations: $\dfrac{3x-12}{x^2-4} \cdot \dfrac{x^2+6x+8}{x-4} = \dfrac{3(x-4)}{(x+2)(x-2)} \cdot \dfrac{(x+4)(x+2)}{x-4} = \dfrac{3(x+4)}{x-2}$

19. Performing the operations: $\dfrac{xy}{xy+1} \div \dfrac{x}{y} = \dfrac{xy}{xy+1} \cdot \dfrac{y}{x} = \dfrac{y^2}{xy+1}$

21. Performing the operations: $\dfrac{1}{x^2-9} \div \dfrac{1}{x^2+9} = \dfrac{1}{x^2-9} \cdot \dfrac{x^2+9}{1} = \dfrac{x^2+9}{x^2-9}$

23. Performing the operations: $\dfrac{y-3}{y^2-6y+9} \cdot \dfrac{y-3}{4} = \dfrac{y-3}{(y-3)^2} \cdot \dfrac{y-3}{4} = \dfrac{1}{4}$

25. Performing the operations: $\dfrac{5x+2y}{25x^2-5xy-6y^2} \cdot \dfrac{20x^2-7xy-3y^2}{4x+y} = \dfrac{5x+2y}{(5x+2y)(5x-3y)} \cdot \dfrac{(5x-3y)(4x+y)}{4x+y} = 1$

27. Performing the operations:
$\dfrac{a^2-5a+6}{a^2-2a-3} \div \dfrac{a-5}{a^2+3a+2} = \dfrac{a^2-5a+6}{a^2-2a-3} \cdot \dfrac{a^2+3a+2}{a-5} = \dfrac{(a-3)(a-2)}{(a-3)(a+1)} \cdot \dfrac{(a+2)(a+1)}{a-5} = \dfrac{(a-2)(a+2)}{a-5}$

29. Performing the operations:
$\dfrac{4t^2-1}{6t^2+t-2} \div \dfrac{8t^3+1}{27t^3+8} = \dfrac{4t^2-1}{6t^2+t-2} \cdot \dfrac{27t^3+8}{8t^3+1} = \dfrac{(2t+1)(2t-1)}{(3t+2)(2t-1)} \cdot \dfrac{(3t+2)(9t^2-6t+4)}{(2t+1)(4t^2-2t+1)} = \dfrac{9t^2-6t+4}{4t^2-2t+1}$

31. Performing the operations:
$\dfrac{2x^2-5x-12}{4x^2+8x+3} \div \dfrac{x^2-16}{2x^2+7x+3} = \dfrac{2x^2-5x-12}{4x^2+8x+3} \cdot \dfrac{2x^2+7x+3}{x^2-16} = \dfrac{(2x+3)(x-4)}{(2x+1)(2x+3)} \cdot \dfrac{(2x+1)(x+3)}{(x+4)(x-4)} = \dfrac{x+3}{x+4}$

33. Performing the operations:
$\dfrac{2a^2-21ab-36b^2}{a^2-11ab-12b^2} \div \dfrac{10a+15b}{a^2-b^2} = \dfrac{2a^2-21ab-36b^2}{a^2-11ab-12b^2} \cdot \dfrac{a^2-b^2}{10a+15b} = \dfrac{(2a+3b)(a-12b)}{(a-12b)(a+b)} \cdot \dfrac{(a+b)(a-b)}{5(2a+3b)} = \dfrac{a-b}{5}$

35. Performing the operations: $\dfrac{6c^2-c-15}{9c^2-25} \cdot \dfrac{15c^2+22c-5}{6c^2+5c-6} = \dfrac{(3c-5)(2c+3)}{(3c+5)(3c-5)} \cdot \dfrac{(3c+5)(5c-1)}{(3c-2)(2c+3)} = \dfrac{5c-1}{3c-2}$

37. Performing the operations:
$$\frac{6a^2b+2ab^2-20b^3}{4a^2b-16b^3} \cdot \frac{10a^2-22ab+4b^2}{27a^3-125b^3} = \frac{2b(3a^2+ab-10b^2)}{4b(a^2-4b^2)} \cdot \frac{2(5a^2-11ab+2b^2)}{27a^3-125b^3}$$
$$= \frac{2b(3a-5b)(a+2b)}{4b(a+2b)(a-2b)} \cdot \frac{2(5a-b)(a-2b)}{(3a-5b)(9a^2+15ab+25b^2)}$$
$$= \frac{5a-b}{9a^2+15ab+25b^2}$$

39. Performing the operations:
$$\frac{360x^3-490x}{36x^2+84x+49} \cdot \frac{30x^2+83x+56}{150x^3+65x^2-280x} = \frac{10x(36x^2-49)}{(6x+7)^2} \cdot \frac{(6x+7)(5x+8)}{5x(30x^2+13x-56)}$$
$$= \frac{10x(6x+7)(6x-7)}{(6x+7)^2} \cdot \frac{(6x+7)(5x+8)}{5x(6x-7)(5x+8)}$$
$$= 2$$

41. Performing the operations:
$$\frac{x^5-x^2}{5x^2-5x} \cdot \frac{10x^4-10x^2}{2x^4+2x^3+2x^2} = \frac{x^2(x^3-1)}{5x(x-1)} \cdot \frac{10x^2(x^2-1)}{2x^2(x^2+x+1)}$$
$$= \frac{x^2(x-1)(x^2+x+1)}{5x(x-1)} \cdot \frac{10x^2(x+1)(x-1)}{2x^2(x^2+x+1)}$$
$$= x(x+1)(x-1)$$

43. Performing the operations:
$$\frac{a^2-16b^2}{a^2-8ab+16b^2} \cdot \frac{a^2-9ab+20b^2}{a^2-7ab+12b^2} \div \frac{a^2-25b^2}{a^2-6ab+9b^2}$$
$$= \frac{a^2-16b^2}{a^2-8ab+16b^2} \cdot \frac{a^2-9ab+20b^2}{a^2-7ab+12b^2} \cdot \frac{a^2-6ab+9b^2}{a^2-25b^2}$$
$$= \frac{(a+4b)(a-4b)}{(a-4b)^2} \cdot \frac{(a-5b)(a-4b)}{(a-3b)(a-4b)} \cdot \frac{(a-3b)^2}{(a+5b)(a-5b)}$$
$$= \frac{(a+4b)(a-3b)}{(a-4b)(a+5b)}$$

45. Performing the operations:
$$\frac{2y^2-7y-15}{42y^2-29y-5} \cdot \frac{12y^2-16y+5}{7y^2-36y+5} \div \frac{4y^2-9}{49y^2-1} = \frac{2y^2-7y-15}{42y^2-29y-5} \cdot \frac{12y^2-16y+5}{7y^2-36y+5} \cdot \frac{49y^2-1}{4y^2-9}$$
$$= \frac{(2y+3)(y-5)}{(6y-5)(7y+1)} \cdot \frac{(6y-5)(2y-1)}{(7y-1)(y-5)} \cdot \frac{(7y+1)(7y-1)}{(2y+3)(2y-3)}$$
$$= \frac{2y-1}{2y-3}$$

47. Performing the operations:
$$\frac{xy-2x+3y-6}{xy+2x-4y-8} \cdot \frac{xy+x-4y-4}{xy-x+3y-3} = \frac{x(y-2)+3(y-2)}{x(y+2)-4(y+2)} \cdot \frac{x(y+1)-4(y+1)}{x(y-1)+3(y-1)}$$
$$= \frac{(y-2)(x+3)}{(y+2)(x-4)} \cdot \frac{(y+1)(x-4)}{(y-1)(x+3)}$$
$$= \frac{(y-2)(y+1)}{(y+2)(y-1)}$$

49. Performing the operations:
$$\frac{xy^2-y^2+4xy-4y}{xy-3y+4x-12} \div \frac{xy^3+2xy^2+y^3+2y^2}{xy^2-3y^2+2xy-6y} = \frac{xy^2-y^2+4xy-4y}{xy-3y+4x-12} \cdot \frac{xy^2-3y^2+2xy-6y}{xy^3+2xy^2+y^3+2y^2}$$
$$= \frac{y^2(x-1)+4y(x-1)}{y(x-3)+4(x-3)} \cdot \frac{y^2(x-3)+2y(x-3)}{xy^2(y+2)+y^2(y+2)}$$
$$= \frac{y(x-1)(y+4)}{(x-3)(y+4)} \cdot \frac{y(x-3)(y+2)}{y^2(y+2)(x+1)}$$
$$= \frac{x-1}{x+1}$$

51. Performing the operations:
$$\frac{2x^3+10x^2-8x-40}{x^3+4x^2-9x-36} \cdot \frac{x^2+x-12}{2x^2+14x+20} = \frac{2x^2(x+5)-8(x+5)}{x^2(x+4)-9(x+4)} \cdot \frac{(x+4)(x-3)}{2(x^2+7x+10)}$$
$$= \frac{2(x+5)(x^2-4)}{(x+4)(x^2-9)} \cdot \frac{(x+4)(x-3)}{2(x+5)(x+2)}$$
$$= \frac{2(x+5)(x+2)(x-2)}{(x+4)(x+3)(x-3)} \cdot \frac{(x+4)(x-3)}{2(x+5)(x+2)}$$
$$= \frac{x-2}{x+3}$$

53. Performing the operations: $\frac{w^3-w^2x}{wy-w} \div \left(\frac{x-w}{y-1}\right)^2 = \frac{w^3-w^2x}{wy-w} \cdot \left(\frac{y-1}{x-w}\right)^2 = \frac{-w^2(x-w)}{w(y-1)} \cdot \frac{(y-1)^2}{(x-w)^2} = -\frac{w(y-1)}{x-w} = \frac{w(y-1)}{w-x}$

55. Performing the operations:
$$\frac{mx+my+2x+2y}{6x^2-5xy-4y^2} \div \frac{2mx-4x+my-2y}{3mx-6x-4my+8y} = \frac{mx+my+2x+2y}{6x^2-5xy-4y^2} \cdot \frac{3mx-6x-4my+8y}{2mx-4x+my-2y}$$
$$= \frac{m(x+y)+2(x+y)}{(3x-4y)(2x+y)} \cdot \frac{3x(m-2)-4y(m-2)}{2x(m-2)+y(m-2)}$$
$$= \frac{(x+y)(m+2)}{(3x-4y)(2x+y)} \cdot \frac{(m-2)(3x-4y)}{(m-2)(2x+y)}$$
$$= \frac{(x+y)(m+2)}{(2x+y)^2}$$

57. Finding the product: $(3x-6) \cdot \frac{x}{x-2} = \frac{3(x-2)}{1} \cdot \frac{x}{x-2} = 3x$

59. Finding the product: $(x^2-25) \cdot \frac{2}{x-5} = \frac{(x+5)(x-5)}{1} \cdot \frac{2}{x-5} = 2(x+5)$

61. Finding the product: $(x^2-3x+2) \cdot \frac{3}{3x-3} = \frac{(x-2)(x-1)}{1} \cdot \frac{3}{3(x-1)} = x-2$

63. Finding the product: $(y-3)(y-4)(y+3) \cdot \frac{-1}{y^2-9} = \frac{(y-3)(y-4)(y+3)}{1} \cdot \frac{-1}{(y+3)(y-3)} = -(y-4) = 4-y$

65. Finding the product: $a(a+5)(a-5) \cdot \frac{a+1}{a^2+5a} = \frac{a(a+5)(a-5)}{1} \cdot \frac{a+1}{a(a+5)} = (a-5)(a+1)$

67. a. Simplifying: $\dfrac{16-1}{64-1} = \dfrac{15}{63} = \dfrac{5}{21}$

   b. Reducing: $\dfrac{25x^2-9}{125x^3-27} = \dfrac{(5x-3)(5x+3)}{(5x-3)(25x^2+15x+9)} = \dfrac{5x+3}{25x^2+15x+9}$

   c. Multiplying: $\dfrac{25x^2-9}{125x^3-27} \cdot \dfrac{5x-3}{5x+3} = \dfrac{(5x-3)(5x+3)}{(5x-3)(25x^2+15x+9)} \cdot \dfrac{5x-3}{5x+3} = \dfrac{5x-3}{25x^2+15x+9}$

   d. Dividing: $\dfrac{25x^2-9}{125x^3-27} \div \dfrac{5x-3}{25x^2+15x+9} = \dfrac{(5x-3)(5x+3)}{(5x-3)(25x^2+15x+9)} \cdot \dfrac{25x^2+15x+9}{5x-3} = \dfrac{5x+3}{5x-3}$

69. Combining: $\dfrac{4}{9} + \dfrac{2}{9} = \dfrac{6}{9} = \dfrac{2 \cdot 3}{3 \cdot 3} = \dfrac{2}{3}$

71. Combining: $\dfrac{3}{14} + \dfrac{7}{30} = \dfrac{3}{14} \cdot \dfrac{15}{15} + \dfrac{7}{30} \cdot \dfrac{7}{7} = \dfrac{45}{210} + \dfrac{49}{210} = \dfrac{94}{210} = \dfrac{47}{105}$

73. Multiplying: $-1(7-x) = -7 + x = x - 7$

75. Factoring: $x^2 - 1 = (x+1)(x-1)$

77. Factoring: $2x + 10 = 2(x+5)$

79. Factoring: $a^3 - b^3 = (a-b)(a^2 + ab + b^2)$

7.3 Addition and Subtraction of Rational Expressions

1. Combining the fractions: $\dfrac{3}{4} + \dfrac{1}{2} = \dfrac{3}{4} + \dfrac{1}{2} \cdot \dfrac{2}{2} = \dfrac{3}{4} + \dfrac{2}{4} = \dfrac{5}{4}$

3. Combining the fractions: $\dfrac{2}{5} - \dfrac{1}{15} = \dfrac{2}{5} \cdot \dfrac{3}{3} - \dfrac{1}{15} = \dfrac{6}{15} - \dfrac{1}{15} = \dfrac{5}{15} = \dfrac{1}{3}$

5. Combining the fractions: $\dfrac{5}{6} + \dfrac{7}{8} = \dfrac{5}{6} \cdot \dfrac{4}{4} + \dfrac{7}{8} \cdot \dfrac{3}{3} = \dfrac{20}{24} + \dfrac{21}{24} = \dfrac{41}{24}$

7. Combining the fractions: $\dfrac{9}{48} - \dfrac{3}{54} = \dfrac{9}{48} \cdot \dfrac{9}{9} - \dfrac{3}{54} \cdot \dfrac{8}{8} = \dfrac{81}{432} - \dfrac{24}{432} = \dfrac{57}{432} = \dfrac{19}{144}$

9. Combining the fractions: $\dfrac{3}{4} - \dfrac{1}{8} + \dfrac{2}{3} = \dfrac{3}{4} \cdot \dfrac{6}{6} - \dfrac{1}{8} \cdot \dfrac{3}{3} + \dfrac{2}{3} \cdot \dfrac{8}{8} = \dfrac{18}{24} - \dfrac{3}{24} + \dfrac{16}{24} = \dfrac{31}{24}$

11. Combining the rational expressions: $\dfrac{x}{x+3} + \dfrac{3}{x+3} = \dfrac{x+3}{x+3} = 1$

13. Combining the rational expressions: $\dfrac{4}{y-4} - \dfrac{y}{y-4} = \dfrac{4-y}{y-4} = \dfrac{-1(y-4)}{y-4} = -1$

15. Combining the rational expressions: $\dfrac{x}{x^2-y^2} - \dfrac{y}{x^2-y^2} = \dfrac{x-y}{x^2-y^2} = \dfrac{x-y}{(x+y)(x-y)} = \dfrac{1}{x+y}$

17. Combining the rational expressions: $\dfrac{2x-3}{x-2} - \dfrac{x-1}{x-2} = \dfrac{2x-3-x+1}{x-2} = \dfrac{x-2}{x-2} = 1$

19. Combining the rational expressions: $\dfrac{1}{a} + \dfrac{2}{a^2} - \dfrac{3}{a^3} = \dfrac{1}{a} \cdot \dfrac{a^2}{a^2} + \dfrac{2}{a^2} \cdot \dfrac{a}{a} - \dfrac{3}{a^3} = \dfrac{a^2 + 2a - 3}{a^3}$

21. Combining the rational expressions: $\dfrac{7x-2}{2x+1} - \dfrac{5x-3}{2x+1} = \dfrac{7x-2-5x+3}{2x+1} = \dfrac{2x+1}{2x+1} = 1$

23.

a. Multiplying: $\dfrac{3}{8} \cdot \dfrac{1}{6} = \dfrac{3}{8} \cdot \dfrac{1}{2 \cdot 3} = \dfrac{1}{16}$

b. Dividing: $\dfrac{3}{8} \div \dfrac{1}{6} = \dfrac{3}{8} \cdot \dfrac{6}{1} = \dfrac{3}{2 \cdot 4} \cdot \dfrac{2 \cdot 3}{1} = \dfrac{9}{4}$

c. Adding: $\dfrac{3}{8} + \dfrac{1}{6} = \dfrac{3}{8} \cdot \dfrac{3}{3} + \dfrac{1}{6} \cdot \dfrac{4}{4} = \dfrac{9}{24} + \dfrac{4}{24} = \dfrac{13}{24}$

d. Multiplying: $\dfrac{x+3}{x-3} \cdot \dfrac{5x+15}{x^2-9} = \dfrac{x+3}{x-3} \cdot \dfrac{5(x+3)}{(x+3)(x-3)} = \dfrac{5(x+3)}{(x-3)^2} = \dfrac{5x+15}{(x-3)^2}$

e. Dividing: $\dfrac{x+3}{x-3} \div \dfrac{5x+15}{x^2-9} = \dfrac{x+3}{x-3} \cdot \dfrac{(x+3)(x-3)}{5(x+3)} = \dfrac{x+3}{5}$

f. Subtracting:
$$\dfrac{x+3}{x-3} - \dfrac{5x+15}{x^2-9} = \dfrac{x+3}{x-3} \cdot \dfrac{x+3}{x+3} - \dfrac{5x+15}{(x+3)(x-3)}$$
$$= \dfrac{x^2+6x+9}{(x+3)(x-3)} - \dfrac{5x+15}{(x+3)(x-3)}$$
$$= \dfrac{x^2+x-6}{(x+3)(x-3)}$$
$$= \dfrac{(x+3)(x-2)}{(x+3)(x-3)}$$
$$= \dfrac{x-2}{x-3}$$

25. Combining the rational expressions:
$$\dfrac{3x+1}{2x-6} - \dfrac{x+2}{x-3} = \dfrac{3x+1}{2(x-3)} - \dfrac{x+2}{x-3} \cdot \dfrac{2}{2} = \dfrac{3x+1}{2(x-3)} - \dfrac{2x+4}{2(x-3)} = \dfrac{3x+1-2x-4}{2(x-3)} = \dfrac{x-3}{2(x-3)} = \dfrac{1}{2}$$

27. Combining the rational expressions:
$$\dfrac{6x+5}{5x-25} - \dfrac{x+2}{x-5} = \dfrac{6x+5}{5(x-5)} - \dfrac{x+2}{x-5} \cdot \dfrac{5}{5} = \dfrac{6x+5}{5(x-5)} - \dfrac{5x+10}{5(x-5)} = \dfrac{6x+5-5x-10}{5(x-5)} = \dfrac{x-5}{5(x-5)} = \dfrac{1}{5}$$

29. Combining the rational expressions:
$$\dfrac{x+1}{2x-2} - \dfrac{2}{x^2-1} = \dfrac{x+1}{2(x-1)} \cdot \dfrac{x+1}{x+1} - \dfrac{2}{(x+1)(x-1)} \cdot \dfrac{2}{2}$$
$$= \dfrac{x^2+2x+1}{2(x+1)(x-1)} - \dfrac{4}{2(x+1)(x-1)}$$
$$= \dfrac{x^2+2x-3}{2(x+1)(x-1)}$$
$$= \dfrac{(x+3)(x-1)}{2(x+1)(x-1)}$$
$$= \dfrac{x+3}{2(x+1)}$$

31. Combining the rational expressions:
$$\frac{1}{a-b} - \frac{3ab}{a^3-b^3} = \frac{1}{a-b} \cdot \frac{a^2+ab+b^2}{a^2+ab+b^2} - \frac{3ab}{a^3-b^3}$$
$$= \frac{a^2+ab+b^2}{a^3-b^3} - \frac{3ab}{a^3-b^3}$$
$$= \frac{a^2-2ab+b^2}{a^3-b^3}$$
$$= \frac{(a-b)^2}{(a-b)(a^2+ab+b^2)}$$
$$= \frac{a-b}{a^2+ab+b^2}$$

33. Combining the rational expressions:
$$\frac{1}{2y-3} - \frac{18y}{8y^3-27} = \frac{1}{2y-3} \cdot \frac{4y^2+6y+9}{4y^2+6y+9} - \frac{18y}{8y^3-27}$$
$$= \frac{4y^2+6y+9}{8y^3-27} - \frac{18y}{8y^3-27}$$
$$= \frac{4y^2-12y+9}{8y^3-27}$$
$$= \frac{(2y-3)^2}{(2y-3)(4y^2+6y+9)}$$
$$= \frac{2y-3}{4y^2+6y+9}$$

35. Combining the rational expressions:
$$\frac{x}{x^2-5x+6} - \frac{3}{3-x} = \frac{x}{(x-2)(x-3)} + \frac{3}{x-3} \cdot \frac{x-2}{x-2}$$
$$= \frac{x}{(x-2)(x-3)} + \frac{3x-6}{(x-2)(x-3)}$$
$$= \frac{4x-6}{(x-2)(x-3)}$$
$$= \frac{2(2x-3)}{(x-3)(x-2)}$$

37. Combining the rational expressions:
$$\frac{2}{4t-5} + \frac{9}{8t^2-38t+35} = \frac{2}{4t-5} \cdot \frac{2t-7}{2t-7} + \frac{9}{(4t-5)(2t-7)}$$
$$= \frac{4t-14}{(4t-5)(2t-7)} + \frac{9}{(4t-5)(2t-7)}$$
$$= \frac{4t-5}{(4t-5)(2t-7)}$$
$$= \frac{1}{2t-7}$$

39. Combining the rational expressions:
$$\frac{1}{a^2-5a+6}+\frac{3}{a^2-a-2}=\frac{1}{(a-2)(a-3)}\cdot\frac{a+1}{a+1}+\frac{3}{(a-2)(a+1)}\cdot\frac{a-3}{a-3}$$
$$=\frac{a+1}{(a-2)(a-3)(a+1)}+\frac{3a-9}{(a-2)(a-3)(a+1)}$$
$$=\frac{4a-8}{(a-2)(a-3)(a+1)}$$
$$=\frac{4(a-2)}{(a-2)(a-3)(a+1)}$$
$$=\frac{4}{(a-3)(a+1)}$$

41. Combining the rational expressions:
$$\frac{1}{8x^3-1}-\frac{1}{4x^2-1}=\frac{1}{(2x-1)(4x^2+2x+1)}\cdot\frac{2x+1}{2x+1}-\frac{1}{(2x+1)(2x-1)}\cdot\frac{4x^2+2x+1}{4x^2+2x+1}$$
$$=\frac{2x+1}{(2x+1)(2x-1)(4x^2+2x+1)}-\frac{4x^2+2x+1}{(2x+1)(2x-1)(4x^2+2x+1)}$$
$$=\frac{2x+1-4x^2-2x-1}{(2x+1)(2x-1)(4x^2+2x+1)}$$
$$=\frac{-4x^2}{(2x+1)(2x-1)(4x^2+2x+1)}$$

43. Combining the rational expressions:
$$\frac{4}{4x^2-9}-\frac{6}{8x^2-6x-9}=\frac{4}{(2x+3)(2x-3)}\cdot\frac{4x+3}{4x+3}-\frac{6}{(2x-3)(4x+3)}\cdot\frac{2x+3}{2x+3}$$
$$=\frac{16x+12}{(2x+3)(2x-3)(4x+3)}-\frac{12x+18}{(2x+3)(2x-3)(4x+3)}$$
$$=\frac{16x+12-12x-18}{(2x+3)(2x-3)(4x+3)}$$
$$=\frac{4x-6}{(2x+3)(2x-3)(4x+3)}$$
$$=\frac{2(2x-3)}{(2x+3)(2x-3)(4x+3)}$$
$$=\frac{2}{(2x+3)(4x+3)}$$

45. Combining the rational expressions:
$$\frac{4a}{a^2+6a+5} - \frac{3a}{a^2+5a+4} = \frac{4a}{(a+5)(a+1)} \cdot \frac{a+4}{a+4} - \frac{3a}{(a+4)(a+1)} \cdot \frac{a+5}{a+5}$$
$$= \frac{4a^2+16a}{(a+4)(a+5)(a+1)} - \frac{3a^2+15a}{(a+4)(a+5)(a+1)}$$
$$= \frac{4a^2+16a-3a^2-15a}{(a+4)(a+5)(a+1)}$$
$$= \frac{a^2+a}{(a+4)(a+5)(a+1)}$$
$$= \frac{a(a+1)}{(a+4)(a+5)(a+1)}$$
$$= \frac{a}{(a+4)(a+5)}$$

47. Combining the rational expressions:
$$\frac{2x-1}{x^2+x-6} - \frac{x+2}{x^2+5x+6} = \frac{2x-1}{(x+3)(x-2)} \cdot \frac{x+2}{x+2} - \frac{x+2}{(x+3)(x+2)} \cdot \frac{x-2}{x-2}$$
$$= \frac{2x^2+3x-2}{(x+3)(x+2)(x-2)} - \frac{x^2-4}{(x+3)(x+2)(x-2)}$$
$$= \frac{2x^2+3x-2-x^2+4}{(x+3)(x+2)(x-2)}$$
$$= \frac{x^2+3x+2}{(x+3)(x+2)(x-2)}$$
$$= \frac{(x+2)(x+1)}{(x+3)(x+2)(x-2)}$$
$$= \frac{x+1}{(x-2)(x+3)}$$

49. Combining the rational expressions:
$$\frac{2x-8}{3x^2+8x+4} + \frac{x+3}{3x^2+5x+2} = \frac{2x-8}{(3x+2)(x+2)} + \frac{x+3}{(3x+2)(x+1)}$$
$$= \frac{2x-8}{(3x+2)(x+2)} \cdot \frac{x+1}{x+1} + \frac{x+3}{(3x+2)(x+1)} \cdot \frac{x+2}{x+2}$$
$$= \frac{2x^2-6x-8}{(3x+2)(x+2)(x+1)} + \frac{x^2+5x+6}{(3x+2)(x+2)(x+1)}$$
$$= \frac{3x^2-x-2}{(3x+2)(x+2)(x+1)}$$
$$= \frac{(3x+2)(x-1)}{(3x+2)(x+2)(x+1)}$$
$$= \frac{x-1}{(x+1)(x+2)}$$

51. Combining the rational expressions:
$$\frac{2}{x^2+5x+6} - \frac{4}{x^2+4x+3} + \frac{3}{x^2+3x+2} = \frac{2}{(x+3)(x+2)} - \frac{4}{(x+3)(x+1)} + \frac{3}{(x+2)(x+1)}$$
$$= \frac{2}{(x+3)(x+2)} \cdot \frac{x+1}{x+1} - \frac{4}{(x+3)(x+1)} \cdot \frac{x+2}{x+2} + \frac{3}{(x+2)(x+1)} \cdot \frac{x+3}{x+3}$$
$$= \frac{2x+2}{(x+3)(x+2)(x+1)} - \frac{4x+8}{(x+3)(x+2)(x+1)} + \frac{3x+9}{(x+3)(x+2)(x+1)}$$
$$= \frac{2x+2-4x-8+3x+9}{(x+3)(x+2)(x+1)}$$
$$= \frac{x+3}{(x+3)(x+2)(x+1)}$$
$$= \frac{1}{(x+2)(x+1)}$$

53. Combining the rational expressions:
$$\frac{2x+8}{x^2+5x+6} - \frac{x+5}{x^2+4x+3} - \frac{x-1}{x^2+3x+2} = \frac{2x+8}{(x+3)(x+2)} - \frac{x+5}{(x+3)(x+1)} - \frac{x-1}{(x+2)(x+1)}$$
$$= \frac{2x+8}{(x+3)(x+2)} \cdot \frac{x+1}{x+1} - \frac{x+5}{(x+3)(x+1)} \cdot \frac{x+2}{x+2} - \frac{x-1}{(x+2)(x+1)} \cdot \frac{x+3}{x+3}$$
$$= \frac{2x^2+10x+8}{(x+3)(x+2)(x+1)} - \frac{x^2+7x+10}{(x+3)(x+2)(x+1)} - \frac{x^2+2x-3}{(x+3)(x+2)(x+1)}$$
$$= \frac{2x^2+10x+8-x^2-7x-10-x^2-2x+3}{(x+3)(x+2)(x+1)}$$
$$= \frac{x+1}{(x+3)(x+2)(x+1)}$$
$$= \frac{1}{(x+2)(x+3)}$$

55. Combining the rational expressions: $2 + \frac{3}{2x+1} = \frac{2}{1} \cdot \frac{2x+1}{2x+1} + \frac{3}{2x+1} = \frac{4x+2}{2x+1} + \frac{3}{2x+1} = \frac{4x+5}{2x+1}$

57. Combining the rational expressions: $5 + \frac{2}{4-t} = \frac{5}{1} \cdot \frac{4-t}{4-t} + \frac{2}{4-t} = \frac{20-5t}{4-t} + \frac{2}{4-t} = \frac{22-5t}{4-t}$

59. Combining the rational expressions: $x - \frac{4}{2x+3} = \frac{x}{1} \cdot \frac{2x+3}{2x+3} - \frac{4}{2x+3} = \frac{2x^2+3x}{2x+3} - \frac{4}{2x+3} = \frac{2x^2+3x-4}{2x+3}$

61. Combining the rational expressions:
$$\frac{x}{x+2} + \frac{1}{2x+4} - \frac{3}{x^2+2x} = \frac{x}{x+2} \cdot \frac{2x}{2x} + \frac{1}{2(x+2)} \cdot \frac{x}{x} - \frac{3}{x(x+2)} \cdot \frac{2}{2}$$
$$= \frac{2x^2}{2x(x+2)} + \frac{x}{2x(x+2)} - \frac{6}{2x(x+2)}$$
$$= \frac{2x^2+x-6}{2x(x+2)}$$
$$= \frac{(2x-3)(x+2)}{2x(x+2)}$$
$$= \frac{2x-3}{2x}$$

63. Combining the rational expressions:
$$\frac{1}{x} + \frac{x}{2x+4} - \frac{2}{x^2+2x} = \frac{1}{x} \cdot \frac{2(x+2)}{2(x+2)} + \frac{x}{2(x+2)} \cdot \frac{x}{x} - \frac{2}{x(x+2)} \cdot \frac{2}{2}$$
$$= \frac{2x+4}{2x(x+2)} + \frac{x^2}{2x(x+2)} - \frac{4}{2x(x+2)}$$
$$= \frac{x^2+2x}{2x(x+2)}$$
$$= \frac{x(x+2)}{2x(x+2)}$$
$$= \frac{1}{2}$$

65. Finding the sum:
$$f(x)+g(x) = \frac{2}{x+4} + \frac{x-1}{x^2+3x-4}$$
$$= \frac{2}{x+4} \cdot \frac{x-1}{x-1} + \frac{x-1}{(x+4)(x-1)}$$
$$= \frac{2x-2}{(x+4)(x-1)} + \frac{x-1}{(x+4)(x-1)}$$
$$= \frac{3x-3}{(x+4)(x-1)}$$
$$= \frac{3(x-1)}{(x+4)(x-1)}$$
$$= \frac{3}{x+4}$$

67. Finding the sum:
$$f(x)+g(x) = \frac{2x}{x^2-x-2} + \frac{5}{x^2+x-6}$$
$$= \frac{2x}{(x-2)(x+1)} \cdot \frac{x+3}{x+3} + \frac{5}{(x+3)(x-2)} \cdot \frac{x+1}{x+1}$$
$$= \frac{2x^2+6x}{(x-2)(x+1)(x+3)} + \frac{5x+5}{(x-2)(x+1)(x+3)}$$
$$= \frac{2x^2+11x+5}{(x-2)(x+1)(x+3)}$$
$$= \frac{(2x+1)(x+5)}{(x-2)(x+1)(x+3)}$$

69. Substituting the values: $P = \frac{1}{10} + \frac{1}{0.2} = 0.1 + 5 = 5.1 = \frac{51}{10}$

71. Writing the expression and simplifying: $x + \frac{4}{x} = \frac{x^2+4}{x}$

73. Writing the expression and simplifying: $\frac{1}{x} + \frac{1}{x+1} = \frac{1}{x} \cdot \frac{x+1}{x+1} + \frac{1}{x+1} \cdot \frac{x}{x} = \frac{x+1}{x(x+1)} + \frac{x}{x(x+1)} = \frac{2x+1}{x(x+1)}$

75. Dividing: $\dfrac{3}{4} \div \dfrac{5}{8} = \dfrac{3}{4} \cdot \dfrac{8}{5} = \dfrac{24}{20} = \dfrac{4 \cdot 6}{4 \cdot 5} = \dfrac{6}{5}$

77. Multiplying: $x\left(1 + \dfrac{2}{x}\right) = x \cdot 1 + x \cdot \dfrac{2}{x} = x + 2$

79. Multiplying: $3x\left(\dfrac{1}{x} - \dfrac{1}{3}\right) = 3x \cdot \dfrac{1}{x} - 3x \cdot \dfrac{1}{3} = 3 - x$

81. Factoring: $x^2 - 4 = (x+2)(x-2)$

7.4 Complex Fractions

1. Simplifying the complex fraction: $\dfrac{\frac{3}{4}}{\frac{2}{3}} = \dfrac{\frac{3}{4} \cdot 12}{\frac{2}{3} \cdot 12} = \dfrac{9}{8}$

3. Simplifying the complex fraction: $\dfrac{\frac{1}{3} - \frac{1}{4}}{\frac{1}{2} + \frac{1}{8}} = \dfrac{\left(\frac{1}{3} - \frac{1}{4}\right) \cdot 24}{\left(\frac{1}{2} + \frac{1}{8}\right) \cdot 24} = \dfrac{8 - 6}{12 + 3} = \dfrac{2}{15}$

5. Simplifying the complex fraction: $\dfrac{3 + \frac{2}{5}}{1 - \frac{3}{7}} = \dfrac{\left(3 + \frac{2}{5}\right) \cdot 35}{\left(1 - \frac{3}{7}\right) \cdot 35} = \dfrac{105 + 14}{35 - 15} = \dfrac{119}{20}$

7. Simplifying the complex fraction: $\dfrac{\frac{1}{x}}{1 + \frac{1}{x}} = \dfrac{\left(\frac{1}{x}\right) \cdot x}{\left(1 + \frac{1}{x}\right) \cdot x} = \dfrac{1}{x+1}$

9. Simplifying the complex fraction: $\dfrac{1 + \frac{1}{a}}{1 - \frac{1}{a}} = \dfrac{\left(1 + \frac{1}{a}\right) \cdot a}{\left(1 - \frac{1}{a}\right) \cdot a} = \dfrac{a+1}{a-1}$

11. Simplifying the complex fraction: $\dfrac{\frac{1}{x} - \frac{1}{y}}{\frac{1}{x} + \frac{1}{y}} = \dfrac{\left(\frac{1}{x} - \frac{1}{y}\right) \cdot xy}{\left(\frac{1}{x} + \frac{1}{y}\right) \cdot xy} = \dfrac{y - x}{y + x}$

13. Simplifying the complex fraction: $\dfrac{\frac{x-5}{x^2-4}}{\frac{x^2-25}{x+2}} = \dfrac{\frac{x-5}{(x+2)(x-2)} \cdot (x+2)(x-2)}{\frac{(x+5)(x-5)}{x+2} \cdot (x+2)(x-2)} = \dfrac{x-5}{(x+5)(x-5)(x-2)} = \dfrac{1}{(x+5)(x-2)}$

15. Simplifying the complex fraction: $\dfrac{\frac{4a}{2a^3+2}}{\frac{8a}{4a+4}} = \dfrac{\frac{4a}{2(a+1)(a^2-a+1)} \cdot 2(a+1)(a^2-a+1)}{\frac{8a}{4(a+1)} \cdot 2(a+1)(a^2-a+1)} = \dfrac{4a}{4a(a^2-a+1)} = \dfrac{1}{a^2-a+1}$

17. Simplifying the complex fraction: $\dfrac{1 - \frac{9}{x^2}}{1 - \frac{1}{x} - \frac{6}{x^2}} = \dfrac{\left(1 - \frac{9}{x^2}\right) \cdot x^2}{\left(1 - \frac{1}{x} - \frac{6}{x^2}\right) \cdot x^2} = \dfrac{x^2 - 9}{x^2 - x - 6} = \dfrac{(x+3)(x-3)}{(x+2)(x-3)} = \dfrac{x+3}{x+2}$

19. Simplifying the complex fraction: $\dfrac{2+\dfrac{5}{a}-\dfrac{3}{a^2}}{2-\dfrac{5}{a}+\dfrac{2}{a^2}} = \dfrac{\left(2+\dfrac{5}{a}-\dfrac{3}{a^2}\right)\cdot a^2}{\left(2-\dfrac{5}{a}+\dfrac{2}{a^2}\right)\cdot a^2} = \dfrac{2a^2+5a-3}{2a^2-5a+2} = \dfrac{(2a-1)(a+3)}{(2a-1)(a-2)} = \dfrac{a+3}{a-2}$

21. Simplifying the complex fraction:

$$\dfrac{2+\dfrac{3}{x}-\dfrac{18}{x^2}-\dfrac{27}{x^3}}{2+\dfrac{9}{x}+\dfrac{9}{x^2}} = \dfrac{\left(2+\dfrac{3}{x}-\dfrac{18}{x^2}-\dfrac{27}{x^3}\right)\cdot x^3}{\left(2+\dfrac{9}{x}+\dfrac{9}{x^2}\right)\cdot x^3}$$

$$= \dfrac{2x^3+3x^2-18x-27}{2x^3+9x^2+9x}$$

$$= \dfrac{x^2(2x+3)-9(2x+3)}{x(2x^2+9x+9)}$$

$$= \dfrac{(2x+3)(x^2-9)}{x(2x+3)(x+3)}$$

$$= \dfrac{(2x+3)(x+3)(x-3)}{x(2x+3)(x+3)}$$

$$= \dfrac{x-3}{x}$$

23. Simplifying the complex fraction: $\dfrac{1+\dfrac{1}{x+3}}{1-\dfrac{1}{x+3}} = \dfrac{\left(1+\dfrac{1}{x+3}\right)\cdot(x+3)}{\left(1-\dfrac{1}{x+3}\right)\cdot(x+3)} = \dfrac{x+3+1}{x+3-1} = \dfrac{x+4}{x+2}$

25. Simplifying the complex fraction:

$$\dfrac{1+\dfrac{1}{x+3}}{1+\dfrac{7}{x-3}} = \dfrac{\left(1+\dfrac{1}{x+3}\right)\cdot(x+3)(x-3)}{\left(1+\dfrac{7}{x-3}\right)\cdot(x+3)(x-3)}$$

$$= \dfrac{(x+3)(x-3)+x-3}{(x+3)(x-3)+7(x+3)}$$

$$= \dfrac{x^2-9+x-3}{x^2-9+7x+21}$$

$$= \dfrac{x^2+x-12}{x^2+7x+12}$$

$$= \dfrac{(x+4)(x-3)}{(x+4)(x+3)}$$

$$= \dfrac{x-3}{x+3}$$

27. Simplifying the complex fraction:

$$\dfrac{1-\dfrac{1}{a+1}}{1+\dfrac{1}{a-1}} = \dfrac{\left(1-\dfrac{1}{a+1}\right)\cdot(a+1)(a-1)}{\left(1+\dfrac{1}{a-1}\right)\cdot(a+1)(a-1)}$$

$$= \dfrac{(a+1)(a-1)-(a-1)}{(a+1)(a-1)+(a+1)}$$

$$= \dfrac{(a-1)(a+1-1)}{(a+1)(a-1+1)}$$

$$= \dfrac{a(a-1)}{a(a+1)}$$

$$= \dfrac{a-1}{a+1}$$

29. Simplifying the complex fraction:

$$\frac{\frac{1}{x+3}+\frac{1}{x-3}}{\frac{1}{x+3}-\frac{1}{x-3}} = \frac{\left(\frac{1}{x+3}+\frac{1}{x-3}\right)\cdot(x+3)(x-3)}{\left(\frac{1}{x+3}-\frac{1}{x-3}\right)\cdot(x+3)(x-3)}$$

$$= \frac{(x-3)+(x+3)}{(x-3)-(x+3)}$$

$$= \frac{2x}{-6}$$

$$= -\frac{x}{3}$$

31. Simplifying the complex fraction:

$$\frac{\frac{y+1}{y-1}+\frac{y-1}{y+1}}{\frac{y+1}{y-1}-\frac{y-1}{y+1}} = \frac{\left(\frac{y+1}{y-1}+\frac{y-1}{y+1}\right)\cdot(y+1)(y-1)}{\left(\frac{y+1}{y-1}-\frac{y-1}{y+1}\right)\cdot(y+1)(y-1)}$$

$$= \frac{(y+1)^2+(y-1)^2}{(y+1)^2-(y-1)^2}$$

$$= \frac{y^2+2y+1+y^2-2y+1}{y^2+2y+1-y^2+2y-1}$$

$$= \frac{2y^2+2}{4y}$$

$$= \frac{2(y^2+1)}{4y}$$

$$= \frac{y^2+1}{2y}$$

33. Simplifying the complex fraction: $1-\dfrac{x}{1-\dfrac{1}{x}} = 1-\dfrac{x\bullet x}{\left(1-\dfrac{1}{x}\right)\bullet x} = 1-\dfrac{x^2}{x-1} = \dfrac{x-1-x^2}{x-1} = \dfrac{-x^2+x-1}{x-1}$

35. Simplifying the complex fraction: $1+\dfrac{1}{1+\dfrac{1}{1+1}} = 1+\dfrac{1}{1+\dfrac{1}{2}} = 1+\dfrac{1}{\dfrac{3}{2}} = 1+\dfrac{2}{3} = \dfrac{5}{3}$

37. Simplifying the complex fraction:

$$\frac{1-\dfrac{1}{x+\dfrac{1}{2}}}{1+\dfrac{1}{x+\dfrac{1}{2}}} = \frac{1-\dfrac{1\bullet 2}{\left(x+\dfrac{1}{2}\right)\bullet 2}}{1+\dfrac{1\bullet 2}{\left(x+\dfrac{1}{2}\right)\bullet 2}} = \frac{1-\dfrac{2}{2x+1}}{1+\dfrac{2}{2x+1}} = \frac{\left(1-\dfrac{2}{2x+1}\right)(2x+1)}{\left(1+\dfrac{2}{2x+1}\right)(2x+1)} = \frac{2x+1-2}{2x+1+2} = \frac{2x-1}{2x+3}$$

39. Simplifying the complex fraction: $\dfrac{\dfrac{1}{x+h}-\dfrac{1}{x}}{h} = \dfrac{\left(\dfrac{1}{x+h}-\dfrac{1}{x}\right)\bullet x(x+h)}{h\bullet x(x+h)} = \dfrac{x-(x+h)}{hx(x+h)} = \dfrac{x-x-h}{hx(x+h)} = \dfrac{-h}{hx(x+h)} = -\dfrac{1}{x(x+h)}$

41. Simplifying the complex fraction: $\dfrac{\dfrac{3}{ab}+\dfrac{4}{bc}-\dfrac{2}{ac}}{\dfrac{5}{abc}} = \dfrac{\left(\dfrac{3}{ab}+\dfrac{4}{bc}-\dfrac{2}{ac}\right)\bullet abc}{\left(\dfrac{5}{abc}\right)\bullet abc} = \dfrac{3c+4a-2b}{5}$

43. Simplifying the complex fraction: $\dfrac{\dfrac{t^2-2t-8}{t^2+7t+6}}{\dfrac{t^2-t-6}{t^2+2t+1}} = \dfrac{\dfrac{(t-4)(t+2)}{(t+6)(t+1)}\bullet(t+6)(t+1)^2}{\dfrac{(t-3)(t+2)}{(t+1)^2}\bullet(t+6)(t+1)^2} = \dfrac{(t-4)(t+2)(t+1)}{(t-3)(t+2)(t+6)} = \dfrac{(t-4)(t+1)}{(t+6)(t-3)}$

45. Simplifying the complex fraction:

$$\frac{5+\frac{4}{b-1}}{\frac{7}{b+5}-\frac{3}{b-1}} = \frac{\left(5+\frac{4}{b-1}\right)\cdot(b+5)(b-1)}{\left(\frac{7}{b+5}-\frac{3}{b-1}\right)\cdot(b+5)(b-1)}$$

$$= \frac{5(b+5)(b-1)+4(b+5)}{7(b-1)-3(b+5)}$$

$$= \frac{(b+5)(5b-5+4)}{7b-7-3b-15}$$

$$= \frac{(b+5)(5b-1)}{4b-22}$$

$$= \frac{(5b-1)(b+5)}{2(2b-11)}$$

47. Simplifying the complex fraction:

$$\frac{\frac{3}{x^2-x-6}}{\frac{2}{x+2}-\frac{4}{x-3}} = \frac{\frac{3}{(x-3)(x+2)}\cdot(x-3)(x+2)}{\left(\frac{2}{x+2}-\frac{4}{x-3}\right)\cdot(x-3)(x+2)}$$

$$= \frac{3}{2(x-3)-4(x+2)}$$

$$= \frac{3}{2x-6-4x-8}$$

$$= \frac{3}{-2x-14}$$

$$= -\frac{3}{2x+14}$$

49. Simplifying the complex fraction: $\dfrac{\frac{1}{m-4}+\frac{1}{m-5}}{\frac{1}{m^2-9m+20}} = \dfrac{\left(\frac{1}{m-4}+\frac{1}{m-5}\right)\cdot(m-4)(m-5)}{\frac{1}{(m-4)(m-5)}\cdot(m-4)(m-5)} = \dfrac{(m-5)+(m-4)}{1} = 2m-9$

51. a. Simplifying the difference quotient: $\dfrac{f(x)-f(a)}{x-a} = \dfrac{\frac{4}{x}-\frac{4}{a}}{x-a} = \dfrac{\left(\frac{4}{x}-\frac{4}{a}\right)ax}{(x-a)ax} = \dfrac{4a-4x}{ax(x-a)} = \dfrac{-4(x-a)}{ax(x-a)} = -\dfrac{4}{ax}$

b. Simplifying the difference quotient:

$$\frac{f(x)-f(a)}{x-a} = \frac{\frac{1}{x+1}-\frac{1}{a+1}}{x-a}$$

$$= \frac{\left(\frac{1}{x+1}-\frac{1}{a+1}\right)(x+1)(a+1)}{(x-a)(x+1)(a+1)}$$

$$= \frac{a+1-x-1}{(x-a)(x+1)(a+1)}$$

$$= \frac{a-x}{(x-a)(x+1)(a+1)}$$

$$= -\frac{1}{(x+1)(a+1)}$$

    c. Simplifying the difference quotient:

$$\frac{f(x)-f(a)}{x-a} = \frac{\frac{1}{x^2}-\frac{1}{a^2}}{x-a} = \frac{\left(\frac{1}{x^2}-\frac{1}{a^2}\right)a^2x^2}{a^2x^2(x-a)} = \frac{a^2-x^2}{a^2x^2(x-a)} = \frac{(a+x)(a-x)}{a^2x^2(x-a)} = -\frac{a+x}{a^2x^2}$$

53. **a.** As v approaches 0, the denominator approaches 1.

      b. Solving for v:

$$h = \frac{f}{1+\frac{v}{s}}$$

$$h = \frac{f \cdot s}{\left(1+\frac{v}{s}\right)s}$$

$$h = \frac{fs}{s+v}$$

$$h(s+v) = fs$$

$$s+v = \frac{fs}{h}$$

$$v = \frac{fs}{h} - s$$

$$v = \frac{fs-sh}{h} = \frac{fs}{h} - s$$

55. Multiplying: $x(y-2) = xy - 2x$

57. Multiplying: $6\left(\frac{x}{2}-3\right) = 6 \cdot \frac{x}{2} - 6 \cdot 3 = 3x - 18$

59. Multiplying: $xab \cdot \frac{1}{x} = ab$

61. Factoring: $y^2 - 25 = (y+5)(y-5)$

63. Factoring: $xa + xb = x(a+b)$

65. Solving the equation:
$$5x - 4 = 6$$
$$5x = 10$$
$$x = 2$$

7.5 Equations With Rational Expressions

1. Solving the equation:
$$\frac{x}{5} + 4 = \frac{5}{3}$$
$$15\left(\frac{x}{5} + 4\right) = 15\left(\frac{5}{3}\right)$$
$$3x + 60 = 25$$
$$3x = -35$$
$$x = -\frac{35}{3}$$

3. Solving the equation:
$$\frac{a}{3} + 2 = \frac{4}{5}$$
$$15\left(\frac{a}{3} + 2\right) = 15\left(\frac{4}{5}\right)$$
$$5a + 30 = 12$$
$$5a = -18$$
$$a = -\frac{18}{5}$$

5. Solving the equation:
$$\frac{y}{2} + \frac{y}{4} + \frac{y}{6} = 3$$
$$12\left(\frac{y}{2} + \frac{y}{4} + \frac{y}{6}\right) = 12(3)$$
$$6y + 3y + 2y = 36$$
$$11y = 36$$
$$y = \frac{36}{11}$$

7. Solving the equation:
$$\frac{5}{2x} = \frac{1}{x} + \frac{3}{4}$$
$$4x\left(\frac{5}{2x}\right) = 4x\left(\frac{1}{x} + \frac{3}{4}\right)$$
$$10 = 4 + 3x$$
$$3x = 6$$
$$x = 2$$

9. Solving the equation:
$$\frac{1}{x} = \frac{1}{3} - \frac{2}{3x}$$
$$3x\left(\frac{1}{x}\right) = 3x\left(\frac{1}{3} - \frac{2}{3x}\right)$$
$$3 = x - 2$$
$$x = 5$$

11. Solving the equation:
$$\frac{2x}{x-3} + 2 = \frac{2}{x-3}$$
$$(x-3)\left(\frac{2x}{x-3} + 2\right) = (x-3)\left(\frac{2}{x-3}\right)$$
$$2x + 2(x-3) = 2$$
$$2x + 2x - 6 = 2$$
$$4x = 8$$
$$x = 2$$

13. Solving the equation:
$$1 - \frac{1}{x} = \frac{12}{x^2}$$
$$x^2\left(1 - \frac{1}{x}\right) = x^2\left(\frac{12}{x^2}\right)$$
$$x^2 - x = 12$$
$$x^2 - x - 12 = 0$$
$$(x+3)(x-4) = 0$$
$$x = -3, 4$$

15. Solving the equation:
$$y - \frac{4}{3y} = -\frac{1}{3}$$
$$3y\left(y - \frac{4}{3y}\right) = 3y\left(-\frac{1}{3}\right)$$
$$3y^2 - 4 = -y$$
$$3y^2 + y - 4 = 0$$
$$(3y+4)(y-1) = 0$$
$$x = -\frac{4}{3}, 1$$

17. Solving the equation:
$$\frac{x+2}{x+1} = \frac{1}{x+1} + 2$$
$$(x+1)\left(\frac{x+2}{x+1}\right) = (x+1)\left(\frac{1}{x+1} + 2\right)$$
$$x+2 = 1 + 2(x+1)$$
$$x+2 = 1 + 2x + 2$$
$$x+2 = 2x + 3$$
$$x = -1 \quad \text{(does not check)}$$
There is no solution (-1 does not check).

19. Solving the equation:
$$\frac{3}{a-2} = \frac{2}{a-3}$$
$$(a-2)(a-3)\left(\frac{3}{a-2}\right) = (a-2)(a-3)\left(\frac{2}{a-3}\right)$$
$$3(a-3) = 2(a-2)$$
$$3a - 9 = 2a - 4$$
$$a = 5$$

21. Solving the equation:
$$6 - \frac{5}{x^2} = \frac{7}{x}$$
$$x^2\left(6 - \frac{5}{x^2}\right) = x^2\left(\frac{7}{x}\right)$$
$$6x^2 - 5 = 7x$$
$$6x^2 - 7x - 5 = 0$$
$$(2x+1)(3x-5) = 0$$
$$x = -\frac{1}{2}, \frac{5}{3}$$

23. Solving the equation:
$$\frac{1}{x-1} - \frac{1}{x+1} = \frac{3x}{x^2-1}$$
$$(x+1)(x-1)\left(\frac{1}{x-1} - \frac{1}{x+1}\right) = (x+1)(x-1)\left(\frac{3x}{(x+1)(x-1)}\right)$$
$$(x+1) - (x-1) = 3x$$
$$x + 1 - x + 1 = 3x$$
$$3x = 2$$
$$x = \frac{2}{3}$$

25. Solving the equation:
$$\frac{2}{x-3} + \frac{x}{x^2-9} = \frac{4}{x+3}$$
$$(x+3)(x-3)\left(\frac{2}{x-3} + \frac{x}{(x+3)(x-3)}\right) = (x+3)(x-3)\left(\frac{4}{x+3}\right)$$
$$2(x+3) + x = 4(x-3)$$
$$2x + 6 + x = 4x - 12$$
$$3x + 6 = 4x - 12$$
$$-x = -18$$
$$x = 18$$

27. Solving the equation:
$$\frac{3}{2} - \frac{1}{x-4} = \frac{-2}{2x-8}$$
$$2(x-4)\left(\frac{3}{2} - \frac{1}{x-4}\right) = 2(x-4)\left(\frac{-2}{2(x-4)}\right)$$
$$3(x-4) - 2 = -2$$
$$3x - 12 - 2 = -2$$
$$3x - 14 = -2$$
$$3x = 12$$
$$x = 4 \quad \text{(does not check)}$$

There is no solution (4 does not check).

29. Solving the equation:
$$\frac{t-4}{t^2 - 3t} = \frac{-2}{t^2 - 9}$$
$$t(t+3)(t-3) \cdot \frac{t-4}{t(t-3)} = t(t+3)(t-3) \cdot \frac{-2}{(t+3)(t-3)}$$
$$(t+3)(t-4) = -2t$$
$$t^2 - t - 12 = -2t$$
$$t^2 + t - 12 = 0$$
$$(t+4)(t-3) = 0$$
$$t = -4 \quad (t = 3 \text{ does not check})$$

31. Solving the equation:
$$\frac{3}{y-4} - \frac{2}{y+1} = \frac{5}{y^2 - 3y - 4}$$
$$(y-4)(y+1)\left(\frac{3}{y-4} - \frac{2}{y+1}\right) = (y-4)(y+1)\left(\frac{5}{(y-4)(y+1)}\right)$$
$$3(y+1) - 2(y-4) = 5$$
$$3y + 3 - 2y + 8 = 5$$
$$y + 11 = 5$$
$$y = -6$$

33. Solving the equation:
$$\frac{2}{1+a} = \frac{3}{1-a} + \frac{5}{a}$$
$$a(1+a)(1-a)\left(\frac{2}{1+a}\right) = a(1+a)(1-a)\left(\frac{3}{1-a} + \frac{5}{a}\right)$$
$$2a(1-a) = 3a(1+a) + 5(1+a)(1-a)$$
$$2a - 2a^2 = 3a + 3a^2 + 5 - 5a^2$$
$$-2a^2 + 2a = -2a^2 + 3a + 5$$
$$2a = 3a + 5$$
$$-a = 5$$
$$a = -5$$

35. Solving the equation:
$$\frac{3}{2x-6} - \frac{x+1}{4x-12} = 4$$
$$4(x-3)\left(\frac{3}{2(x-3)} - \frac{x+1}{4(x-3)}\right) = 4(x-3)(4)$$
$$6 - (x+1) = 16x - 48$$
$$5 - x = 16x - 48$$
$$-17x = -53$$
$$x = \frac{53}{17}$$

37. Solving the equation:
$$\frac{y+2}{y^2-y} - \frac{6}{y^2-1} = 0$$
$$y(y+1)(y-1)\left(\frac{y+2}{y(y-1)} - \frac{6}{(y+1)(y-1)}\right) = y(y+1)(y-1)(0)$$
$$(y+1)(y+2) - 6y = 0$$
$$y^2 + 3y + 2 - 6y = 0$$
$$y^2 - 3y + 2 = 0$$
$$(y-1)(y-2) = 0$$
$$y = 2 \quad (y = 1 \text{ does not check})$$

39. Solving the equation:
$$\frac{4}{2x-6} - \frac{12}{4x+12} = \frac{12}{x^2-9}$$
$$4(x+3)(x-3)\left(\frac{4}{2(x-3)} - \frac{12}{4(x+3)}\right) = 4(x+3)(x-3)\left(\frac{12}{(x+3)(x-3)}\right)$$
$$8(x+3) - 12(x-3) = 48$$
$$8x + 24 - 12x + 36 = 48$$
$$-4x + 60 = 48$$
$$-4x = -12$$
$$x = 3 \quad (x = 3 \text{ does not check})$$

There is no solution (3 does not check).

41. Solving the equation:
$$\frac{2}{y^2-7y+12} - \frac{1}{y^2-9} = \frac{4}{y^2-y-12}$$
$$(y+3)(y-3)(y-4)\left(\frac{2}{(y-3)(y-4)} - \frac{1}{(y+3)(y-3)}\right) = (y+3)(y-3)(y-4)\left(\frac{4}{(y-4)(y+3)}\right)$$
$$2(y+3) - (y-4) = 4(y-3)$$
$$2y + 6 - y + 4 = 4y - 12$$
$$y + 10 = 4y - 12$$
$$-3y = -22$$
$$y = \frac{22}{3}$$

43. **a.** Solving the equation:
$$f(x) + g(x) = \frac{5}{8}$$
$$\frac{1}{x-3} + \frac{1}{x+3} = \frac{5}{8}$$
$$8(x-3)(x+3)\left(\frac{1}{x-3} + \frac{1}{x+3}\right) = 8(x-3)(x+3)\left(\frac{5}{8}\right)$$
$$8(x+3) + 8(x-3) = 5(x-3)(x+3)$$
$$8x + 24 + 8x - 24 = 5x^2 - 45$$
$$16x = 5x^2 - 45$$
$$0 = 5x^2 - 16x - 45$$
$$0 = (5x+9)(x-5)$$
$$x = -\frac{9}{5}, 5$$

b. Solving the equation:
$$\frac{f(x)}{g(x)} = 5$$
$$\frac{\frac{1}{x-3}}{\frac{1}{x+3}} = 5$$
$$\frac{x+3}{x-3} = 5$$
$$x + 3 = 5(x-3)$$
$$x + 3 = 5x - 15$$
$$18 = 4x$$
$$x = \frac{9}{2}$$

c. Solving the equation:
$$f(x) = g(x)$$
$$\frac{1}{x-3} = \frac{1}{x+3}$$
$$x + 3 = x - 3$$
$$3 = -3 \text{ (false)}$$
There is no solution $(\varnothing)$.

45. **a.** Solving the equation:
$$6x - 2 = 0$$
$$6x = 2$$
$$x = \frac{1}{3}$$

b. Solving the equation:
$$\frac{6}{x} - 2 = 0$$
$$x\left(\frac{6}{x} - 2\right) = x(0)$$
$$6 - 2x = 0$$
$$6 = 2x$$
$$x = 3$$

c. Solving the equation:
$$\frac{x}{6} - 2 = -\frac{1}{2}$$
$$6\left(\frac{x}{6} - 2\right) = 6\left(-\frac{1}{2}\right)$$
$$x - 12 = -3$$
$$x = 9$$

d. Solving the equation:
$$\frac{6}{x} - 2 = -\frac{1}{2}$$
$$2x\left(\frac{6}{x} - 2\right) = 2x\left(-\frac{1}{2}\right)$$
$$12 - 4x = -x$$
$$12 = 3x$$
$$x = 4$$

e. Solving the equation:
$$\frac{6}{x^2} + 6 = \frac{20}{x}$$
$$x^2\left(\frac{6}{x^2} + 6\right) = x^2\left(\frac{20}{x}\right)$$
$$6 + 6x^2 = 20x$$
$$6x^2 - 20x + 6 = 0$$
$$3x^2 - 10x + 3 = 0$$
$$(3x - 1)(x - 3) = 0$$
$$x = \frac{1}{3}, 3$$

47. a. Dividing: $\dfrac{6}{x^2 - 2x - 8} \div \dfrac{x+3}{x+2} = \dfrac{6}{(x-4)(x+2)} \cdot \dfrac{x+2}{x+3} = \dfrac{6}{(x-4)(x+3)}$

   b. Adding:
$$\frac{6}{x^2 - 2x - 8} + \frac{x+3}{x+2} = \frac{6}{(x-4)(x+2)} + \frac{x+3}{x+2} \cdot \frac{x-4}{x-4}$$
$$= \frac{6}{(x-4)(x+2)} + \frac{x^2 - x - 12}{(x-4)(x+2)}$$
$$= \frac{x^2 - x - 6}{(x-4)(x+2)}$$
$$= \frac{(x-3)(x+2)}{(x-4)(x+2)}$$
$$= \frac{x-3}{x-4}$$

   c. Solving the equation:
$$\frac{6}{x^2 - 2x - 8} + \frac{x+3}{x+2} = 2$$
$$(x-4)(x+2)\left(\frac{6}{(x-4)(x+2)} + \frac{x+3}{x+2}\right) = (x-4)(x+2)(2)$$
$$6 + (x-4)(x+3) = 2(x-4)(x+2)$$
$$6 + x^2 - x - 12 = 2x^2 - 4x - 16$$
$$0 = x^2 - 3x - 10$$
$$0 = (x-5)(x+2)$$
$$x = 5 \qquad (x = -2 \text{ does not check})$$

49. Solving for x:
$$\frac{1}{x} = \frac{1}{b} - \frac{1}{a}$$
$$abx\left(\frac{1}{x}\right) = abx\left(\frac{1}{b} - \frac{1}{a}\right)$$
$$ab = ax - bx$$
$$ab = x(a-b)$$
$$x = \frac{ab}{a-b}$$

51. Solving for y:
$$x = \frac{y-3}{y-1}$$
$$x(y-1) = y-3$$
$$xy - x = y - 3$$
$$xy - y = x - 3$$
$$y(x-1) = x-3$$
$$y = \frac{x-3}{x-1}$$

53. Solving for y:
$$x = \frac{2y+1}{3y+1}$$
$$x(3y+1) = 2y+1$$
$$3xy + x = 2y+1$$
$$3xy - 2y = -x+1$$
$$y(3x-2) = -x+1$$
$$y = \frac{1-x}{3x-2}$$

55. Graphing the function:

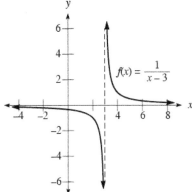

57. Graphing the function:

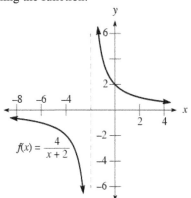

59. Graphing the function:

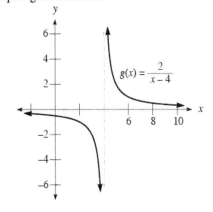

61. Graphing the function:

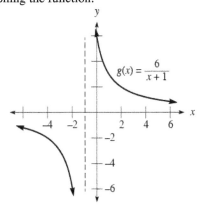

63. Substituting $y_1 = 12$ and $y_2 = 8$:
$$\frac{1}{h} = \frac{1}{12} + \frac{1}{8} = \frac{2}{24} + \frac{3}{24} = \frac{5}{24}$$
$$h = \frac{24}{5} \text{ feet}$$

65. Multiplying: $39.3 \cdot 60 = 2{,}358$

67. Dividing: $65{,}000 \div 5{,}280 \approx 12.3$

69. Multiplying: $2x\left(\frac{1}{x} + \frac{1}{2x}\right) = 2x \cdot \frac{1}{x} + 2x \cdot \frac{1}{2x} = 2 + 1 = 3$

71. Solving the equation:
$$12(x+3) + 12(x-3) = 3(x^2 - 9)$$
$$12x + 36 + 12x - 36 = 3x^2 - 27$$
$$24x = 3x^2 - 27$$
$$3x^2 - 24x - 27 = 0$$
$$3(x^2 - 8x - 9) = 0$$
$$3(x-9)(x+1) = 0$$
$$x = -1, 9$$

73. Solving the equation:
$$\frac{1}{10} - \frac{1}{12} = \frac{1}{x}$$
$$60x\left(\frac{1}{10} - \frac{1}{12}\right) = 60x\left(\frac{1}{x}\right)$$
$$6x - 5x = 60$$
$$x = 60$$

7.6 Applications

1. Let x and $3x$ represent the two numbers. The equation is:
$$\frac{1}{x} + \frac{1}{3x} = \frac{20}{3}$$
$$3x\left(\frac{1}{x} + \frac{1}{3x}\right) = 3x\left(\frac{20}{3}\right)$$
$$3 + 1 = 20x$$
$$20x = 4$$
$$x = \frac{1}{5}$$
$$3x = \frac{3}{5}$$
The numbers are $\frac{1}{5}$ and $\frac{3}{5}$.

3. Let x represent the number. The equation is:
$$x + \frac{1}{x} = \frac{10}{3}$$
$$3x\left(x + \frac{1}{x}\right) = 3x\left(\frac{10}{3}\right)$$
$$3x^2 + 3 = 10x$$
$$3x^2 - 10x + 3 = 0$$
$$(3x-1)(x-3) = 0$$
$$x = \frac{1}{3}, 3$$
The number is either 3 or $\frac{1}{3}$.

5. Let x and $x + 1$ represent the two integers. The equation is:
$$\frac{1}{x} + \frac{1}{x+1} = \frac{7}{12}$$
$$12x(x+1)\left(\frac{1}{x} + \frac{1}{x+1}\right) = 12x(x+1)\left(\frac{7}{12}\right)$$
$$12(x+1) + 12x = 7x(x+1)$$
$$12x + 12 + 12x = 7x^2 + 7x$$
$$0 = 7x^2 - 17x - 12$$
$$0 = (7x+4)(x-3)$$
$$x = 3 \quad \left(x = -\frac{4}{7} \text{ is not an integer}\right)$$

The two integers are 3 and 4.

7. Let x represent the number. The equation is:
$$\frac{7+x}{9+x} = \frac{5}{6}$$
$$6(9+x)\left(\frac{7+x}{9+x}\right) = 6(9+x)\left(\frac{5}{6}\right)$$
$$6(7+x) = 5(9+x)$$
$$42 + 6x = 45 + 5x$$
$$x = 3$$

The number is 3.

9. a-b. Completing the table.

	d (miles)	r (mph)	t (hours)
Upstream	1.5	$5-x$	$\frac{1.5}{5-x}$
Downstream	3	$5+x$	$\frac{3}{5+x}$

   c. The times are the same, so the equation is $\frac{3}{5+x} = \frac{1.5}{5-x}$.

   d. Setting the times equal:
$$\frac{3}{5+x} = \frac{1.5}{5-x}$$
$$3(5-x) = 1.5(5+x)$$
$$15 - 3x = 7.5 + 1.5x$$
$$7.5 = 4.5x$$
$$x = \frac{75}{45} = \frac{5}{3}$$

The speed of the current is $\frac{5}{3}$ mph.

11. Let x represent the speed of the boat. Since the total time is 3 hours:
$$\frac{8}{x-2}+\frac{8}{x+2}=3$$
$$(x+2)(x-2)\left(\frac{8}{x-2}+\frac{8}{x+2}\right)=3(x+2)(x-2)$$
$$8(x+2)+8(x-2)=3x^2-12$$
$$16x=3x^2-12$$
$$0=3x^2-16x-12$$
$$0=(3x+2)(x-6)$$
$$x=6 \quad \left(x=-\frac{2}{3} \text{ is impossible}\right)$$
The speed of the boat is 6 mph.

13. **a-b.** Completing the table.

	d (miles)	r (mph)	t (hours)
Train A	150	$x+15$	$\frac{150}{x+15}$
Train B	120	x	$\frac{120}{x}$

   c. The times are the same, so the equation is $\frac{150}{x+15}=\frac{120}{x}$.

   d. Setting the times equal:
   $$\frac{150}{x+15}=\frac{120}{x}$$
   $$150x=120(x+15)$$
   $$150x=120x+1800$$
   $$30x=1800$$
   $$x=60$$
   The speed of train A is 75 mph and the speed of train B is 60 mph.

15. The smaller plane makes the trip in 3 hours, so the 747 must take $1\frac{1}{2}$ hours to complete the trip. Thus the average speed is given by: $\dfrac{810 \text{ miles}}{1\frac{1}{2} \text{ hours}} = 540$ miles per hour

17. Let r represent the bus's usual speed. The difference of the two times is $\frac{1}{2}$ hour, therefore:
$$\frac{270}{r}-\frac{270}{r+6}=\frac{1}{2}$$
$$2r(r+6)\left(\frac{270}{r}-\frac{270}{r+6}\right)=2r(r+6)\left(\frac{1}{2}\right)$$
$$540(r+6)-540(r)=r(r+6)$$
$$540r+3240-540r=r^2+6r$$
$$0=r^2+6r-3240$$
$$0=(r-54)(r+60)$$
$$r=54 \quad (r=-60 \text{ is impossible})$$
The usual speed is 54 mph.

19. Let x represent the time to fill the tank if both pipes are open. The rate equation is:
$$\frac{1}{8} - \frac{1}{16} = \frac{1}{x}$$
$$16x\left(\frac{1}{8} - \frac{1}{16}\right) = 16x\left(\frac{1}{x}\right)$$
$$2x - x = 16$$
$$x = 16$$
It will take 16 hours to fill the tank if both pipes are open.

21. Let x represent the time to fill the pool with both pipes open. The rate equation is:
$$\frac{1}{10} - \frac{1}{15} = \frac{1}{2} \cdot \frac{1}{x}$$
$$30x\left(\frac{1}{10} - \frac{1}{15}\right) = 30x\left(\frac{1}{2x}\right)$$
$$3x - 2x = 15$$
$$x = 15$$
It will take 15 hours to fill the pool with both pipes open.

23. Let x represent the time to fill the sink with the hot water faucet. The rate equation is:
$$\frac{1}{3.5} + \frac{1}{x} = \frac{1}{2.1}$$
$$7.35x\left(\frac{1}{3.5} + \frac{1}{x}\right) = 7.35x\left(\frac{1}{2.1}\right)$$
$$2.1x + 7.35 = 3.5x$$
$$7.35 = 1.4x$$
$$x = 5.25$$
It will take 5.25 minutes to fill the sink with the hot water faucet alone.

25. Solving the equation:
$$\frac{1}{3}\left[\left(x + \frac{2}{3}x\right) + \frac{1}{3}\left(x + \frac{2}{3}x\right)\right] = 10$$
$$\left(x + \frac{2}{3}x\right) + \frac{1}{3}\left(x + \frac{2}{3}x\right) = 30$$
$$x + \frac{2}{3}x + \frac{1}{3}x + \frac{2}{9}x = 30$$
$$\frac{20}{9}x = 30$$
$$20x = 270$$
$$x = \frac{27}{2}$$

27. a. Finding the grams of carbon: $(2.5 \text{ moles})\left(\frac{12.01 \text{ grams}}{1 \text{ mole}}\right) \approx 30$ grams

    b. Finding the moles of carbon: $(39 \text{ grams})\left(\frac{1 \text{ mole}}{12.01 \text{ grams}}\right) \approx 3.25$ moles

29. Sketching the graph:

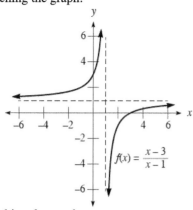

31. Sketching the graph:

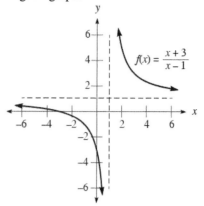

33. Sketching the graph:

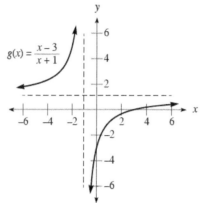

35. Dividing: $\dfrac{10x^2}{5x^2} = 2x^0 = 2$

37. Dividing: $\dfrac{4x^4 y^3}{-2x^2 y} = -2x^{4-2} y^{3-1} = -2x^2 y^2$

39. Dividing: $4{,}628 \div 25 = 185.12$

41. Multiplying: $2x^2 (2x-4) = 4x^3 - 8x^2$

43. Multiplying:
$$(2x-4)(2x^2 + 4x + 5) = 2x(2x^2 + 4x + 5) - 4(2x^2 + 4x + 5)$$
$$= 4x^3 + 8x^2 + 10x - 8x^2 - 16x - 20$$
$$= 4x^3 - 6x - 20$$

45. Subtracting: $(2x^2 - 7x + 9) - (2x^2 - 4x) = 2x^2 - 7x + 9 - 2x^2 + 4x = -3x + 9$

47. Factoring: $x^2 - a^2 = (x+a)(x-a)$

49. Factoring: $x^2 - 6xy - 7y^2 = (x-7y)(x+y)$

7.7 Division of Polynomials

1. Dividing: $\dfrac{4x^3 - 8x^2 + 6x}{2x} = \dfrac{4x^3}{2x} - \dfrac{8x^2}{2x} + \dfrac{6x}{2x} = 2x^2 - 4x + 3$

3. Dividing: $\dfrac{10x^4 + 15x^3 - 20x^2}{-5x^2} = \dfrac{10x^4}{-5x^2} + \dfrac{15x^3}{-5x^2} - \dfrac{20x^2}{-5x^2} = -2x^2 - 3x + 4$

5. Dividing: $\dfrac{8y^5 + 10y^3 - 6y}{4y^3} = \dfrac{8y^5}{4y^3} + \dfrac{10y^3}{4y^3} - \dfrac{6y}{4y^3} = 2y^2 + \dfrac{5}{2} - \dfrac{3}{2y^2}$

7. Dividing: $\dfrac{5x^3 - 8x^2 - 6x}{-2x^2} = \dfrac{5x^3}{-2x^2} - \dfrac{8x^2}{-2x^2} - \dfrac{6x}{-2x^2} = -\dfrac{5}{2}x + 4 + \dfrac{3}{x}$

9. Dividing: $\dfrac{28a^3b^5 + 42a^4b^3}{7a^2b^2} = \dfrac{28a^3b^5}{7a^2b^2} + \dfrac{42a^4b^3}{7a^2b^2} = 4ab^3 + 6a^2b$

11. Dividing: $\dfrac{10x^3y^2 - 20x^2y^3 - 30x^3y^3}{-10x^2y} = \dfrac{10x^3y^2}{-10x^2y} - \dfrac{20x^2y^3}{-10x^2y} - \dfrac{30x^3y^3}{-10x^2y} = -xy + 2y^2 + 3xy^2$

13. Dividing by factoring: $\dfrac{x^2 - x - 6}{x - 3} = \dfrac{(x-3)(x+2)}{x-3} = x + 2$

15. Dividing by factoring: $\dfrac{2a^2 - 3a - 9}{2a + 3} = \dfrac{(2a+3)(a-3)}{2a+3} = a - 3$

17. Dividing by factoring: $\dfrac{5x^2 - 14xy - 24y^2}{x - 4y} = \dfrac{(5x+6y)(x-4y)}{x-4y} = 5x + 6y$

19. Dividing by factoring: $\dfrac{x^3 - y^3}{x - y} = \dfrac{(x-y)(x^2+xy+y^2)}{x-y} = x^2 + xy + y^2$

21. Dividing by factoring: $\dfrac{y^4 - 16}{y - 2} = \dfrac{(y^2+4)(y^2-4)}{y-2} = \dfrac{(y^2+4)(y+2)(y-2)}{y-2} = (y^2+4)(y+2)$

23. Dividing by factoring:

$\dfrac{x^3 + 2x^2 - 25x - 50}{x - 5} = \dfrac{x^2(x+2) - 25(x+2)}{x-5} = \dfrac{(x+2)(x^2-25)}{x-5} = \dfrac{(x+2)(x+5)(x-5)}{x-5} = (x+2)(x+5)$

25. Dividing by factoring:

$\dfrac{4x^3 + 12x^2 - 9x - 27}{x + 3} = \dfrac{4x^2(x+3) - 9(x+3)}{x+3} = \dfrac{(x+3)(4x^2-9)}{x+3} = \dfrac{(x+3)(2x+3)(2x-3)}{x+3} = (2x+3)(2x-3)$

27. Dividing using long division:

$$\begin{array}{r} x - 7 \\ x+2 \overline{\smash{)}x^2 - 5x - 7} \\ \underline{x^2 + 2x} \\ -7x - 7 \\ \underline{-7x - 14} \\ 7 \end{array}$$

The quotient is $x - 7 + \dfrac{7}{x+2}$.

29. Dividing using long division:

$$\begin{array}{r} 2x + 5 \\ 3x-4 \overline{\smash{)}6x^2 + 7x - 18} \\ \underline{6x^2 - 8x} \\ 15x - 18 \\ \underline{15x - 20} \\ 2 \end{array}$$

The quotient is $2x + 5 + \dfrac{2}{3x-4}$.

31. Finding the quotient using long division:

$$\begin{array}{r} 2x^2 - 5x + 1 \\ x+1 \overline{) 2x^3 - 3x^2 - 4x + 5} \\ \underline{2x^3 + 2x^2} \\ -5x^2 - 4x \\ \underline{-5x^2 - 5x} \\ x + 5 \\ \underline{x + 1} \\ 4 \end{array}$$

The quotient is $2x^2 - 5x + 1 + \dfrac{4}{x+1}$.

33. Finding the quotient using long division:

$$\begin{array}{r} y^2 - 3y - 13 \\ 2y-3 \overline{) 2y^3 - 9y^2 - 17y + 39} \\ \underline{2y^3 - 3y^2} \\ -6y^2 - 17y \\ \underline{-6y^2 + 9y} \\ -26y + 39 \\ \underline{-26y + 39} \\ 0 \end{array}$$

The quotient is $y^2 - 3y - 13$.

35. Dividing using long division:

$$\begin{array}{r} x - 3 \\ 2x^2 - 3x + 2 \overline{) 2x^3 - 9x^2 + 11x - 6} \\ \underline{2x^3 - 3x^2 + 2x} \\ -6x^2 + 9x - 6 \\ \underline{-6x^2 + 9x - 6} \\ 0 \end{array}$$

The quotient is $x - 3$.

37. Dividing using long division:

$$\begin{array}{r} 3y^2 + 6y + 8 \\ 2y-4 \overline{) 6y^3 + 0y^2 - 8y + 5} \\ \underline{6y^3 - 12y^2} \\ 12y^2 - 8y \\ \underline{12y^2 - 24y} \\ 16y + 5 \\ \underline{16y - 32} \\ 37 \end{array}$$

The quotient is $3y^2 + 6y + 8 + \dfrac{37}{2y-4}$.

39. Dividing using long division:

$$\begin{array}{r} a^3 + 2a^2 + 4a + 6 \\ a-2 \overline{) a^4 + 0a^3 + 0a^2 - 2a + 5} \\ \underline{a^4 - 2a^3} \\ 2a^3 + 0a^2 \\ \underline{2a^3 - 4a^2} \\ 4a^2 - 2a \\ \underline{4a^2 - 8a} \\ 6a + 5 \\ \underline{6a - 12} \\ 17 \end{array}$$

The quotient is $a^3 + 2a^2 + 4a + 6 + \dfrac{17}{a-2}$.

41. Dividing using long division:

$$\begin{array}{r} y^3 + 2y^2 + 4y + 8 \\ y-2 \overline{) y^4 + 0y^3 + 0y^2 + 0y - 16} \\ \underline{y^4 - 2y^3} \\ 2y^3 + 0y^2 \\ \underline{2y^3 - 4y^2} \\ 4y^2 + 0y \\ \underline{4y^2 - 8y} \\ 8y - 16 \\ \underline{8y - 16} \\ 0 \end{array}$$

The quotient is $y^3 + 2y^2 + 4y + 8$.

43. Finding the quotient using long division:

$$\begin{array}{r} x^2-2x+1 \\ x^2+3x+2{\overline{\smash{\big)}\,x^4+x^3-3x^2-x+2}} \\ \underline{x^4+3x^3+2x^2} \\ -2x^3-5x^2-x \\ \underline{-2x^3-6x^2-4x} \\ x^2+3x+2 \\ \underline{x^2+3x+2} \\ 0 \end{array}$$

The quotient is x^2-2x+1.

45. Using long division:

$$\begin{array}{r} x^2+3x+2 \\ x+3{\overline{\smash{\big)}\,x^3+6x^2+11x+6}} \\ \underline{x^3+3x^2} \\ 3x^2+11x \\ \underline{3x^2+9x} \\ 2x+6 \\ \underline{2x+6} \\ 0 \end{array}$$

So $x^3+6x^2+11x+6=(x+3)(x^2+3x+2)=(x+3)(x+2)(x+1)$.

47. Using long division:

$$\begin{array}{r} x^2+2x-8 \\ x+3{\overline{\smash{\big)}\,x^3+5x^2-2x-24}} \\ \underline{x^3+3x^2} \\ 2x^2-2x \\ \underline{2x^2+6x} \\ -8x-24 \\ \underline{-8x-24} \\ 0 \end{array}$$

So $x^3+5x^2-2x-24=(x+3)(x^2+2x-8)=(x+3)(x+4)(x-2)$.

49. Yes, the two answers are equivalent.

51. Evaluating the function: $P(-2)=(-2)^2-5(-2)-7=4+10-7=7$

The remainder is the same (7).

53. **a.** Using long division:

$$\begin{array}{r}x^2-x+3\\x-2\overline{)x^3-3x^2+5x-6}\\\underline{x^3-2x^2}\\-x^2+5x\\\underline{-x^2+2x}\\3x-6\\\underline{3x-6}\\0\end{array}$$

Since the remainder is 0, $x-2$ is a factor of x^3-3x^2+5x-6.

Also note that: $P(2)=(2)^3-3(2)^2+5(2)-6=8-12+10-6=0$

b. Using long division:

$$\begin{array}{r}x^3-x+1\\x-5\overline{)x^4-5x^3-x^2+6x-5}\\\underline{x^4-5x^3}\\-x^2+6x\\\underline{-x^2+5x}\\x-5\\\underline{x-5}\\0\end{array}$$

Since the remainder is 0, $x-5$ is a factor of $x^4-5x^3-x^2+6x-5$.

Also note that: $P(5)=(5)^4-5(5)^3-(5)^2+6(5)-5=625-625-25+30-5=0$

55. **a.** Completing the table:

x	1	5	10	15	20
$C(x)$	2.15	2.75	3.50	4.25	5.00

b. The average cost function is $\bar{C}(x)=\dfrac{2}{x}+0.15$.

c. Completing the table:

x	1	5	10	15	20
$\bar{C}(x)$	2.15	0.55	0.35	0.28	0.25

d. The average cost function decreases.

e. For $C(x)$, the domain is $\{x\,|\,1\le x\le 20\}$ and the range is $\{y\,|\,2.15\le y\le 5.00\}$.

For $\bar{C}(x)$, the domain is $\{x\,|\,1\le x\le 20\}$ and the range is $\{y\,|\,0.25\le y\le 2.15\}$.

57. **a.** Evaluating the total cost:
$T(100)=4.95+0.07(100)=\$11.95$
$T(400)=4.95+0.07(400)=\$32.95$
$T(500)=4.95+0.07(500)=\$39.95$

b. The average cost function is given by: $\bar{T}(m)=\dfrac{4.95+0.07m}{m}=0.07+\dfrac{4.95}{m}$

c. Evaluating the average cost:
$\bar{T}(100)=0.07+\dfrac{4.95}{100}=\$0.1195$
$\bar{T}(400)=0.07+\dfrac{4.95}{400}\approx\$0.0824$
$\bar{T}(500)=0.07+\dfrac{4.95}{500}=\$0.0799$

59. Performing the operations: $\dfrac{2a+10}{a^3} \cdot \dfrac{a^2}{3a+15} = \dfrac{2(a+5)}{a^3} \cdot \dfrac{a^2}{3(a+5)} = \dfrac{2}{3a}$

61. Performing the operations: $(x^2-9)\left(\dfrac{x+2}{x+3}\right) = (x+3)(x-3)\left(\dfrac{x+2}{x+3}\right) = (x-3)(x+2)$

63. Performing the operations: $\dfrac{2x-7}{x-2} - \dfrac{x-5}{x-2} = \dfrac{2x-7-x+5}{x-2} = \dfrac{x-2}{x-2} = 1$

65. Simplifying the expression: $\dfrac{\dfrac{1}{x}-\dfrac{1}{3}}{\dfrac{1}{x}+\dfrac{1}{3}} = \dfrac{\left(\dfrac{1}{x}-\dfrac{1}{3}\right)\cdot 3x}{\left(\dfrac{1}{x}+\dfrac{1}{3}\right)\cdot 3x} = \dfrac{3-x}{3+x}$

67. Solving the equation:
$$\dfrac{x}{x-3} + \dfrac{3}{2} = \dfrac{3}{x-3}$$
$$2(x-3)\left(\dfrac{x}{x-3}+\dfrac{3}{2}\right) = 2(x-3)\left(\dfrac{3}{x-3}\right)$$
$$2x + 3(x-3) = 6$$
$$2x + 3x - 9 = 6$$
$$5x = 15$$
$$x = 3 \quad \text{(does not check)}$$
There is no solution (3 does not check).

Chapter 7 Test

1. Reducing the fraction: $\dfrac{x^2-y^2}{x-y} = \dfrac{(x+y)(x-y)}{x-y} = x+y$

2. Reducing the fraction: $\dfrac{2x^2-5x+3}{2x^2-x-3} = \dfrac{(2x-3)(x-1)}{(2x-3)(x+1)} = \dfrac{x-1}{x+1}$

3. Performing the operations: $\dfrac{a^2-16}{5a-15} \cdot \dfrac{10(a-3)^2}{a^2-7a+12} = \dfrac{(a+4)(a-4)}{5(a-3)} \cdot \dfrac{10(a-3)^2}{(a-4)(a-3)} = 2(a+4)$

4. Performing the operations:
$$\dfrac{a^4-81}{a^2+9} \div \dfrac{a^2-8a+15}{4a-20} = \dfrac{a^4-81}{a^2+9} \cdot \dfrac{4a-20}{a^2-8a+15} = \dfrac{(a^2+9)(a+3)(a-3)}{a^2+9} \cdot \dfrac{4(a-5)}{(a-5)(a-3)} = 4(a+3)$$

5. Performing the operations:
$$\dfrac{x^3-8}{2x^2-9x+10} \div \dfrac{x^2+2x+4}{2x^2+x-15} = \dfrac{x^3-8}{2x^2-9x+10} \cdot \dfrac{2x^2+x-15}{x^2+2x+4} = \dfrac{(x-2)(x^2+2x+4)}{(2x-5)(x-2)} \cdot \dfrac{(2x-5)(x+3)}{x^2+2x+4} = x+3$$

6. Performing the operations: $\dfrac{4}{21} + \dfrac{6}{35} = \dfrac{4}{21} \cdot \dfrac{5}{5} + \dfrac{6}{35} \cdot \dfrac{3}{3} = \dfrac{20}{105} + \dfrac{18}{105} = \dfrac{38}{105}$

7. Performing the operations: $\dfrac{3}{4} - \dfrac{1}{2} + \dfrac{5}{8} = \dfrac{3}{4} \cdot \dfrac{2}{2} - \dfrac{1}{2} \cdot \dfrac{4}{4} + \dfrac{5}{8} = \dfrac{6}{8} - \dfrac{4}{8} + \dfrac{5}{8} = \dfrac{7}{8}$

8. Performing the operations: $\dfrac{a}{a^2-9} + \dfrac{3}{a^2-9} = \dfrac{a+3}{a^2-9} = \dfrac{a+3}{(a+3)(a-3)} = \dfrac{1}{a-3}$

9. Performing the operations: $\dfrac{1}{x} + \dfrac{2}{x-3} = \dfrac{1}{x} \cdot \dfrac{x-3}{x-3} + \dfrac{2}{x-3} \cdot \dfrac{x}{x} = \dfrac{x-3}{x(x-3)} + \dfrac{2x}{x(x-3)} = \dfrac{3x-3}{x(x-3)} = \dfrac{3(x-1)}{x(x-3)}$

10. Performing the operations:

$$\frac{4x}{x^2+6x+5} - \frac{3x}{x^2+5x+4} = \frac{4x}{(x+5)(x+1)} - \frac{3x}{(x+4)(x+1)}$$

$$= \frac{4x}{(x+5)(x+1)} \cdot \frac{x+4}{x+4} - \frac{3x}{(x+4)(x+1)} \cdot \frac{x+5}{x+5}$$

$$= \frac{4x^2+16x}{(x+5)(x+1)(x+4)} - \frac{3x^2+15x}{(x+5)(x+1)(x+4)}$$

$$= \frac{4x^2+16x-3x^2-15x}{(x+5)(x+1)(x+4)}$$

$$= \frac{x^2+x}{(x+5)(x+1)(x+4)}$$

$$= \frac{x(x+1)}{(x+5)(x+1)(x+4)}$$

$$= \frac{x}{(x+4)(x+5)}$$

11. Performing the operations:

$$\frac{2x+8}{x^2+4x+3} - \frac{x+4}{x^2+5x+6} = \frac{2x+8}{(x+3)(x+1)} - \frac{x+4}{(x+3)(x+2)}$$

$$= \frac{2x+8}{(x+3)(x+1)} \cdot \frac{x+2}{x+2} - \frac{x+4}{(x+3)(x+2)} \cdot \frac{x+1}{x+1}$$

$$= \frac{2x^2+12x+16}{(x+1)(x+2)(x+3)} - \frac{x^2+5x+4}{(x+1)(x+2)(x+3)}$$

$$= \frac{2x^2+12x+16-x^2-5x-4}{(x+1)(x+2)(x+3)}$$

$$= \frac{x^2+7x+12}{(x+1)(x+2)(x+3)}$$

$$= \frac{(x+3)(x+4)}{(x+1)(x+2)(x+3)}$$

$$= \frac{x+4}{(x+1)(x+2)}$$

12. Simplifying the complex fraction: $\dfrac{3-\dfrac{1}{a+3}}{3+\dfrac{1}{a+3}} = \dfrac{\left(3-\dfrac{1}{a+3}\right)(a+3)}{\left(3+\dfrac{1}{a+3}\right)(a+3)} = \dfrac{3(a+3)-1}{3(a+3)+1} = \dfrac{3a+9-1}{3a+9+1} = \dfrac{3a+8}{3a+10}$

13. Simplifying the complex fraction: $\dfrac{1-\dfrac{9}{x^2}}{1+\dfrac{1}{x}-\dfrac{6}{x^2}} = \dfrac{\left(1-\dfrac{9}{x^2}\right)\cdot x^2}{\left(1+\dfrac{1}{x}-\dfrac{6}{x^2}\right)\cdot x^2} = \dfrac{x^2-9}{x^2+x-6} = \dfrac{(x+3)(x-3)}{(x+3)(x-2)} = \dfrac{x-3}{x-2}$

14. Solving the equation:
$$\frac{1}{x} + 3 = \frac{4}{3}$$
$$3x\left(\frac{1}{x} + 3\right) = 3x\left(\frac{4}{3}\right)$$
$$3 + 9x = 4x$$
$$5x = -3$$
$$x = -\frac{3}{5}$$

15. Solving the equation:
$$\frac{x}{x-3} + 3 = \frac{3}{x-3}$$
$$(x-3)\left(\frac{x}{x-3} + 3\right) = (x-3)\left(\frac{3}{x-3}\right)$$
$$x + 3(x-3) = 3$$
$$x + 3x - 9 = 3$$
$$4x = 12$$
$$x = 3 \quad (x = 3 \text{ does not check})$$
There is no solution (3 does not check).

16. Solving the equation:
$$\frac{y+3}{2y} + \frac{5}{y-1} = \frac{1}{2}$$
$$2y(y-1)\left(\frac{y+3}{2y} + \frac{5}{y-1}\right) = 2y(y-1)\left(\frac{1}{2}\right)$$
$$(y-1)(y+3) + 10y = y(y-1)$$
$$y^2 + 2y - 3 + 10y = y^2 - y$$
$$12y - 3 = -y$$
$$13y = 3$$
$$y = \frac{3}{13}$$

17. Solving the equation:
$$1 - \frac{1}{x} = \frac{6}{x^2}$$
$$x^2\left(1 - \frac{1}{x}\right) = x^2\left(\frac{6}{x^2}\right)$$
$$x^2 - x = 6$$
$$x^2 - x - 6 = 0$$
$$(x+2)(x-3) = 0$$
$$x = -2, 3$$

18. Graphing the function:

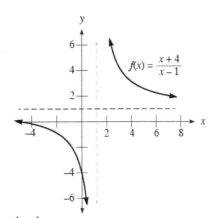

19. Let x represent the number. The equation is:
$$\frac{10}{23-x} = \frac{1}{3}$$
$$30 = 23 - x$$
$$x = -7$$
The number is -7.

20. Let x represent the speed of the boat. Since the total time is 3 hours:
$$\frac{8}{x-2} + \frac{8}{x+2} = 3$$
$$3(x-2)(x+2)\left(\frac{8}{x-2} + \frac{8}{x+2}\right) = 3(x-2)(x+2) \cdot 3$$
$$24(x+2) + 24(x-2) = 9(x^2 - 4)$$
$$48x = 9x^2 - 36$$
$$0 = 9x^2 - 48x - 36$$
$$0 = 3(3x^2 - 16x - 12)$$
$$0 = 3(3x+2)(x-6)$$
$$x = 6 \quad \left(x = -\frac{2}{3} \text{ is impossible}\right)$$

The speed of the boat is 6 mph.

21. Let x represent the time to fill the pool with both the pipe and drain open. The rate equation is:
$$\frac{1}{10} - \frac{1}{15} = \frac{1}{2} \cdot \frac{1}{x}$$
$$30x\left(\frac{1}{10} - \frac{1}{15}\right) = 30x\left(\frac{1}{2x}\right)$$
$$3x - 2x = 15$$
$$x = 15$$

The pool can be filled in 15 hours with both the pipe and drain open.

22. Converting the units: $14{,}494 \text{ feet} \cdot \dfrac{1 \text{ mile}}{5{,}280 \text{ feet}} \approx 2.7$ miles

23. Finding the speed: $\dfrac{4{,}750 \text{ feet}}{3.2 \text{ seconds}} \cdot \dfrac{1 \text{ mile}}{5{,}280 \text{ feet}} \cdot \dfrac{3{,}600 \text{ seconds}}{1 \text{ hour}} \approx 1{,}012$ miles/hour

24. Dividing: $\dfrac{24x^3y + 12x^2y^2 - 16xy^3}{4xy} = \dfrac{24x^3y}{4xy} + \dfrac{12x^2y^2}{4xy} - \dfrac{16xy^3}{4xy} = 6x^2 + 3xy - 4y^2$

25. Dividing using long division:

$$\begin{array}{r}
x^2 - 4x - 2 \\
2x-1 \overline{)2x^3 - 9x^2 + 0x + 10} \\
\underline{2x^3 - x^2} \\
-8x^2 + 0x \\
\underline{-8x^2 + 4x} \\
-4x + 10 \\
\underline{-4x + 2} \\
8
\end{array}$$

The quotient is $x^2 - 4x - 2 + \dfrac{8}{2x-1}$.

Chapter 8
Rational Exponents and Radicals

8.1 Rational Exponents

1. Finding the root: $\sqrt{144} = 12$
3. Finding the root: $\sqrt{-144}$ is not a real number
5. Finding the root: $-\sqrt{49} = -7$
7. Finding the root: $\sqrt[3]{-27} = -3$
9. Finding the root: $\sqrt[4]{16} = 2$
11. Finding the root: $\sqrt[4]{-16}$ is not a real number
13. Finding the root: $\sqrt{0.04} = 0.2$
15. Finding the root: $\sqrt[3]{0.008} = 0.2$
17. Simplifying: $\sqrt{36a^8} = 6a^4$
19. Simplifying: $\sqrt[3]{27a^{12}} = 3a^4$
21. Simplifying: $\sqrt[3]{x^3 y^6} = xy^2$
23. Simplifying: $\sqrt[5]{32x^{10}y^5} = 2x^2 y$
25. Simplifying: $\sqrt[4]{16a^{12}b^{20}} = 2a^3 b^5$
27. Writing as a root and simplifying: $36^{1/2} = \sqrt{36} = 6$
29. Writing as a root and simplifying: $-9^{1/2} = -\sqrt{9} = -3$
31. Writing as a root and simplifying: $8^{1/3} = \sqrt[3]{8} = 2$
33. Writing as a root and simplifying: $(-8)^{1/3} = \sqrt[3]{-8} = -2$
35. Writing as a root and simplifying: $32^{1/5} = \sqrt[5]{32} = 2$
37. Writing as a root and simplifying: $\left(\frac{81}{25}\right)^{1/2} = \sqrt{\frac{81}{25}} = \frac{9}{5}$
39. Writing as a root and simplifying: $\left(\frac{64}{125}\right)^{1/3} = \sqrt[3]{\frac{64}{125}} = \frac{4}{5}$
41. Simplifying: $27^{2/3} = \left(27^{1/3}\right)^2 = 3^2 = 9$
43. Simplifying: $25^{3/2} = \left(25^{1/2}\right)^3 = 5^3 = 125$
45. Simplifying: $16^{3/4} = \left(16^{1/4}\right)^3 = 2^3 = 8$
47. Simplifying: $27^{-1/3} = \left(27^{1/3}\right)^{-1} = 3^{-1} = \frac{1}{3}$
49. Simplifying: $81^{-3/4} = \left(81^{1/4}\right)^{-3} = 3^{-3} = \frac{1}{3^3} = \frac{1}{27}$
51. Simplifying: $\left(\frac{25}{36}\right)^{-1/2} = \left(\frac{36}{25}\right)^{1/2} = \frac{6}{5}$
53. Simplifying: $\left(\frac{81}{16}\right)^{-3/4} = \left(\frac{16}{81}\right)^{3/4} = \left[\left(\frac{16}{81}\right)^{1/4}\right]^3 = \left(\frac{2}{3}\right)^3 = \frac{8}{27}$
55. Simplifying: $16^{1/2} + 27^{1/3} = 4 + 3 = 7$
57. Simplifying: $8^{-2/3} + 4^{-1/2} = \left(8^{1/3}\right)^{-2} + \left(4^{1/2}\right)^{-1} = 2^{-2} + 2^{-1} = \frac{1}{4} + \frac{1}{2} = \frac{3}{4}$
59. Using properties of exponents: $x^{3/5} \cdot x^{1/5} = x^{3/5+1/5} = x^{4/5}$
61. Using properties of exponents: $\left(a^{3/4}\right)^{4/3} = a^{3/4 \cdot 4/3} = a$
63. Using properties of exponents: $\frac{x^{1/5}}{x^{3/5}} = x^{1/5-3/5} = x^{-2/5} = \frac{1}{x^{2/5}}$

65. Using properties of exponents: $\dfrac{x^{5/6}}{x^{2/3}} = x^{5/6-2/3} = x^{5/6-4/6} = x^{1/6}$

67. Using properties of exponents: $\left(x^{3/5}y^{5/6}z^{1/3}\right)^{3/5} = x^{3/5 \cdot 3/5}y^{5/6 \cdot 3/5}z^{1/3 \cdot 3/5} = x^{9/25}y^{1/2}z^{1/5}$

69. Using properties of exponents: $\dfrac{a^{3/4}b^2}{a^{7/8}b^{1/4}} = a^{3/4-7/8}b^{2-1/4} = a^{6/8-7/8}b^{8/4-1/4} = a^{-1/8}b^{7/4} = \dfrac{b^{7/4}}{a^{1/8}}$

71. Using properties of exponents: $\dfrac{\left(y^{2/3}\right)^{3/4}}{\left(y^{1/3}\right)^{3/5}} = \dfrac{y^{1/2}}{y^{1/5}} = y^{1/2-1/5} = y^{5/10-2/10} = y^{3/10}$

73. Using properties of exponents: $\left(\dfrac{a^{-1/4}}{b^{1/2}}\right)^8 = \dfrac{a^{-1/4 \cdot 8}}{b^{1/2 \cdot 8}} = \dfrac{a^{-2}}{b^4} = \dfrac{1}{a^2 b^4}$

75. a. Simplifying: $\sqrt{25} = \sqrt{5^2} = 5$ b. Simplifying: $\sqrt{0.25} = \sqrt{0.5^2} = 0.5$
    c. Simplifying: $\sqrt{2{,}500} = \sqrt{50^2} = 50$ d. Simplifying: $\sqrt{0.0025} = \sqrt{0.05^2} = 0.05$

77. a. Simplifying: $\sqrt{16a^4 b^8} = \sqrt{\left(4a^2 b^4\right)^2} = 4a^2 b^4$

    b. Simplifying: $\sqrt[3]{16a^4 b^8} = \sqrt[3]{\left(2ab^2\right)^3 \left(2ab^2\right)} = 2ab^2 \sqrt[3]{2ab^2}$

    c. Simplifying: $\sqrt[4]{16a^4 b^8} = \sqrt[4]{\left(2ab^2\right)^4} = 2ab^2$

79. Simplifying each expression:
$\left(9^{1/2} + 4^{1/2}\right)^2 = (3+2)^2 = 5^2 = 25$
$9 + 4 = 13$
Note that the values are not equal.

81. Rewriting with exponents: $\sqrt{\sqrt{a}} = \sqrt{a^{1/2}} = \left(a^{1/2}\right)^{1/2} = a^{1/4} = \sqrt[4]{a}$

83. Substituting $r = 250$: $v = \left(\dfrac{5 \cdot 250}{2}\right)^{1/2} = 625^{1/2} = 25$. The maximum speed is 25 mph.

85. Using a calculator: $\dfrac{1+\sqrt{5}}{2} \approx 1.618$

87. The pattern is to add the previous two numerators (for the numerator), and to add the previous two denominators (for the denominator). So the next term is $\dfrac{5+8}{3+5} = \dfrac{13}{8}$. This is a similar pattern to the Fibonacci sequence, except in fractional form.

89. a. Each side of the square is: 60 + 300 + 60 = 420 picometers
    b. Let d represent the length of the diagonal. Using the Pythagorean theorem:
    $420^2 + 420^2 = d^2$
    $d^2 = 352{,}800$
    $d = \sqrt{352{,}800} \approx 594$ picometers
    c. Converting to meters: $594 \text{ pm} \cdot \dfrac{1 \text{ m}}{10^{12} \text{ pm}} = 5.94 \times 10^{-10}$ m

91. Simplifying: $\sqrt{25} = 5$

93. Simplifying: $\sqrt{6^2} = 6$

95. Simplifying: $\sqrt{16x^4y^2} = 4x^2y$

97. Simplifying: $\sqrt{(5y)^2} = 5y$

99. Simplifying: $\sqrt[3]{27} = 3$

101. Simplifying: $\sqrt[3]{2^3} = 2$

103. Simplifying: $\sqrt[3]{8a^3b^3} = 2ab$

105. Filling in the blank: $50 = 25 \cdot 2$

107. Filling in the blank: $48x^4y^3 = 48x^4y^2 \cdot y$

109. Filling in the blank: $12x^7y^6 = 4x^6y^6 \cdot 3x$

8.2 Simplified Form for Radicals

1. Simplifying the radical: $\sqrt{8} = \sqrt{4 \cdot 2} = 2\sqrt{2}$

3. Simplifying the radical: $\sqrt{98} = \sqrt{49 \cdot 2} = 7\sqrt{2}$

5. Simplifying the radical: $\sqrt{288} = \sqrt{144 \cdot 2} = 12\sqrt{2}$

7. Simplifying the radical: $\sqrt{80} = \sqrt{16 \cdot 5} = 4\sqrt{5}$

9. Simplifying the radical: $\sqrt{48} = \sqrt{16 \cdot 3} = 4\sqrt{3}$

11. Simplifying the radical: $\sqrt{675} = \sqrt{225 \cdot 3} = 15\sqrt{3}$

13. Simplifying the radical: $\sqrt[3]{54} = \sqrt[3]{27 \cdot 2} = 3\sqrt[3]{2}$

15. Simplifying the radical: $\sqrt[3]{128} = \sqrt[3]{64 \cdot 2} = 4\sqrt[3]{2}$

17. Simplifying the radical: $\sqrt[3]{432} = \sqrt[3]{216 \cdot 2} = 6\sqrt[3]{2}$

19. Simplifying the radical: $\sqrt[5]{64} = \sqrt[5]{32 \cdot 2} = 2\sqrt[5]{2}$

21. Simplifying the radical: $\sqrt{18x^3} = \sqrt{9x^2 \cdot 2x} = 3x\sqrt{2x}$

23. Simplifying the radical: $\sqrt[4]{32y^7} = \sqrt[4]{16y^4 \cdot 2y^3} = 2y\sqrt[4]{2y^3}$

25. Simplifying the radical: $\sqrt[3]{40x^4y^7} = \sqrt[3]{8x^3y^6 \cdot 5xy} = 2xy^2\sqrt[3]{5xy}$

27. Simplifying the radical: $\sqrt{48a^2b^3c^4} = \sqrt{16a^2b^2c^4 \cdot 3b} = 4abc^2\sqrt{3b}$

29. Simplifying the radical: $\sqrt[3]{48a^2b^3c^4} = \sqrt[3]{8b^3c^3 \cdot 6a^2c} = 2bc\sqrt[3]{6a^2c}$

31. Simplifying the radical: $\sqrt[5]{64x^8y^{12}} = \sqrt[5]{32x^5y^{10} \cdot 2x^3y^2} = 2xy^2\sqrt[5]{2x^3y^2}$

33. Simplifying the radical: $\sqrt[5]{243x^7y^{10}z^5} = \sqrt[5]{243x^5y^{10}z^5 \cdot x^2} = 3xy^2z\sqrt[5]{x^2}$

35. Substituting into the expression: $\sqrt{b^2 - 4ac} = \sqrt{(-6)^2 - 4(2)(3)} = \sqrt{36 - 24} = \sqrt{12} = 2\sqrt{3}$

37. Substituting into the expression: $\sqrt{b^2 - 4ac} = \sqrt{(2)^2 - 4(1)(6)} = \sqrt{4 - 24} = \sqrt{-20}$, which is not a real number

39. Substituting into the expression: $\sqrt{b^2 - 4ac} = \sqrt{\left(-\frac{1}{2}\right)^2 - 4\left(\frac{1}{2}\right)\left(-\frac{5}{4}\right)} = \sqrt{\frac{1}{4} + \frac{5}{2}} = \sqrt{\frac{11}{4}} = \frac{\sqrt{11}}{2}$

41. a. Simplifying: $\dfrac{\sqrt{20}}{4} = \dfrac{\sqrt{4 \cdot 5}}{4} = \dfrac{2\sqrt{5}}{4} = \dfrac{\sqrt{5}}{2}$

   b. Simplifying: $\dfrac{3\sqrt{20}}{15} = \dfrac{3\sqrt{4 \cdot 5}}{15} = \dfrac{6\sqrt{5}}{15} = \dfrac{2\sqrt{5}}{5}$

   c. Simplifying: $\dfrac{4 + \sqrt{12}}{2} = \dfrac{4 + \sqrt{4 \cdot 3}}{2} = \dfrac{4 + 2\sqrt{3}}{2} = \dfrac{2(2 + \sqrt{3})}{2} = 2 + \sqrt{3}$

   d. Simplifying: $\dfrac{2 + \sqrt{9}}{5} = \dfrac{2 + 3}{5} = \dfrac{5}{5} = 1$

43.
  a. Simplifying: $\dfrac{10+\sqrt{75}}{5} = \dfrac{10+\sqrt{25 \cdot 3}}{5} = \dfrac{10+5\sqrt{3}}{5} = \dfrac{5(2+\sqrt{3})}{5} = 2+\sqrt{3}$

  b. Simplifying: $\dfrac{-6+\sqrt{45}}{3} = \dfrac{-6+\sqrt{9 \cdot 5}}{3} = \dfrac{-6+3\sqrt{5}}{3} = \dfrac{3(-2+\sqrt{5})}{3} = -2+\sqrt{5}$

  c. Simplifying: $\dfrac{-2-\sqrt{27}}{6} = \dfrac{-2-\sqrt{9 \cdot 3}}{6} = \dfrac{-2-3\sqrt{3}}{6}$

45. Rationalizing the denominator: $\dfrac{2}{\sqrt{3}} = \dfrac{2}{\sqrt{3}} \cdot \dfrac{\sqrt{3}}{\sqrt{3}} = \dfrac{2\sqrt{3}}{3}$

47. Rationalizing the denominator: $\dfrac{5}{\sqrt{6}} = \dfrac{5}{\sqrt{6}} \cdot \dfrac{\sqrt{6}}{\sqrt{6}} = \dfrac{5\sqrt{6}}{6}$

49. Rationalizing the denominator: $\sqrt{\dfrac{1}{2}} = \dfrac{1}{\sqrt{2}} \cdot \dfrac{\sqrt{2}}{\sqrt{2}} = \dfrac{\sqrt{2}}{2}$

51. Rationalizing the denominator: $\sqrt{\dfrac{1}{5}} = \dfrac{1}{\sqrt{5}} \cdot \dfrac{\sqrt{5}}{\sqrt{5}} = \dfrac{\sqrt{5}}{5}$

53. Rationalizing the denominator: $\dfrac{4}{\sqrt[3]{2}} = \dfrac{4}{\sqrt[3]{2}} \cdot \dfrac{\sqrt[3]{4}}{\sqrt[3]{4}} = \dfrac{4\sqrt[3]{4}}{2} = 2\sqrt[3]{4}$

55. Rationalizing the denominator: $\dfrac{2}{\sqrt[3]{9}} = \dfrac{2}{\sqrt[3]{9}} \cdot \dfrac{\sqrt[3]{3}}{\sqrt[3]{3}} = \dfrac{2\sqrt[3]{3}}{3}$

57. Rationalizing the denominator: $\sqrt[4]{\dfrac{3}{2x^2}} = \dfrac{\sqrt[4]{3}}{\sqrt[4]{2x^2}} \cdot \dfrac{\sqrt[4]{8x^2}}{\sqrt[4]{8x^2}} = \dfrac{\sqrt[4]{24x^2}}{2x}$

59. Rationalizing the denominator: $\sqrt[4]{\dfrac{8}{y}} = \dfrac{\sqrt[4]{8}}{\sqrt[4]{y}} \cdot \dfrac{\sqrt[4]{y^3}}{\sqrt[4]{y^3}} = \dfrac{\sqrt[4]{8y^3}}{y}$

61. Rationalizing the denominator: $\sqrt[3]{\dfrac{4x}{3y}} = \dfrac{\sqrt[3]{4x}}{\sqrt[3]{3y}} \cdot \dfrac{\sqrt[3]{9y^2}}{\sqrt[3]{9y^2}} = \dfrac{\sqrt[3]{36xy^2}}{3y}$

63. Rationalizing the denominator: $\sqrt[3]{\dfrac{2x}{9y}} = \dfrac{\sqrt[3]{2x}}{\sqrt[3]{9y}} \cdot \dfrac{\sqrt[3]{3y^2}}{\sqrt[3]{3y^2}} = \dfrac{\sqrt[3]{6xy^2}}{3y}$

65. Simplifying: $\sqrt{\dfrac{27x^3}{5y}} = \dfrac{\sqrt{27x^3}}{\sqrt{5y}} \cdot \dfrac{\sqrt{5y}}{\sqrt{5y}} = \dfrac{\sqrt{135x^3 y}}{5y} = \dfrac{3x\sqrt{15xy}}{5y}$

67. Simplifying: $\sqrt{\dfrac{75x^3 y^2}{2z}} = \dfrac{\sqrt{75x^3 y^2}}{\sqrt{2z}} \cdot \dfrac{\sqrt{2z}}{\sqrt{2z}} = \dfrac{\sqrt{150x^3 y^2 z}}{2z} = \dfrac{5xy\sqrt{6xz}}{2z}$

69.
  a. Rationalizing the denominator: $\dfrac{1}{\sqrt{2}} = \dfrac{1}{\sqrt{2}} \cdot \dfrac{\sqrt{2}}{\sqrt{2}} = \dfrac{\sqrt{2}}{2}$

  b. Rationalizing the denominator: $\dfrac{1}{\sqrt[3]{2}} = \dfrac{1}{\sqrt[3]{2}} \cdot \dfrac{\sqrt[3]{4}}{\sqrt[3]{4}} = \dfrac{\sqrt[3]{4}}{2}$

  c. Rationalizing the denominator: $\dfrac{1}{\sqrt[4]{2}} = \dfrac{1}{\sqrt[4]{2}} \cdot \dfrac{\sqrt[4]{8}}{\sqrt[4]{8}} = \dfrac{\sqrt[4]{8}}{2}$

71. Simplifying: $\sqrt{25x^2} = 5|x|$

73. Simplifying: $\sqrt{27x^3y^2} = \sqrt{9x^2y^2 \cdot 3x} = 3|xy|\sqrt{3x}$

75. Simplifying: $\sqrt{x^2 - 10x + 25} = \sqrt{(x-5)^2} = |x-5|$

77. Simplifying: $\sqrt{4x^2 + 12x + 9} = \sqrt{(2x+3)^2} = |2x+3|$

79. Simplifying: $\sqrt{4a^4 + 16a^3 + 16a^2} = \sqrt{4a^2(a^2 + 4a + 4)} = \sqrt{4a^2(a+2)^2} = 2|a(a+2)|$

81. Simplifying: $\sqrt{4x^3 - 8x^2} = \sqrt{4x^2(x-2)} = 2|x|\sqrt{x-2}$

83. Substituting $a = 9$ and $b = 16$:
$\sqrt{a+b} = \sqrt{9+16} = \sqrt{25} = 5$
$\sqrt{a} + \sqrt{b} = \sqrt{9} + \sqrt{16} = 3 + 4 = 7$
Thus $\sqrt{a+b} \neq \sqrt{a} + \sqrt{b}$.

85. Substituting $w = 10$ and $l = 15$: $d = \sqrt{l^2 + w^2} = \sqrt{15^2 + 10^2} = \sqrt{225 + 100} = \sqrt{325} = \sqrt{25 \cdot 13} = 5\sqrt{13}$ feet

87. a. Substituting $k = 1$: $d = \sqrt{8000k + k^2} = \sqrt{8000(1) + (1)^2} = \sqrt{8001} \approx 89.4$ miles

   b. Substituting $k = 2$: $d = \sqrt{8000k + k^2} = \sqrt{8000(2) + (2)^2} = \sqrt{16,004} \approx 126.5$ miles

   c. Substituting $k = 3$: $d = \sqrt{8000k + k^2} = \sqrt{8000(3) + (3)^2} = \sqrt{24,009} \approx 154.9$ miles

89. Answers will vary.

91. The terms are $f(1) = \sqrt{2}, f(\sqrt{2}) = \sqrt{3}, \ldots$ So $f(a_{10}) = \sqrt{11}$ and $f(a_{100}) = \sqrt{101}$.

93. Simplifying: $5x - 4x + 6x = 7x$

95. Simplifying: $35xy^2 - 8xy^2 = 27xy^2$

97. Simplifying: $\frac{1}{2}x + \frac{1}{3}x = \frac{3}{6}x + \frac{2}{6}x = \frac{5}{6}x$

99. Simplifying: $\sqrt{18} = \sqrt{9 \cdot 2} = 3\sqrt{2}$

101. Simplifying: $\sqrt{75xy^3} = \sqrt{25y^2 \cdot 3xy} = 5y\sqrt{3xy}$

103. Simplifying: $\sqrt[3]{8a^4b^2} = \sqrt[3]{8a^3 \cdot ab^2} = 2a\sqrt[3]{ab^2}$

8.3 Addition and Subtraction of Radical Expressions

1. Combining radicals: $3\sqrt{5} + 4\sqrt{5} = 7\sqrt{5}$

3. Combining radicals: $3x\sqrt{7} - 4x\sqrt{7} = -x\sqrt{7}$

5. Combining radicals: $5\sqrt[3]{10} - 4\sqrt[3]{10} = \sqrt[3]{10}$

7. Combining radicals: $8\sqrt[5]{6} - 2\sqrt[5]{6} + 3\sqrt[5]{6} = 9\sqrt[5]{6}$

9. Combining radicals: $3x\sqrt{2} - 4x\sqrt{2} + x\sqrt{2} = 0$

11. Combining radicals: $\sqrt{20} - \sqrt{80} + \sqrt{45} = 2\sqrt{5} - 4\sqrt{5} + 3\sqrt{5} = \sqrt{5}$

13. Combining radicals: $4\sqrt{8} - 2\sqrt{50} - 5\sqrt{72} = 8\sqrt{2} - 10\sqrt{2} - 30\sqrt{2} = -32\sqrt{2}$

15. Combining radicals: $5x\sqrt{8} + 3\sqrt{32x^2} - 5\sqrt{50x^2} = 10x\sqrt{2} + 12x\sqrt{2} - 25x\sqrt{2} = -3x\sqrt{2}$

17. Combining radicals: $5\sqrt[3]{16} - 4\sqrt[3]{54} = 10\sqrt[3]{2} - 12\sqrt[3]{2} = -2\sqrt[3]{2}$

19. Combining radicals: $\sqrt[3]{x^4y^2} + 7x\sqrt[3]{xy^2} = x\sqrt[3]{xy^2} + 7x\sqrt[3]{xy^2} = 8x\sqrt[3]{xy^2}$

21. Combining radicals: $5a^2\sqrt{27ab^3} - 6b\sqrt{12a^5b} = 15a^2b\sqrt{3ab} - 12a^2b\sqrt{3ab} = 3a^2b\sqrt{3ab}$

23. Combining radicals: $b\sqrt[3]{24a^5b} + 3a\sqrt[3]{81a^2b^4} = 2ab\sqrt[3]{3a^2b} + 9ab\sqrt[3]{3a^2b} = 11ab\sqrt[3]{3a^2b}$

25. Combining radicals: $5x\sqrt[4]{3y^5} + y\sqrt[4]{243x^4y} + \sqrt[4]{48x^4y^5} = 5xy\sqrt[4]{3y} + 3xy\sqrt[4]{3y} + 2xy\sqrt[4]{3y} = 10xy\sqrt[4]{3y}$

27. Combining radicals: $\frac{\sqrt{2}}{2} + \frac{1}{\sqrt{2}} = \frac{\sqrt{2}}{2} + \frac{1}{\sqrt{2}} \cdot \frac{\sqrt{2}}{\sqrt{2}} = \frac{\sqrt{2}}{2} + \frac{\sqrt{2}}{2} = \sqrt{2}$

29. Combining radicals: $\frac{\sqrt{5}}{3} + \frac{1}{\sqrt{5}} = \frac{\sqrt{5}}{3} + \frac{1}{\sqrt{5}} \cdot \frac{\sqrt{5}}{\sqrt{5}} = \frac{\sqrt{5}}{3} + \frac{\sqrt{5}}{5} = \frac{5\sqrt{5}}{15} + \frac{3\sqrt{5}}{15} = \frac{8\sqrt{5}}{15}$

31. Combining radicals: $\sqrt{x} - \dfrac{1}{\sqrt{x}} = \sqrt{x} - \dfrac{1}{\sqrt{x}} \cdot \dfrac{\sqrt{x}}{\sqrt{x}} = \sqrt{x} - \dfrac{\sqrt{x}}{x} = \dfrac{x\sqrt{x}}{x} - \dfrac{\sqrt{x}}{x} = \dfrac{(x-1)\sqrt{x}}{x}$

33. Combining radicals: $\dfrac{\sqrt{18}}{6} + \sqrt{\dfrac{1}{2}} + \dfrac{\sqrt{2}}{2} = \dfrac{3\sqrt{2}}{6} + \dfrac{1}{\sqrt{2}} \cdot \dfrac{\sqrt{2}}{\sqrt{2}} + \dfrac{\sqrt{2}}{2} = \dfrac{\sqrt{2}}{2} + \dfrac{\sqrt{2}}{2} + \dfrac{\sqrt{2}}{2} = \dfrac{3\sqrt{2}}{2}$

35. Combining radicals: $\sqrt{6} - \sqrt{\dfrac{2}{3}} + \sqrt{\dfrac{1}{6}} = \sqrt{6} - \dfrac{\sqrt{2}}{\sqrt{3}} \cdot \dfrac{\sqrt{3}}{\sqrt{3}} + \dfrac{1}{\sqrt{6}} \cdot \dfrac{\sqrt{6}}{\sqrt{6}} = \sqrt{6} - \dfrac{\sqrt{6}}{3} + \dfrac{\sqrt{6}}{6} = \dfrac{6\sqrt{6}}{6} - \dfrac{2\sqrt{6}}{6} + \dfrac{\sqrt{6}}{6} = \dfrac{5\sqrt{6}}{6}$

37. Combining radicals: $\sqrt[3]{25} + \dfrac{3}{\sqrt[3]{5}} = \sqrt[3]{25} + \dfrac{3}{\sqrt[3]{5}} \cdot \dfrac{\sqrt[3]{25}}{\sqrt[3]{25}} = \sqrt[3]{25} + \dfrac{3\sqrt[3]{25}}{5} = \dfrac{5\sqrt[3]{25}}{5} + \dfrac{3\sqrt[3]{25}}{5} = \dfrac{8\sqrt[3]{25}}{5}$

39. Using a calculator:
$\sqrt{12} \approx 3.464 \qquad 2\sqrt{3} \approx 3.464$

41. It is equal to the decimal approximation for $\sqrt{50}$:
$\sqrt{8} + \sqrt{18} \approx 7.071 \approx \sqrt{50} \qquad \sqrt{26} \approx 5.099$

43. Correcting the right side: $3\sqrt{2x} + 5\sqrt{2x} = 8\sqrt{2x}$

45. Correcting the right side: $\sqrt{9+16} = \sqrt{25} = 5$

47. Answers will vary.

49. Answers will vary.

51. Answers will vary.

53. Using x as the length of the sides, the hypotenuse d is given by:
$$x^2 + x^2 = d^2$$
$$d^2 = 2x^2$$
$$d = x\sqrt{2}$$
Therefore the ratio is: $\dfrac{x\sqrt{2}}{x} = \sqrt{2}$

55. a. Since the diagonal of the base is $5\sqrt{2}$, the ratio is: $\dfrac{5\sqrt{2}}{5} = \sqrt{2}$

    b. Since the area of the base is 25, the ratio is: $\dfrac{25}{5\sqrt{2}} = \dfrac{5}{\sqrt{2}}$

    c. Since the area of the base is 25 and the perimeter is 20, the ratio is: $\dfrac{25}{20} = \dfrac{5}{4}$

57. Simplifying: $3 \cdot 2 = 6$

59. Simplifying: $(x+y)(4x-y) = 4x^2 - xy + 4xy - y^2 = 4x^2 + 3xy - y^2$

61. Simplifying: $(x+3)^2 = x^2 + 2(3x) + 3^2 = x^2 + 6x + 9$

63. Simplifying: $(x-2)(x+2) = x^2 - 2^2 = x^2 - 4$

65. Simplifying: $2\sqrt{18} = 2\sqrt{9 \cdot 2} = 2 \cdot 3\sqrt{2} = 6\sqrt{2}$

67. Simplifying: $\left(\sqrt{6}\right)^2 = 6$

69. Simplifying: $\left(3\sqrt{x}\right)^2 = 9x$

71. Rationalizing the denominator: $\dfrac{\sqrt{3}}{\sqrt{2}} = \dfrac{\sqrt{3}}{\sqrt{2}} \cdot \dfrac{\sqrt{2}}{\sqrt{2}} = \dfrac{\sqrt{6}}{2}$

8.4 Multiplication and Division of Radical Expressions

1. Multiplying: $\sqrt{6}\sqrt{3} = \sqrt{18} = 3\sqrt{2}$
3. Multiplying: $(2\sqrt{3})(5\sqrt{7}) = 10\sqrt{21}$
5. Multiplying: $(4\sqrt{6})(2\sqrt{15})(3\sqrt{10}) = 24\sqrt{900} = 24 \cdot 30 = 720$
7. Multiplying: $(3\sqrt[3]{3})(6\sqrt[3]{9}) = 18\sqrt[3]{27} = 18 \cdot 3 = 54$
9. Multiplying: $\sqrt{3}(\sqrt{2} - 3\sqrt{3}) = \sqrt{6} - 3\sqrt{9} = \sqrt{6} - 9$
11. Multiplying: $6\sqrt[3]{4}(2\sqrt[3]{2} + 1) = 12\sqrt[3]{8} + 6\sqrt[3]{4} = 24 + 6\sqrt[3]{4}$
13. Multiplying: $(\sqrt{3} + \sqrt{2})(3\sqrt{3} - \sqrt{2}) = 3\sqrt{9} - \sqrt{6} + 3\sqrt{6} - \sqrt{4} = 9 + 2\sqrt{6} - 2 = 7 + 2\sqrt{6}$
15. Multiplying: $(\sqrt{x} + 5)(\sqrt{x} - 3) = x - 3\sqrt{x} + 5\sqrt{x} - 15 = x + 2\sqrt{x} - 15$
17. Multiplying: $(3\sqrt{6} + 4\sqrt{2})(\sqrt{6} + 2\sqrt{2}) = 3\sqrt{36} + 4\sqrt{12} + 6\sqrt{12} + 8\sqrt{4} = 18 + 8\sqrt{3} + 12\sqrt{3} + 16 = 34 + 20\sqrt{3}$
19. Multiplying: $(\sqrt{3} + 4)^2 = (\sqrt{3} + 4)(\sqrt{3} + 4) = \sqrt{9} + 4\sqrt{3} + 4\sqrt{3} + 16 = 19 + 8\sqrt{3}$
21. Multiplying: $(\sqrt{x} - 3)^2 = (\sqrt{x} - 3)(\sqrt{x} - 3) = x - 3\sqrt{x} - 3\sqrt{x} + 9 = x - 6\sqrt{x} + 9$
23. Multiplying: $(2\sqrt{a} - 3\sqrt{b})^2 = (2\sqrt{a} - 3\sqrt{b})(2\sqrt{a} - 3\sqrt{b}) = 4a - 6\sqrt{ab} - 6\sqrt{ab} + 9b = 4a - 12\sqrt{ab} + 9b$
25. Multiplying: $(\sqrt{x-4} + 2)^2 = (\sqrt{x-4} + 2)(\sqrt{x-4} + 2) = x - 4 + 2\sqrt{x-4} + 2\sqrt{x-4} + 4 = x + 4\sqrt{x-4}$
27. Multiplying: $(\sqrt{x-5} - 3)^2 = (\sqrt{x-5} - 3)(\sqrt{x-5} - 3) = x - 5 - 3\sqrt{x-5} - 3\sqrt{x-5} + 9 = x - 6\sqrt{x-5} + 4$
29. Multiplying: $(\sqrt{3} - \sqrt{2})(\sqrt{3} + \sqrt{2}) = (\sqrt{3})^2 - (\sqrt{2})^2 = 3 - 2 = 1$
31. Multiplying: $(\sqrt{a} + 7)(\sqrt{a} - 7) = (\sqrt{a})^2 - (7)^2 = a - 49$
33. Multiplying: $(5 - \sqrt{x})(5 + \sqrt{x}) = (5)^2 - (\sqrt{x})^2 = 25 - x$
35. Multiplying: $(\sqrt{x-4} + 2)(\sqrt{x-4} - 2) = (\sqrt{x-4})^2 - (2)^2 = x - 4 - 4 = x - 8$
37. Multiplying: $(\sqrt{3} + 1)^3 = (\sqrt{3} + 1)(3 + 2\sqrt{3} + 1) = (\sqrt{3} + 1)(4 + 2\sqrt{3}) = 4\sqrt{3} + 4 + 6 + 2\sqrt{3} = 10 + 6\sqrt{3}$
39. Rationalizing the denominator: $\dfrac{\sqrt{2}}{\sqrt{6} - \sqrt{2}} = \dfrac{\sqrt{2}}{\sqrt{6} - \sqrt{2}} \cdot \dfrac{\sqrt{6} + \sqrt{2}}{\sqrt{6} + \sqrt{2}} = \dfrac{\sqrt{12} + 2}{6 - 2} = \dfrac{2\sqrt{3} + 2}{4} = \dfrac{1 + \sqrt{3}}{2}$
41. Rationalizing the denominator: $\dfrac{\sqrt{5}}{\sqrt{5} + 1} = \dfrac{\sqrt{5}}{\sqrt{5} + 1} \cdot \dfrac{\sqrt{5} - 1}{\sqrt{5} - 1} = \dfrac{5 - \sqrt{5}}{5 - 1} = \dfrac{5 - \sqrt{5}}{4}$
43. Rationalizing the denominator: $\dfrac{\sqrt{x}}{\sqrt{x} - 3} = \dfrac{\sqrt{x}}{\sqrt{x} - 3} \cdot \dfrac{\sqrt{x} + 3}{\sqrt{x} + 3} = \dfrac{x + 3\sqrt{x}}{x - 9}$
45. Rationalizing the denominator: $\dfrac{\sqrt{5}}{2\sqrt{5} - 3} = \dfrac{\sqrt{5}}{2\sqrt{5} - 3} \cdot \dfrac{2\sqrt{5} + 3}{2\sqrt{5} + 3} = \dfrac{2\sqrt{25} + 3\sqrt{5}}{20 - 9} = \dfrac{10 + 3\sqrt{5}}{11}$
47. Rationalizing the denominator: $\dfrac{3}{\sqrt{x} - \sqrt{y}} = \dfrac{3}{\sqrt{x} - \sqrt{y}} \cdot \dfrac{\sqrt{x} + \sqrt{y}}{\sqrt{x} + \sqrt{y}} = \dfrac{3\sqrt{x} + 3\sqrt{y}}{x - y}$
49. Rationalizing the denominator: $\dfrac{\sqrt{6} + \sqrt{2}}{\sqrt{6} - \sqrt{2}} = \dfrac{\sqrt{6} + \sqrt{2}}{\sqrt{6} - \sqrt{2}} \cdot \dfrac{\sqrt{6} + \sqrt{2}}{\sqrt{6} + \sqrt{2}} = \dfrac{6 + 2\sqrt{12} + 2}{6 - 2} = \dfrac{8 + 4\sqrt{3}}{4} = 2 + \sqrt{3}$

51. Rationalizing the denominator: $\dfrac{\sqrt{7}-2}{\sqrt{7}+2} = \dfrac{\sqrt{7}-2}{\sqrt{7}+2} \cdot \dfrac{\sqrt{7}-2}{\sqrt{7}-2} = \dfrac{7-4\sqrt{7}+4}{7-4} = \dfrac{11-4\sqrt{7}}{3}$

53.
   a. Adding: $\left(\sqrt{x}+2\right)+\left(\sqrt{x}-2\right) = \sqrt{x}+2+\sqrt{x}-2 = 2\sqrt{x}$

   b. Multiplying: $\left(\sqrt{x}+2\right)\left(\sqrt{x}-2\right) = x+2\sqrt{x}-2\sqrt{x}-4 = x-4$

   c. Squaring: $\left(\sqrt{x}+2\right)^2 = \left(\sqrt{x}+2\right)\left(\sqrt{x}+2\right) = x+2\sqrt{x}+2\sqrt{x}+4 = x+4\sqrt{x}+4$

   d. Dividing: $\dfrac{\sqrt{x}+2}{\sqrt{x}-2} = \dfrac{\sqrt{x}+2}{\sqrt{x}-2} \cdot \dfrac{\sqrt{x}+2}{\sqrt{x}+2} = \dfrac{x+4\sqrt{x}+4}{x-4}$

55.
   a. Adding: $\left(5+\sqrt{2}\right)+\left(5-\sqrt{2}\right) = 5+\sqrt{2}+5-\sqrt{2} = 10$

   b. Multiplying: $\left(5+\sqrt{2}\right)\left(5-\sqrt{2}\right) = 25+5\sqrt{2}-5\sqrt{2}-2 = 23$

   c. Squaring: $\left(5+\sqrt{2}\right)^2 = \left(5+\sqrt{2}\right)\left(5+\sqrt{2}\right) = 25+5\sqrt{2}+5\sqrt{2}+2 = 27+10\sqrt{2}$

   d. Dividing: $\dfrac{5+\sqrt{2}}{5-\sqrt{2}} = \dfrac{5+\sqrt{2}}{5-\sqrt{2}} \cdot \dfrac{5+\sqrt{2}}{5+\sqrt{2}} = \dfrac{25+5\sqrt{2}+5\sqrt{2}+2}{25-2} = \dfrac{27+10\sqrt{2}}{23}$

57.
   a. Adding: $\sqrt{2}+\left(\sqrt{6}+\sqrt{2}\right) = \sqrt{2}+\sqrt{6}+\sqrt{2} = \sqrt{6}+2\sqrt{2}$

   b. Multiplying: $\sqrt{2}\left(\sqrt{6}+\sqrt{2}\right) = \sqrt{12}+\sqrt{4} = 2+2\sqrt{3}$

   c. Dividing: $\dfrac{\sqrt{6}+\sqrt{2}}{\sqrt{2}} = \dfrac{\sqrt{6}+\sqrt{2}}{\sqrt{2}} \cdot \dfrac{\sqrt{2}}{\sqrt{2}} = \dfrac{\sqrt{12}+2}{2} = \dfrac{2+2\sqrt{3}}{2} = 1+\sqrt{3}$

   d. Dividing: $\dfrac{\sqrt{2}}{\sqrt{6}+\sqrt{2}} = \dfrac{\sqrt{2}}{\sqrt{6}+\sqrt{2}} \cdot \dfrac{\sqrt{6}-\sqrt{2}}{\sqrt{6}-\sqrt{2}} = \dfrac{\sqrt{12}-2}{6-2} = \dfrac{-2+2\sqrt{3}}{4} = \dfrac{-1+\sqrt{3}}{2}$

59.
   a. Adding: $\left(\dfrac{1+\sqrt{5}}{2}\right)+\left(\dfrac{1-\sqrt{5}}{2}\right) = \dfrac{1+\sqrt{5}}{2}+\dfrac{1-\sqrt{5}}{2} = \dfrac{2}{2} = 1$

   b. Multiplying: $\left(\dfrac{1+\sqrt{5}}{2}\right)\left(\dfrac{1-\sqrt{5}}{2}\right) = \dfrac{1-5}{4} = \dfrac{-4}{4} = -1$

61. Simplifying the product: $\left(\sqrt[3]{2}+\sqrt[3]{3}\right)\left(\sqrt[3]{4}-\sqrt[3]{6}+\sqrt[3]{9}\right) = \sqrt[3]{8}-\sqrt[3]{12}+\sqrt[3]{18}+\sqrt[3]{12}-\sqrt[3]{18}+\sqrt[3]{27} = 2+3 = 5$

63. The correct statement is: $5\left(2\sqrt{3}\right) = 10\sqrt{3}$

65. The correct statement is: $\left(\sqrt{x}+3\right)^2 = \left(\sqrt{x}+3\right)\left(\sqrt{x}+3\right) = x+6\sqrt{x}+9$

67. The correct statement is: $\left(5\sqrt{3}\right)^2 = \left(5\sqrt{3}\right)\left(5\sqrt{3}\right) = 25 \cdot 3 = 75$

69. Substituting $h = 50$: $t = \dfrac{\sqrt{100-50}}{4} = \dfrac{\sqrt{50}}{4} = \dfrac{5\sqrt{2}}{4}$ seconds

    Substituting $h = 0$: $t = \dfrac{\sqrt{100-0}}{4} = \dfrac{\sqrt{100}}{4} = \dfrac{10}{4} = \dfrac{5}{2}$ seconds

71. Since the large rectangle is a golden rectangle and $AC = 6$, then $CE = 6\left(\dfrac{1+\sqrt{5}}{2}\right) = 3+3\sqrt{5}$. Since $CD = 6$, then

    $DE = 3+3\sqrt{5}-6 = 3\sqrt{5}-3$. Now computing the ratio:

    $\dfrac{EF}{DE} = \dfrac{6}{3\sqrt{5}-3} \cdot \dfrac{3\sqrt{5}+3}{3\sqrt{5}+3} = \dfrac{18\left(\sqrt{5}+1\right)}{45-9} = \dfrac{18\left(\sqrt{5}+1\right)}{36} = \dfrac{1+\sqrt{5}}{2}$

    Therefore the smaller rectangle BDEF is also a golden rectangle.

73. Since the large rectangle is a golden rectangle and $AC = 2x$, then $CE = 2x\left(\dfrac{1+\sqrt{5}}{2}\right) = x\left(1+\sqrt{5}\right)$. Since $CD = 2x$, then
$DE = x\left(1+\sqrt{5}\right) - 2x = x\left(-1+\sqrt{5}\right)$. Now computing the ratio:
$$\dfrac{EF}{DE} = \dfrac{2x}{x\left(-1+\sqrt{5}\right)} = \dfrac{2}{-1+\sqrt{5}} \cdot \dfrac{-1-\sqrt{5}}{-1-\sqrt{5}} = \dfrac{-2\left(\sqrt{5}+1\right)}{1-5} = \dfrac{-2\left(\sqrt{5}+1\right)}{-4} = \dfrac{1+\sqrt{5}}{2}$$
Therefore the smaller rectangle BDEF is also a golden rectangle.

75. Simplifying: $(t+5)^2 = t^2 + 2(5t) + 5^2 = t^2 + 10t + 25$

77. Simplifying: $\sqrt{x} \cdot \sqrt{x} = \sqrt{x^2} = x$

79. Solving the equation:
$$3x + 4 = 5^2$$
$$3x + 4 = 25$$
$$3x = 21$$
$$x = 7$$

81. Solving the equation:
$$t^2 + 7t + 12 = 0$$
$$(t+4)(t+3) = 0$$
$$t = -4, -3$$

83. Solving the equation:
$$t^2 + 10t + 25 = t + 7$$
$$t^2 + 9t + 18 = 0$$
$$(t+6)(t+3) = 0$$
$$t = -6, -3$$

85. Solving the equation:
$$(x+4)^2 = x + 6$$
$$x^2 + 8x + 16 = x + 6$$
$$x^2 + 7x + 10 = 0$$
$$(x+5)(x+2) = 0$$
$$x = -5, -2$$

87. Substituting $x = 7$: $\sqrt{3(7)+4} = \sqrt{25} = 5$. Yes, it is a solution.

89. Substituting $t = -6$:
$$-6 + 5 = \sqrt{-6+7}$$
$$-1 = 1$$
No, it is not a solution.

8.5 Equations Involving Radicals

1. Solving the equation:
$$\sqrt{2x+1} = 3$$
$$\left(\sqrt{2x+1}\right)^2 = 3^2$$
$$2x + 1 = 9$$
$$2x = 8$$
$$x = 4$$

3. Solving the equation:
$$\sqrt{4x+1} = -5$$
$$\left(\sqrt{4x+1}\right)^2 = (-5)^2$$
$$4x + 1 = 25$$
$$4x = 24$$
$$x = 6$$
Since this value does not check, there is no solution.

5. Solving the equation:
$$\sqrt{2y-1} = 3$$
$$\left(\sqrt{2y-1}\right)^2 = 3^2$$
$$2y - 1 = 9$$
$$2y = 10$$
$$y = 5$$

7. Solving the equation:
$$\sqrt{5x-7} = -1$$
$$\left(\sqrt{5x-7}\right)^2 = (-1)^2$$
$$5x - 7 = 1$$
$$5x = 8$$
$$x = \dfrac{8}{5}$$
Since this value does not check, there is no solution.

9. Solving the equation:
$$\sqrt{2x-3} - 2 = 4$$
$$\sqrt{2x-3} = 6$$
$$\left(\sqrt{2x-3}\right)^2 = 6^2$$
$$2x - 3 = 36$$
$$2x = 39$$
$$x = \frac{39}{2}$$

11. Solving the equation:
$$\sqrt{4a+1} + 3 = 2$$
$$\sqrt{4a+1} = -1$$
$$\left(\sqrt{4a+1}\right)^2 = (-1)^2$$
$$4a + 1 = 1$$
$$4a = 0$$
$$a = 0$$
Since this value does not check, there is no solution.

13. Solving the equation:
$$\sqrt[4]{3x+1} = 2$$
$$\left(\sqrt[4]{3x+1}\right)^4 = 2^4$$
$$3x + 1 = 16$$
$$3x = 15$$
$$x = 5$$

15. Solving the equation:
$$\sqrt[3]{2x-5} = 1$$
$$\left(\sqrt[3]{2x-5}\right)^3 = 1^3$$
$$2x - 5 = 1$$
$$2x = 6$$
$$x = 3$$

17. Solving the equation:
$$\sqrt[3]{3a+5} = -3$$
$$\left(\sqrt[3]{3a+5}\right)^3 = (-3)^3$$
$$3a + 5 = -27$$
$$3a = -32$$
$$a = -\frac{32}{3}$$

19. Solving the equation:
$$\sqrt{y-3} = y - 3$$
$$\left(\sqrt{y-3}\right)^2 = (y-3)^2$$
$$y - 3 = y^2 - 6y + 9$$
$$0 = y^2 - 7y + 12$$
$$0 = (y-3)(y-4)$$
$$y = 3, 4$$

21. Solving the equation:
$$\sqrt{a+2} = a + 2$$
$$\left(\sqrt{a+2}\right)^2 = (a+2)^2$$
$$a + 2 = a^2 + 4a + 4$$
$$0 = a^2 + 3a + 2$$
$$0 = (a+2)(a+1)$$
$$a = -2, -1$$

23. Solving the equation:
$$\sqrt{2x+4} = \sqrt{1-x}$$
$$\left(\sqrt{2x+4}\right)^2 = \left(\sqrt{1-x}\right)^2$$
$$2x + 4 = 1 - x$$
$$3x = -3$$
$$x = -1$$

25. Solving the equation:
$$\sqrt{4a+7} = -\sqrt{a+2}$$
$$\left(\sqrt{4a+7}\right)^2 = \left(-\sqrt{a+2}\right)^2$$
$$4a + 7 = a + 2$$
$$3a = -5$$
$$a = -\frac{5}{3}$$
Since this value does not check, there is no solution.

27. Solving the equation:
$$\sqrt[4]{5x-8} = \sqrt[4]{4x-1}$$
$$\left(\sqrt[4]{5x-8}\right)^4 = \left(\sqrt[4]{4x-1}\right)^4$$
$$5x - 8 = 4x - 1$$
$$x = 7$$

29. Solving the equation:
$$x + 1 = \sqrt{5x+1}$$
$$(x+1)^2 = \left(\sqrt{5x+1}\right)^2$$
$$x^2 + 2x + 1 = 5x + 1$$
$$x^2 - 3x = 0$$
$$x(x-3) = 0$$
$$x = 0, 3$$

31. Solving the equation:
$$t + 5 = \sqrt{2t+9}$$
$$(t+5)^2 = \left(\sqrt{2t+9}\right)^2$$
$$t^2 + 10t + 25 = 2t + 9$$
$$t^2 + 8t + 16 = 0$$
$$(t+4)^2 = 0$$
$$t = -4$$

33. Solving the equation:
$$\sqrt{y-8} = \sqrt{8-y}$$
$$\left(\sqrt{y-8}\right)^2 = \left(\sqrt{8-y}\right)^2$$
$$y-8 = 8-y$$
$$2y = 16$$
$$y = 8$$

35. Solving the equation:
$$\sqrt[3]{3x+5} = \sqrt[3]{5-2x}$$
$$\left(\sqrt[3]{3x+5}\right)^3 = \left(\sqrt[3]{5-2x}\right)^3$$
$$3x+5 = 5-2x$$
$$5x = 0$$
$$x = 0$$

37. Solving the equation:
$$\sqrt{x-8} = \sqrt{x}-2$$
$$\left(\sqrt{x-8}\right)^2 = \left(\sqrt{x}-2\right)^2$$
$$x-8 = x-4\sqrt{x}+4$$
$$-12 = -4\sqrt{x}$$
$$\sqrt{x} = 3$$
$$x = 9$$

39. Solving the equation:
$$\sqrt{x+1} = \sqrt{x}+1$$
$$\left(\sqrt{x+1}\right)^2 = \left(\sqrt{x}+1\right)^2$$
$$x+1 = x+2\sqrt{x}+1$$
$$0 = 2\sqrt{x}$$
$$\sqrt{x} = 0$$
$$x = 0$$

41. Solving the equation:
$$\sqrt{x+8} = \sqrt{x-4}+2$$
$$\left(\sqrt{x+8}\right)^2 = \left(\sqrt{x-4}+2\right)^2$$
$$x+8 = x-4+4\sqrt{x-4}+4$$
$$8 = 4\sqrt{x-4}$$
$$\sqrt{x-4} = 2$$
$$x-4 = 4$$
$$x = 8$$

43. Solving the equation:
$$\sqrt{x-5} - 3 = \sqrt{x-8}$$
$$\left(\sqrt{x-5}-3\right)^2 = \left(\sqrt{x-8}\right)^2$$
$$x-5-6\sqrt{x-5}+9 = x-8$$
$$-6\sqrt{x-5} = -12$$
$$\sqrt{x-5} = 2$$
$$x-5 = 4$$
$$x = 9$$
Since this value does not check, there is no solution.

45. **a.** Solving the equation:
$$\sqrt{y} - 4 = 6$$
$$\sqrt{y} = 10$$
$$\left(\sqrt{y}\right)^2 = 10^2$$
$$y = 100$$

b. Solving the equation:
$$\sqrt{y-4} = 6$$
$$\left(\sqrt{y-4}\right)^2 = 6^2$$
$$y-4 = 36$$
$$y = 40$$

c. Solving the equation:
$$\sqrt{y-4} = -6$$
$$\left(\sqrt{y-4}\right)^2 = (-6)^2$$
$$y-4 = 36$$
$$y = 40$$

There is no solution (40 does not check).

d. Solving the equation:
$$\sqrt{y-4} = y-6$$
$$\left(\sqrt{y-4}\right)^2 = (y-6)^2$$
$$y-4 = y^2 - 12y + 36$$
$$0 = y^2 - 13y + 40$$
$$0 = (y-5)(y-8)$$
$$y = 5, 8$$

The solution is 8 (5 does not check).

47.

a. Solving the equation:
$$x - 3 = 0$$
$$x = 3$$

b. Solving the equation:
$$\sqrt{x} - 3 = 0$$
$$\left(\sqrt{x}\right)^2 = 3^2$$
$$x = 9$$

c. Solving the equation:
$$\sqrt{x-3} = 0$$
$$\left(\sqrt{x-3}\right)^2 = 0^2$$
$$x - 3 = 0$$
$$x = 3$$

d. Solving the equation:
$$\sqrt{x} + 3 = 0$$
$$\sqrt{x} = -3$$
$$\left(\sqrt{x}\right)^2 = (-3)^2$$
$$x = 9$$
There is no solution (9 does not check).

e. Solving the equation:
$$\sqrt{x} + 3 = 5$$
$$\sqrt{x} = 2$$
$$\left(\sqrt{x}\right)^2 = 2^2$$
$$x = 4$$

f. Solving the equation:
$$\sqrt{x} + 3 = -5$$
$$\sqrt{x} = -8$$
$$\left(\sqrt{x}\right)^2 = (-8)^2$$
$$x = 64$$
There is no solution (64 does not check).

g. Solving the equation:
$$x - 3 = \sqrt{5 - x}$$
$$(x-3)^2 = \left(\sqrt{5-x}\right)^2$$
$$x^2 - 6x + 9 = 5 - x$$
$$x^2 - 5x + 4 = 0$$
$$(x-1)(x-4) = 0$$
$$x = 1, 4$$
The solution is 4 (1 does not check).

49. Solving for h:
$$t = \frac{\sqrt{100 - h}}{4}$$
$$4t = \sqrt{100 - h}$$
$$16t^2 = 100 - h$$
$$h = 100 - 16t^2$$

51. Solving for L:
$$2 = 2\left(\frac{22}{7}\right)\sqrt{\frac{L}{32}}$$
$$\frac{7}{22} = \sqrt{\frac{L}{32}}$$
$$\left(\frac{7}{22}\right)^2 = \frac{L}{32}$$
$$L = 32\left(\frac{7}{22}\right)^2 \approx 3.24 \text{ feet}$$

53. Graphing the equation:

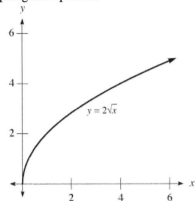

55. Graphing the equation:

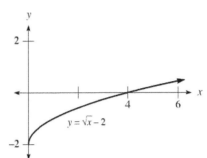

57. Graphing the equation:

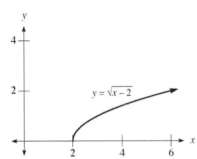

59. Graphing the equation:

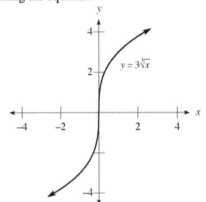

61. Graphing the equation:

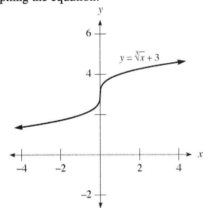

63. Graphing the equation:

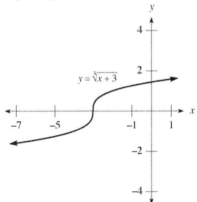

65. Simplifying: $\sqrt{25} = 5$

67. Simplifying: $\sqrt{12} = \sqrt{4 \cdot 3} = 2\sqrt{3}$

69. Simplifying: $(-1)^{15} = -1$

71. Simplifying: $(-1)^{50} = 1$

73. Solving the equation:
$$3x = 12$$
$$x = 4$$

75. Solving the equation:
$$4x - 3 = 5$$
$$4x = 8$$
$$x = 2$$

77. Performing the operations: $(3+4x)+(7-6x)=10-2x$

79. Performing the operations: $(7+3x)-(5+6x)=7+3x-5-6x=2-3x$

81. Performing the operations: $(3-4x)(2+5x)=6+15x-8x-20x^2=6+7x-20x^2$

83. Performing the operations: $2x(4-6x)=8x-12x^2$

85. Performing the operations: $(2+3x)^2=2^2+2(2)(3x)+(3x)^2=4+12x+9x^2$

87. Performing the operations: $(2-3x)(2+3x)=2^2-(3x)^2=4-9x^2$

8.6 Complex Numbers

1. Writing in terms of i: $\sqrt{-36}=6i$

3. Writing in terms of i: $-\sqrt{-25}=-5i$

5. Writing in terms of i: $\sqrt{-72}=6i\sqrt{2}$

7. Writing in terms of i: $-\sqrt{-12}=-2i\sqrt{3}$

9. Rewriting the expression: $i^{28}=\left(i^4\right)^7=(1)^7=1$

11. Rewriting the expression: $i^{26}=i^{24}i^2=\left(i^4\right)^6 i^2=(1)^6(-1)=-1$

13. Rewriting the expression: $i^{75}=i^{72}i^3=\left(i^4\right)^{18}i^2 i=(1)^{18}(-1)i=-i$

15. Setting real and imaginary parts equal:
$2x=6 \qquad 3y=-3$
$x=3 \qquad y=-1$

17. Setting real and imaginary parts equal:
$-x=2 \qquad 10y=-5$
$x=-2 \qquad y=-\dfrac{1}{2}$

19. Setting real and imaginary parts equal:
$2x=-16 \qquad -2y=10$
$x=-8 \qquad y=-5$

21. Setting real and imaginary parts equal:
$2x-4=10 \qquad -6y=-3$
$2x=14 \qquad y=\dfrac{1}{2}$
$x=7$

23. Setting real and imaginary parts equal:
$7x-1=2 \qquad 5y+2=4$
$7x=3 \qquad 5y=2$
$x=\dfrac{3}{7} \qquad y=\dfrac{2}{5}$

25. Combining the numbers: $(2+3i)+(3+6i)=5+9i$

27. Combining the numbers: $(3-5i)+(2+4i)=5-i$

29. Combining the numbers: $(5+2i)-(3+6i)=5+2i-3-6i=2-4i$

31. Combining the numbers: $(3-5i)-(2+i)=3-5i-2-i=1-6i$

33. Combining the numbers: $[(3+2i)-(6+i)]+(5+i)=3+2i-6-i+5+i=2+2i$

35. Combining the numbers: $[(7-i)-(2+4i)]-(6+2i)=7-i-2-4i-6-2i=-1-7i$

37. Combining the numbers:
$(3+2i)-[(3-4i)-(6+2i)]=(3+2i)-(3-4i-6-2i)=(3+2i)-(-3-6i)=3+2i+3+6i=6+8i$

39. Combining the numbers: $(4-9i)+[(2-7i)-(4+8i)]=(4-9i)+(2-7i-4-8i)=(4-9i)+(-2-15i)=2-24i$

41. Finding the product: $3i(4+5i)=12i+15i^2=-15+12i$

43. Finding the product: $6i(4-3i)=24i-18i^2=18+24i$

45. Finding the product: $(3+2i)(4+i)=12+8i+3i+2i^2=12+11i-2=10+11i$

47. Finding the product: $(4+9i)(3-i)=12+27i-4i-9i^2=12+23i+9=21+23i$

49. Finding the product: $(1+i)^3 = (1+i)(1+i)^2 = (1+i)(1+2i-1) = (1+i)(2i) = -2+2i$

51. Finding the product: $(2-i)^3 = (2-i)(2-i)^2 = (2-i)(4-4i-1) = (2-i)(3-4i) = 6-11i-4 = 2-11i$

53. Finding the product: $(2+5i)^2 = (2+5i)(2+5i) = 4+10i+10i-25 = -21+20i$

55. Finding the product: $(1-i)^2 = (1-i)(1-i) = 1-i-i-1 = -2i$

57. Finding the product: $(3-4i)^2 = (3-4i)(3-4i) = 9-12i-12i-16 = -7-24i$

59. Finding the product: $(2+i)(2-i) = 4-i^2 = 4+1 = 5$

61. Finding the product: $(6-2i)(6+2i) = 36-4i^2 = 36+4 = 40$

63. Finding the product: $(2+3i)(2-3i) = 4-9i^2 = 4+9 = 13$

65. Finding the product: $(10+8i)(10-8i) = 100-64i^2 = 100+64 = 164$

67. Finding the quotient: $\dfrac{2-3i}{i} = \dfrac{2-3i}{i} \cdot \dfrac{i}{i} = \dfrac{2i+3}{-1} = -3-2i$

69. Finding the quotient: $\dfrac{5+2i}{-i} = \dfrac{5+2i}{-i} \cdot \dfrac{i}{i} = \dfrac{5i-2}{1} = -2+5i$

71. Finding the quotient: $\dfrac{4}{2-3i} = \dfrac{4}{2-3i} \cdot \dfrac{2+3i}{2+3i} = \dfrac{8+12i}{4+9} = \dfrac{8+12i}{13} = \dfrac{8}{13} + \dfrac{12}{13}i$

73. Finding the quotient: $\dfrac{6}{-3+2i} = \dfrac{6}{-3+2i} \cdot \dfrac{-3-2i}{-3-2i} = \dfrac{-18-12i}{9+4} = \dfrac{-18-12i}{13} = -\dfrac{18}{13} - \dfrac{12}{13}i$

75. Finding the quotient: $\dfrac{2+3i}{2-3i} = \dfrac{2+3i}{2-3i} \cdot \dfrac{2+3i}{2+3i} = \dfrac{4+12i-9}{4+9} = \dfrac{-5+12i}{13} = -\dfrac{5}{13} + \dfrac{12}{13}i$

77. Finding the quotient: $\dfrac{5+4i}{3+6i} = \dfrac{5+4i}{3+6i} \cdot \dfrac{3-6i}{3-6i} = \dfrac{15-18i+24}{9+36} = \dfrac{39-18i}{45} = \dfrac{13}{15} - \dfrac{2}{5}i$

79. Dividing to find R: $R = \dfrac{80+20i}{-6+2i} = \dfrac{80+20i}{-6+2i} \cdot \dfrac{-6-2i}{-6-2i} = \dfrac{-480-280i+40}{36+4} = \dfrac{-440-280i}{40} = (-11-7i)$ ohms

81. Solving the equation:

$$\dfrac{t}{3} - \dfrac{1}{2} = -1$$
$$6\left(\dfrac{t}{3} - \dfrac{1}{2}\right) = 6(-1)$$
$$2t - 3 = -6$$
$$2t = -3$$
$$t = -\dfrac{3}{2}$$

83. Solving the equation:

$$2 + \dfrac{5}{y} = \dfrac{3}{y^2}$$
$$y^2\left(2 + \dfrac{5}{y}\right) = y^2\left(\dfrac{3}{y^2}\right)$$
$$2y^2 + 5y = 3$$
$$2y^2 + 5y - 3 = 0$$
$$(2y-1)(y+3) = 0$$
$$y = -3, \dfrac{1}{2}$$

85. Let x represent the number. The equation is:
$$x + \frac{1}{x} = \frac{41}{20}$$
$$20x\left(x + \frac{1}{x}\right) = 20x\left(\frac{41}{20}\right)$$
$$20x^2 + 20 = 41x$$
$$20x^2 - 41x + 20 = 0$$
$$(5x-4)(4x-5) = 0$$
$$x = \frac{4}{5}, \frac{5}{4}$$
The number is either $\frac{5}{4}$ or $\frac{4}{5}$.

Chapter 8 test

1. Simplifying: $27^{-2/3} = \left(27^{1/3}\right)^{-2} = 3^{-2} = \frac{1}{3^2} = \frac{1}{9}$

2. Simplifying: $\left(\frac{25}{49}\right)^{-1/2} = \left(\frac{49}{25}\right)^{1/2} = \frac{7}{5}$

3. Simplifying: $a^{3/4} \cdot a^{-1/3} = a^{3/4 - 1/3} = a^{9/12 - 4/12} = a^{5/12}$

4. Simplifying: $\frac{\left(x^{2/3}y^{-3}\right)^{1/2}}{\left(x^{3/4}y^{1/2}\right)^{-1}} = \frac{x^{1/3}y^{-3/2}}{x^{-3/4}y^{-1/2}} = x^{1/3+3/4}y^{-3/2+1/2} = x^{4/12+9/12}y^{-3/2+1/2} = x^{13/12}y^{-1} = \frac{x^{13/12}}{y}$

5. Simplifying: $\sqrt{49x^8y^{10}} = 7x^4y^5$

6. Simplifying: $\sqrt[5]{32x^{10}y^{20}} = 2x^2y^4$

7. Simplifying: $\frac{\left(36a^8b^4\right)^{1/2}}{\left(27a^9b^6\right)^{1/3}} = \frac{6a^4b^2}{3a^3b^2} = 2a$

8. Simplifying: $\frac{\left(x^n y^{1/n}\right)^n}{\left(x^{1/n}y^n\right)^{n^2}} = \frac{x^{n^2}y}{x^n y^{n^3}} = x^{n^2-n}y^{1-n^3}$

9. Multiplying: $2a^{1/2}\left(3a^{3/2} - 5a^{1/2}\right) = 6a^{4/2} - 10a^{2/2} = 6a^2 - 10a$

10. Multiplying: $\left(4a^{3/2} - 5\right)^2 = \left(4a^{3/2} - 5\right)\left(4a^{3/2} - 5\right) = 16a^3 - 20a^{3/2} - 20a^{3/2} + 25 = 16a^3 - 40a^{3/2} + 25$

11. Factoring: $3x^{2/3} + 5x^{1/3} - 2 = \left(3x^{1/3} - 1\right)\left(x^{1/3} + 2\right)$

12. Factoring: $9x^{2/3} - 49 = \left(3x^{1/3} + 7\right)\left(3x^{1/3} - 7\right)$

13. Combining: $\frac{4}{x^{1/2}} + x^{1/2} = \frac{4}{x^{1/2}} + x^{1/2} \cdot \frac{x^{1/2}}{x^{1/2}} = \frac{4+x}{x^{1/2}} = \frac{x+4}{x^{1/2}}$

14. Combining: $\frac{x^2}{\left(x^2-3\right)^{1/2}} - \left(x^2-3\right)^{1/2} = \frac{x^2}{\left(x^2-3\right)^{1/2}} - \left(x^2-3\right)^{1/2} \cdot \frac{\left(x^2-3\right)^{1/2}}{\left(x^2-3\right)^{1/2}} = \frac{x^2 - x^2 + 3}{\left(x^2-3\right)^{1/2}} = \frac{3}{\left(x^2-3\right)^{1/2}}$

15. Simplifying: $\sqrt{125x^3y^5} = \sqrt{25x^2y^4 \cdot 5xy} = 5xy^2\sqrt{5xy}$

16. Simplifying: $\sqrt[3]{40x^7y^8} = \sqrt[3]{8x^6y^6 \cdot 5xy^2} = 2x^2y^2\sqrt[3]{5xy^2}$

17. Simplifying: $\sqrt{\frac{2}{3}} = \frac{\sqrt{2}}{\sqrt{3}} \cdot \frac{\sqrt{3}}{\sqrt{3}} = \frac{\sqrt{6}}{3}$

18. Simplifying: $\sqrt{\frac{12a^4b^3}{5c}} = \frac{\sqrt{12a^4b^3}}{\sqrt{5c}} \cdot \frac{\sqrt{5c}}{\sqrt{5c}} = \frac{\sqrt{60a^4b^3c}}{5c} = \frac{2a^2b\sqrt{15bc}}{5c}$

19. Combining: $3\sqrt{12} - 4\sqrt{27} = 6\sqrt{3} - 12\sqrt{3} = -6\sqrt{3}$

20. Combining: $\sqrt[3]{24a^3b^3} - 5a\sqrt[3]{3b^3} = 2ab\sqrt[3]{3} - 5ab\sqrt[3]{3} = -3ab\sqrt[3]{3}$

21. Multiplying: $(\sqrt{x}+7)(\sqrt{x}-4) = x - 4\sqrt{x} + 7\sqrt{x} - 28 = x + 3\sqrt{x} - 28$

22. Multiplying: $(3\sqrt{2}-\sqrt{3})^2 = (3\sqrt{2}-\sqrt{3})(3\sqrt{2}-\sqrt{3}) = 18 - 3\sqrt{6} - 3\sqrt{6} + 3 = 21 - 6\sqrt{6}$

23. Rationalizing the denominator: $\dfrac{5}{\sqrt{3}-1} = \dfrac{5}{\sqrt{3}-1} \cdot \dfrac{\sqrt{3}+1}{\sqrt{3}+1} = \dfrac{5\sqrt{3}+5}{3-1} = \dfrac{5+5\sqrt{3}}{2}$

24. Rationalizing the denominator: $\dfrac{\sqrt{x}-\sqrt{2}}{\sqrt{x}+\sqrt{2}} = \dfrac{\sqrt{x}-\sqrt{2}}{\sqrt{x}+\sqrt{2}} \cdot \dfrac{\sqrt{x}-\sqrt{2}}{\sqrt{x}-\sqrt{2}} = \dfrac{x-\sqrt{2x}-\sqrt{2x}+2}{x-2} = \dfrac{x-2\sqrt{2x}+2}{x-2}$

25. Solving the equation:
$$\sqrt{3x+1} = x-3$$
$$\left(\sqrt{3x+1}\right)^2 = (x-3)^2$$
$$3x+1 = x^2 - 6x + 9$$
$$0 = x^2 - 9x + 8$$
$$0 = (x-1)(x-8)$$
$$x = 1, 8$$
The solution is 8 (1 does not check).

26. Solving the equation:
$$\sqrt[3]{2x+7} = -1$$
$$\left(\sqrt[3]{2x+7}\right)^3 = (-1)^3$$
$$2x+7 = -1$$
$$2x = -8$$
$$x = -4$$

27. Solving the equation:
$$\sqrt{x+3} = \sqrt{x+4} - 1$$
$$\left(\sqrt{x+3}\right)^2 = \left(\sqrt{x+4}-1\right)^2$$
$$x+3 = x+4 - 2\sqrt{x+4} + 1$$
$$-2 = -2\sqrt{x+4}$$
$$\sqrt{x+4} = 1$$
$$x+4 = 1$$
$$x = -3$$

28. Graphing the equation:

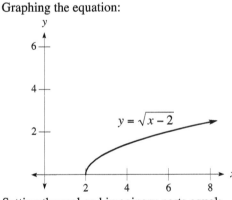

29. Graphing the equation:

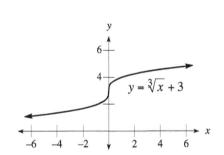

30. Setting the real and imaginary parts equal:
$$2x+5 = 6$$
$$2x = 1$$
$$x = \dfrac{1}{2}$$

$$-(y-3) = -4$$
$$y-3 = 4$$
$$y = 7$$

31. Performing the operations: $(3+2i)-[(7-i)-(4+3i)] = (3+2i)-(7-i-4-3i) = (3+2i)-(3-4i) = 3+2i-3+4i = 6i$

32. Performing the operations: $(2-3i)(4+3i) = 8+6i-12i+9 = 17-6i$

33. Performing the operations: $(5-4i)^2 = (5-4i)(5-4i) = 25-20i-20i-16 = 9-40i$

34. Performing the operations: $\dfrac{2-3i}{2+3i} = \dfrac{2-3i}{2+3i} \cdot \dfrac{2-3i}{2-3i} = \dfrac{4-12i-9}{4+9} = \dfrac{-5-12i}{13} = -\dfrac{5}{13}-\dfrac{12}{13}i$

35. Rewriting the exponent: $i^{38} = (i^2)^{19} = (-1)^{19} = -1$

Chapter 9
Quadratic Equations

9.1 Completing the Square

1. Solving the equation:
$$x^2 = 25$$
$$x = \pm\sqrt{25} = \pm 5$$

3. Solving the equation:
$$a^2 = -9$$
$$a = \pm\sqrt{-9} = \pm 3i$$

5. Solving the equation:
$$y^2 = \frac{3}{4}$$
$$y = \pm\sqrt{\frac{3}{4}} = \pm\frac{\sqrt{3}}{2}$$

7. Solving the equation:
$$x^2 + 12 = 0$$
$$x^2 = -12$$
$$x = \pm\sqrt{-12} = \pm 2i\sqrt{3}$$

9. Solving the equation:
$$4a^2 - 45 = 0$$
$$4a^2 = 45$$
$$a^2 = \frac{45}{4}$$
$$a = \pm\sqrt{\frac{45}{4}} = \pm\frac{3\sqrt{5}}{2}$$

11. Solving the equation:
$$(2y-1)^2 = 25$$
$$2y - 1 = \pm\sqrt{25} = \pm 5$$
$$2y - 1 = -5, 5$$
$$2y = -4, 6$$
$$y = -2, 3$$

13. Solving the equation:
$$(2a+3)^2 = -9$$
$$2a + 3 = \pm\sqrt{-9} = \pm 3i$$
$$2a = -3 \pm 3i$$
$$a = \frac{-3 \pm 3i}{2}$$

15. Solving the equation:
$$(5x+2)^2 = -8$$
$$5x + 2 = \pm\sqrt{-2} = \pm 2i\sqrt{2}$$
$$5x = -2 \pm 2i\sqrt{2}$$
$$x = \frac{-2 \pm 2i\sqrt{2}}{5}$$

17. Solving the equation:
$$x^2 + 8x + 16 = -27$$
$$(x+4)^2 = -27$$
$$x + 4 = \pm\sqrt{-27} = \pm 3i\sqrt{3}$$
$$x = -4 \pm 3i\sqrt{3}$$

19. Solving the equation:
$$4a^2 - 12a + 9 = -4$$
$$(2a-3)^2 = -4$$
$$2a - 3 = \pm\sqrt{-4} = \pm 2i$$
$$2a = 3 \pm 2i$$
$$a = \frac{3 \pm 2i}{2}$$

21. Completing the square: $x^2 + 12x + 36 = (x+6)^2$

23. Completing the square: $x^2 - 4x + 4 = (x-2)^2$

25. Completing the square: $a^2 - 10a + 25 = (a-5)^2$

27. Completing the square: $x^2 + 5x + \frac{25}{4} = \left(x + \frac{5}{2}\right)^2$

29. Completing the square: $y^2 - 7y + \frac{49}{4} = \left(y - \frac{7}{2}\right)^2$

31. Completing the square: $x^2 + \frac{1}{2}x + \frac{1}{16} = \left(x + \frac{1}{4}\right)^2$

33. Completing the square: $x^2 + \frac{2}{3}x + \frac{1}{9} = \left(x + \frac{1}{3}\right)^2$

35. Solving the equation:
$$x^2 + 4x = 12$$
$$x^2 + 4x + 4 = 12 + 4$$
$$(x+2)^2 = 16$$
$$x + 2 = \pm\sqrt{16} = \pm 4$$
$$x + 2 = -4, 4$$
$$x = -6, 2$$

37. Solving the equation:
$$x^2 + 12x = -27$$
$$x^2 + 12x + 36 = -27 + 36$$
$$(x+6)^2 = 9$$
$$x + 6 = \pm\sqrt{9} = \pm 3$$
$$x + 6 = -3, 3$$
$$x = -9, -3$$

39. Solving the equation:
$$a^2 - 2a + 5 = 0$$
$$a^2 - 2a + 1 = -5 + 1$$
$$(a-1)^2 = -4$$
$$a - 1 = \pm\sqrt{-4} = \pm 2i$$
$$a = 1 \pm 2i$$

41. Solving the equation:
$$y^2 - 8y + 1 = 0$$
$$y^2 - 8y + 16 = -1 + 16$$
$$(y-4)^2 = 15$$
$$y - 4 = \pm\sqrt{15}$$
$$y = 4 \pm \sqrt{15}$$

43. Solving the equation:
$$x^2 - 5x - 3 = 0$$
$$x^2 - 5x + \frac{25}{4} = 3 + \frac{25}{4}$$
$$\left(x - \frac{5}{2}\right)^2 = \frac{37}{4}$$
$$x - \frac{5}{2} = \pm\frac{\sqrt{37}}{2}$$
$$x = \frac{5 \pm \sqrt{37}}{2}$$

45. Solving the equation:
$$2x^2 - 4x - 8 = 0$$
$$x^2 - 2x - 4 = 0$$
$$x^2 - 2x + 1 = 4 + 1$$
$$(x-1)^2 = 5$$
$$x - 1 = \pm\sqrt{5}$$
$$x = 1 \pm \sqrt{5}$$

47. Solving the equation:
$$3t^2 - 8t + 1 = 0$$
$$t^2 - \frac{8}{3}t + \frac{1}{3} = 0$$
$$t^2 - \frac{8}{3}t + \frac{16}{9} = -\frac{1}{3} + \frac{16}{9}$$
$$\left(t - \frac{4}{3}\right)^2 = \frac{13}{9}$$
$$t - \frac{4}{3} = \pm\sqrt{\frac{13}{9}} = \pm\frac{\sqrt{13}}{3}$$
$$t = \frac{4 \pm \sqrt{13}}{3}$$

49. Solving the equation:
$$4x^2 - 3x + 5 = 0$$
$$x^2 - \frac{3}{4}x + \frac{5}{4} = 0$$
$$x^2 - \frac{3}{4}x + \frac{9}{64} = -\frac{5}{4} + \frac{9}{64}$$
$$\left(x - \frac{3}{8}\right)^2 = -\frac{71}{64}$$
$$x - \frac{3}{8} = \pm\sqrt{-\frac{71}{64}} = \pm\frac{i\sqrt{71}}{8}$$
$$x = \frac{3 \pm i\sqrt{71}}{8}$$

51. Solving the equation:
$$3x^2 + 4x - 1 = 0$$
$$x^2 + \frac{4}{3}x - \frac{1}{3} = 0$$
$$x^2 + \frac{4}{3}x + \frac{4}{9} = \frac{1}{3} + \frac{4}{9}$$
$$\left(x + \frac{2}{3}\right)^2 = \frac{7}{9}$$
$$x + \frac{2}{3} = \pm\sqrt{\frac{7}{9}} = \pm\frac{\sqrt{7}}{3}$$
$$x = \frac{-2 \pm \sqrt{7}}{3}$$

53. Solving the equation:
$$2x^2 - 10x = 11$$
$$x^2 - 5x = \frac{11}{2}$$
$$x^2 - 5x + \frac{25}{4} = \frac{11}{2} + \frac{25}{4}$$
$$\left(x - \frac{5}{2}\right)^2 = \frac{47}{4}$$
$$x - \frac{5}{2} = \pm\sqrt{\frac{47}{4}} = \pm\frac{\sqrt{47}}{2}$$
$$x = \frac{5 \pm \sqrt{47}}{2}$$

55. Solving the equation:
$$4x^2 - 10x + 11 = 0$$
$$x^2 - \frac{5}{2}x + \frac{11}{4} = 0$$
$$x^2 - \frac{5}{2}x + \frac{25}{16} = -\frac{11}{4} + \frac{25}{16}$$
$$\left(x - \frac{5}{4}\right)^2 = -\frac{19}{16}$$
$$x - \frac{5}{4} = \pm\sqrt{-\frac{19}{16}} = \pm\frac{i\sqrt{19}}{4}$$
$$x = \frac{5 \pm i\sqrt{19}}{4}$$

57. **a.** No, it cannot be solved by factoring.
  b. Solving the equation:
$$x^2 = -9$$
$$x = \pm\sqrt{-9}$$
$$x = \pm 3i$$

59. **a.** Solving by factoring:
$$x^2 - 6x = 0$$
$$x(x - 6) = 0$$
$$x = 0, 6$$

  b. Solving by completing the square:
$$x^2 - 6x = 0$$
$$x^2 - 6x + 9 = 0 + 9$$
$$(x - 3)^2 = 9$$
$$x - 3 = \pm\sqrt{9}$$
$$x - 3 = -3, 3$$
$$x = 0, 6$$

61. **a.** Solving by factoring:
$$x^2 + 2x = 35$$
$$x^2 + 2x - 35 = 0$$
$$(x + 7)(x - 5) = 0$$
$$x = -7, 5$$

  b. Solving by completing the square:
$$x^2 + 2x = 35$$
$$x^2 + 2x + 1 = 35 + 1$$
$$(x + 1)^2 = 36$$
$$x + 1 = \pm\sqrt{36}$$
$$x + 1 = -6, 6$$
$$x = -7, 5$$

63. Substituting: $x^2 - 6x - 7 = \left(-3 + \sqrt{2}\right)^2 - 6\left(-3 + \sqrt{2}\right) - 7 = 9 - 6\sqrt{2} + 2 + 18 - 6\sqrt{2} - 7 = 22 - 12\sqrt{2}$

No, $x = -3 + \sqrt{2}$ is not a solution to the equation.

65. **a.** Solving the equation:
$$5x - 7 = 0$$
$$5x = 7$$
$$x = \frac{7}{5}$$

b. Solving the equation:
$$5x - 7 = 8$$
$$5x = 15$$
$$x = 3$$

c. Solving the equation:
$$(5x - 7)^2 = 8$$
$$5x - 7 = \pm\sqrt{8}$$
$$5x - 7 = \pm 2\sqrt{2}$$
$$5x = 7 \pm 2\sqrt{2}$$
$$x = \frac{7 \pm 2\sqrt{2}}{5}$$

d. Solving the equation:
$$\sqrt{5x - 7} = 8$$
$$\left(\sqrt{5x - 7}\right)^2 = (8)^2$$
$$5x - 7 = 64$$
$$5x = 71$$
$$x = \frac{71}{5}$$

e. Solving the equation:
$$\frac{5}{2} - \frac{7}{2x} = \frac{4}{x}$$
$$2x\left(\frac{5}{2} - \frac{7}{2x}\right) = 2x\left(\frac{4}{x}\right)$$
$$5x - 7 = 8$$
$$5x = 15$$
$$x = 3$$

67. The other two sides are $\frac{\sqrt{3}}{2}$ inch, 1 inch.

69. The hypotenuse is $\sqrt{2}$ inches.

71. Let x represent the horizontal distance. Using the Pythagorean theorem:
$$x^2 + 120^2 = 790^2$$
$$x^2 + 14400 = 624100$$
$$x^2 = 609700$$
$$x = \sqrt{609700} \approx 781 \text{ feet}$$

73. Solving for r:
$$3456 = 3000(1 + r)^2$$
$$(1 + r)^2 = 1.152$$
$$1 + r = \sqrt{1.152}$$
$$r = \sqrt{1.152} - 1 \approx 0.073$$
The annual interest rate is 7.3%.

75. Its length is $20\sqrt{2} \approx 28$ feet.

77. Simplifying: $49 - 4(6)(-5) = 49 + 120 = 169$

79. Simplifying: $(-27)^2 - 4(0.1)(1,700) = 729 - 680 = 49$

81. Simplifying: $-7 + \frac{169}{12} = -\frac{84}{12} + \frac{169}{12} = \frac{85}{12}$

83. Factoring: $27t^3 - 8 = (3t - 2)(9t^2 + 6t + 4)$

9.2 The Quadratic Formula

1. Solving the equation:
$$x^2 + 5x + 6 = 0$$
$$(x+3)(x+2) = 0$$
$$x = -3, -2$$

3. Using the quadratic formula: $x = \dfrac{4 \pm \sqrt{(-4)^2 - 4(1)(1)}}{2(1)} = \dfrac{4 \pm \sqrt{16-4}}{2} = \dfrac{4 \pm \sqrt{12}}{2} = \dfrac{4 \pm 2\sqrt{3}}{2} = 2 \pm \sqrt{3}$

5. Solving the equation:
$$\frac{1}{6}x^2 - \frac{1}{2}x + \frac{1}{3} = 0$$
$$x^2 - 3x + 2 = 0$$
$$(x-1)(x-2) = 0$$
$$x = 1, 2$$

7. Solving the equation:
$$\frac{x^2}{2} + 1 = \frac{2x}{3}$$
$$3x^2 + 6 = 4x$$
$$3x^2 - 4x + 6 = 0$$
$$x = \frac{4 \pm \sqrt{16-72}}{6} = \frac{4 \pm \sqrt{-56}}{6} = \frac{4 \pm 2i\sqrt{14}}{6} = \frac{2 \pm i\sqrt{14}}{3}$$

9. Solving the equation:
$$y^2 - 5y = 0$$
$$y(y-5) = 0$$
$$y = 0, 5$$

11. Solving the equation:
$$30x^2 + 40x = 0$$
$$10x(3x+4) = 0$$
$$x = -\frac{4}{3}, 0$$

13. Solving the equation:
$$\frac{2t^2}{3} - t = -\frac{1}{6}$$
$$4t^2 - 6t = -1$$
$$4t^2 - 6t + 1 = 0$$
$$t = \frac{6 \pm \sqrt{36-16}}{8} = \frac{6 \pm \sqrt{20}}{8} = \frac{6 \pm 2\sqrt{5}}{8} = \frac{3 \pm \sqrt{5}}{4}$$

15. Solving the equation:
$$0.01x^2 + 0.06x - 0.08 = 0$$
$$x^2 + 6x - 8 = 0$$
$$x = \frac{-6 \pm \sqrt{36+32}}{2} = \frac{-6 \pm \sqrt{68}}{2} = \frac{-6 \pm 2\sqrt{17}}{2} = -3 \pm \sqrt{17}$$

17. Solving the equation:
$$2x + 3 = -2x^2$$
$$2x^2 + 2x + 3 = 0$$
$$x = \frac{-2 \pm \sqrt{4-24}}{4} = \frac{-2 \pm \sqrt{-20}}{4} = \frac{-2 \pm 2i\sqrt{5}}{4} = \frac{-1 \pm i\sqrt{5}}{2}$$

19. Solving the equation:
$$100x^2 - 200x + 100 = 0$$
$$100(x^2 - 2x + 1) = 0$$
$$100(x-1)^2 = 0$$
$$x = 1$$

21. Solving the equation:
$$\frac{1}{2}r^2 = \frac{1}{6}r - \frac{2}{3}$$
$$3r^2 = r - 4$$
$$3r^2 - r + 4 = 0$$
$$r = \frac{1 \pm \sqrt{1-48}}{6} = \frac{1 \pm \sqrt{-47}}{6} = \frac{1 \pm i\sqrt{47}}{6}$$

23. Solving the equation:
$$(x-3)(x-5) = 1$$
$$x^2 - 8x + 15 = 1$$
$$x^2 - 8x + 14 = 0$$
$$x = \frac{8 \pm \sqrt{64-56}}{2} = \frac{8 \pm \sqrt{8}}{2} = \frac{8 \pm 2\sqrt{2}}{2} = 4 \pm \sqrt{2}$$

25. Solving the equation:
$$(x+3)^2 + (x-8)(x-1) = 16$$
$$x^2 + 6x + 9 + x^2 - 9x + 8 = 16$$
$$2x^2 - 3x + 1 = 0$$
$$(2x-1)(x-1) = 0$$
$$x = \frac{1}{2}, 1$$

27. Solving the equation:
$$\frac{x^2}{3} - \frac{5x}{6} = \frac{1}{2}$$
$$2x^2 - 5x = 3$$
$$2x^2 - 5x - 3 = 0$$
$$(2x+1)(x-3) = 0$$
$$x = -\frac{1}{2}, 3$$

29. Solving the equation:
$$\frac{1}{x+1} - \frac{1}{x} = \frac{1}{2}$$
$$2x(x+1)\left(\frac{1}{x+1} - \frac{1}{x}\right) = 2x(x+1) \cdot \frac{1}{2}$$
$$2x - (2x+2) = x^2 + x$$
$$2x - 2x - 2 = x^2 + x$$
$$x^2 + x + 2 = 0$$
$$x = \frac{-1 \pm \sqrt{1-8}}{2} = \frac{-1 \pm \sqrt{-7}}{2} = \frac{-1 \pm i\sqrt{7}}{2}$$

31. Solving the equation:
$$\frac{1}{y-1} + \frac{1}{y+1} = 1$$
$$(y+1)(y-1)\left(\frac{1}{y-1} + \frac{1}{y+1}\right) = (y+1)(y-1) \cdot 1$$
$$y + 1 + y - 1 = y^2 - 1$$
$$2y = y^2 - 1$$
$$y^2 - 2y - 1 = 0$$
$$y = \frac{2 \pm \sqrt{4+4}}{2} = \frac{2 \pm \sqrt{8}}{2} = \frac{2 \pm 2\sqrt{2}}{2} = 1 \pm \sqrt{2}$$

33. Solving the equation:
$$\frac{1}{x+2}+\frac{1}{x+3}=1$$
$$(x+2)(x+3)\left(\frac{1}{x+2}+\frac{1}{x+3}\right)=(x+2)(x+3)\cdot 1$$
$$x+3+x+2=x^2+5x+6$$
$$2x+5=x^2+5x+6$$
$$x^2+3x+1=0$$
$$x=\frac{-3\pm\sqrt{9-4}}{2}=\frac{-3\pm\sqrt{5}}{2}$$

35. Solving the equation:
$$\frac{6}{r^2-1}-\frac{1}{2}=\frac{1}{r+1}$$
$$2(r+1)(r-1)\left(\frac{6}{(r+1)(r-1)}-\frac{1}{2}\right)=2(r+1)(r-1)\cdot\frac{1}{r+1}$$
$$12-(r^2-1)=2r-2$$
$$12-r^2+1=2r-2$$
$$r^2+2r-15=0$$
$$(r+5)(r-3)=0$$
$$r=-5,3$$

37. Solving the equation:
$$x^3-8=0$$
$$(x-2)(x^2+2x+4)=0$$
$$x=2 \quad \text{or} \quad x=\frac{-2\pm\sqrt{4-16}}{2}=\frac{-2\pm\sqrt{-12}}{2}=\frac{-2\pm 2i\sqrt{3}}{2}=-1\pm i\sqrt{3}$$
$$x=2,-1\pm i\sqrt{3}$$

39. Solving the equation:
$$8a^3+27=0$$
$$(2a+3)(4a^2-6a+9)=0$$
$$a=-\frac{3}{2} \quad \text{or} \quad a=\frac{6\pm\sqrt{36-144}}{8}=\frac{6\pm\sqrt{-108}}{8}=\frac{6\pm 6i\sqrt{3}}{8}=\frac{3\pm 3i\sqrt{3}}{4}$$
$$a=-\frac{3}{2},\frac{3\pm 3i\sqrt{3}}{4}$$

41. Solving the equation:
$$125t^3-1=0$$
$$(5t-1)(25t^2+5t+1)=0$$
$$t=\frac{1}{5} \quad \text{or} \quad t=\frac{-5\pm\sqrt{25-100}}{50}=\frac{-5\pm\sqrt{-75}}{50}=\frac{-5\pm 5i\sqrt{3}}{50}=\frac{-1\pm i\sqrt{3}}{10}$$
$$t=\frac{1}{5},\frac{-1\pm i\sqrt{3}}{10}$$

43. Solving the equation:
$$2x^3 + 2x^2 + 3x = 0$$
$$x(2x^2 + 2x + 3) = 0$$
$$x = 0 \quad \text{or} \quad x = \frac{-2 \pm \sqrt{4-24}}{4} = \frac{-2 \pm \sqrt{-20}}{4} = \frac{-2 \pm 2i\sqrt{5}}{4} = \frac{-1 \pm i\sqrt{5}}{2}$$
$$x = 0, \frac{-1 \pm i\sqrt{5}}{2}$$

45. Solving the equation:
$$3y^4 = 6y^3 - 6y^2$$
$$3y^4 - 6y^3 + 6y^2 = 0$$
$$3y^2(y^2 - 2y + 2) = 0$$
$$y = 0 \quad \text{or} \quad y = \frac{2 \pm \sqrt{4-8}}{2} = \frac{2 \pm \sqrt{-4}}{2} = \frac{2 \pm 2i}{2} = 1 \pm i$$
$$y = 0, 1 \pm i$$

47. Solving the equation:
$$6t^5 + 4t^4 = -2t^3$$
$$6t^5 + 4t^4 + 2t^3 = 0$$
$$2t^3(3t^2 + 2t + 1) = 0$$
$$t = 0 \quad \text{or} \quad t = \frac{-2 \pm \sqrt{4-12}}{6} = \frac{-2 \pm \sqrt{-8}}{6} = \frac{-2 \pm 2i\sqrt{2}}{6} = \frac{-1 \pm i\sqrt{2}}{3}$$
$$t = 0, \frac{-1 \pm i\sqrt{2}}{3}$$

49. The expressions from **a** and **b** are equivalent, since: $\frac{6 + 2\sqrt{3}}{4} = \frac{2(3 + \sqrt{3})}{4} = \frac{3 + \sqrt{3}}{2}$

51. **a.** Solving by factoring:
$$3x^2 - 5x = 0$$
$$x(3x - 5) = 0$$
$$x = 0, \frac{5}{3}$$

b. Using the quadratic formula: $x = \frac{5 \pm \sqrt{(-5)^2 - 4(3)(0)}}{2(3)} = \frac{5 \pm \sqrt{25-0}}{6} = \frac{5 \pm 5}{6} = 0, \frac{5}{3}$

53. No, it cannot be solved by factoring. Using the quadratic formula:
$$x = \frac{4 \pm \sqrt{(-4)^2 - 4(1)(7)}}{2(1)} = \frac{4 \pm \sqrt{16-28}}{2} = \frac{4 \pm \sqrt{-12}}{2} = \frac{4 \pm 2i\sqrt{3}}{2} = 2 \pm i\sqrt{3}$$

55. Substituting: $x^2 + 2x = (-1+i)^2 + 2(-1+i) = 1 - 2i + i^2 - 2 + 2i = 1 - 2i - 1 - 2 + 2i = -2$
Yes, $x = -1 + i$ is a solution to the equation.

57. Substituting $s = 74$:
$$5t + 16t^2 = 74$$
$$16t^2 + 5t - 74 = 0$$
$$(t-2)(16t + 37) = 0$$
$$t = 2 \quad \left(t = -\frac{37}{16} \text{ is impossible}\right)$$
It will take 2 seconds for the object to fall 74 feet.

59. Since profit is revenue minus the cost, the equation is:
$$100x - 0.5x^2 - (60x + 300) = 300$$
$$100x - 0.5x^2 - 60x - 300 = 300$$
$$-0.5x^2 + 40x - 600 = 0$$
$$x^2 - 80x + 1{,}200 = 0$$
$$(x - 20)(x - 60) = 0$$
$$x = 20, 60$$
The weekly profit is $300 if 20 items or 60 items are sold.

61. Evaluating $b^2 - 4ac$: $b^2 - 4ac = (-3)^2 - 4(1)(-40) = 9 + 160 = 169$

63. Evaluating $b^2 - 4ac$: $b^2 - 4ac = 12^2 - 4(4)(9) = 144 - 144 = 0$

65. Solving the equation:
$$k^2 - 144 = 0$$
$$(k + 12)(k - 12) = 0$$
$$k = -12, 12$$

67. Multiplying: $(x - 3)(x + 2) = x^2 + 2x - 3x - 6 = x^2 - x - 6$

69. Multiplying:
$$(x - 3)(x - 3)(x + 2) = (x^2 - 6x + 9)(x + 2)$$
$$= x^3 - 6x^2 + 9x + 2x^2 - 12x + 18$$
$$= x^3 - 4x^2 - 3x + 18$$

9.3 Additional Items Involving Solutions to Equations

1. Computing the discriminant: $D = (-6)^2 - 4(1)(5) = 36 - 20 = 16$. The equation will have two rational solutions.

3. First write the equation as $4x^2 - 4x + 1 = 0$. Computing the discriminant: $D = (-4)^2 - 4(4)(1) = 16 - 16 = 0$
The equation will have one rational solution.

5. Computing the discriminant: $D = 1^2 - 4(1)(-1) = 1 + 4 = 5$. The equation will have two irrational solutions.

7. First write the equation as $2y^2 - 3y - 1 = 0$. Computing the discriminant: $D = (-3)^2 - 4(2)(-1) = 9 + 8 = 17$
The equation will have two irrational solutions.

9. Computing the discriminant: $D = 0^2 - 4(1)(-9) = 36$. The equation will have two rational solutions.

11. First write the equation as $5a^2 - 4a - 5 = 0$. Computing the discriminant: $D = (-4)^2 - 4(5)(-5) = 16 + 100 = 116$
The equation will have two irrational solutions.

13. Setting the discriminant equal to 0:
$$(-k)^2 - 4(1)(25) = 0$$
$$k^2 - 100 = 0$$
$$k^2 = 100$$
$$k = \pm 10$$

15. First write the equation as $x^2 - kx + 36 = 0$. Setting the discriminant equal to 0:
$$(-k)^2 - 4(1)(36) = 0$$
$$k^2 - 144 = 0$$
$$k^2 = 144$$
$$k = \pm 12$$

17. Setting the discriminant equal to 0:
$$(-12)^2 - 4(4)(k) = 0$$
$$144 - 16k = 0$$
$$16k = 144$$
$$k = 9$$

19. First write the equation as $kx^2 - 40x - 25 = 0$. Setting the discriminant equal to 0:
$$(-40)^2 - 4(k)(-25) = 0$$
$$1600 + 100k = 0$$
$$100k = -1600$$
$$k = -16$$

21. Setting the discriminant equal to 0:
$$(-k)^2 - 4(3)(2) = 0$$
$$k^2 - 24 = 0$$
$$k^2 = 24$$
$$k = \pm\sqrt{24} = \pm 2\sqrt{6}$$

23. Writing the equation:
$$(x-5)(x-2) = 0$$
$$x^2 - 7x + 10 = 0$$

25. Writing the equation:
$$(t+3)(t-6) = 0$$
$$t^2 - 3t - 18 = 0$$

27. Writing the equation:
$$(y-2)(y+2)(y-4) = 0$$
$$(y^2 - 4)(y-4) = 0$$
$$y^3 - 4y^2 - 4y + 16 = 0$$

29. Writing the equation:
$$(2x-1)(x-3) = 0$$
$$2x^2 - 7x + 3 = 0$$

31. Writing the equation:
$$(4t+3)(t-3) = 0$$
$$4t^2 - 9t - 9 = 0$$

33. Writing the equation:
$$(x-3)(x+3)(6x-5) = 0$$
$$(x^2 - 9)(6x-5) = 0$$
$$6x^3 - 5x^2 - 54x + 45 = 0$$

35. Writing the equation:
$$(2a+1)(5a-3) = 0$$
$$10a^2 - a - 3 = 0$$

37. Writing the equation:
$$(3x+2)(3x-2)(x-1) = 0$$
$$(9x^2 - 4)(x-1) = 0$$
$$9x^3 - 9x^2 - 4x + 4 = 0$$

39. Writing the equation:
$$(x-2)(x+2)(x-3)(x+3) = 0$$
$$(x^2 - 4)(x^2 - 9) = 0$$
$$x^4 - 13x^2 + 36 = 0$$

41. Starting with the solutions:
$$x = \pm\sqrt{7}$$
$$x^2 = (\pm\sqrt{7})^2$$
$$x^2 = 7$$
$$x^2 - 7 = 0$$

43. Starting with the solutions:
$$x = \pm 5i$$
$$x^2 = (\pm 5i)^2$$
$$x^2 = -25$$
$$x^2 + 25 = 0$$

45. Starting with the solutions:
$$x = 1 \pm i$$
$$x - 1 = \pm i$$
$$(x-1)^2 = (\pm i)^2$$
$$x^2 - 2x + 1 = -1$$
$$x^2 - 2x + 2 = 0$$

47. Starting with the solutions:
$$x = -2 \pm 3i$$
$$x + 2 = \pm 3i$$
$$(x+2)^2 = (\pm 3i)^2$$
$$x^2 + 4x + 4 = -9$$
$$x^2 + 4x + 13 = 0$$

49. Starting with the solutions:
$$(x-3)(x+5)^2 = 0$$
$$(x-3)(x^2 + 10x + 25) = 0$$
$$x^3 + 10x^2 + 25x - 3x^2 - 30x - 75 = 0$$
$$x^3 + 7x^2 - 5x - 75 = 0$$

51. Starting with the solutions:
$$(x-3)^2 (x+3)^2 = 0$$
$$(x^2 - 6x + 9)(x^2 + 6x + 9) = 0$$
$$x^4 + 6x^3 + 9x^2 - 6x^3 - 36x^2 - 54x + 9x^2 + 54x + 81 = 0$$
$$x^4 - 18x^2 + 81 = 0$$

53. First use long division to find the other factor:

$$\begin{array}{r} x^2 + 3x + 2 \\ x+3 \overline{\smash{\big)}\, x^3 + 6x^2 + 11x + 6} \\ \underline{x^3 + 3x^2} \\ 3x^2 + 11x \\ \underline{3x^2 + 9x} \\ 2x + 6 \\ \underline{2x + 6} \\ 0 \end{array}$$

So $x^3 + 6x^2 + 11x + 6 = (x+3)(x^2 + 3x + 2) = (x+3)(x+2)(x+1)$. Therefore the solutions are $-3, -2,$ and -1.

55. First use long division to find the other factor:

$$\begin{array}{r} y^2 + 2y - 8 \\ y+3 \overline{\smash{\big)}\, y^3 + 5y^2 - 2y - 24} \\ \underline{y^3 + 3y^2} \\ 2y^2 - 2y \\ \underline{2y^2 + 6y} \\ -8y - 24 \\ \underline{-8y - 24} \\ 0 \end{array}$$

So $y^3 + 5y^2 - 2y - 24 = (y+3)(y^2 + 2y - 8) = (y+3)(y+4)(y-2)$. Therefore the solutions are $-4, -3,$ and 2.

57. First write the equation as $x^3 - 5x^2 + 8x - 6 = 0$. Using long division to find the other factor:

$$\begin{array}{r}
x^2 - 2x + 2 \\
x-3 \overline{\smash{\big)} x^3 - 5x^2 + 8x - 6} \\
\underline{x^3 - 3x^2} \\
-2x^2 + 8x \\
\underline{-2x^2 + 6x} \\
2x - 6 \\
\underline{2x - 6} \\
0
\end{array}$$

So $x^3 - 5x^2 + 8x - 6 = (x-3)(x^2 - 2x - 2)$. One solution is 3, and the others can be found using the quadratic

formula: $x = \dfrac{2 \pm \sqrt{(-2)^2 - 4(1)(2)}}{2(1)} = \dfrac{2 \pm \sqrt{4-8}}{2} = \dfrac{2 \pm \sqrt{-4}}{2} = \dfrac{2 \pm 2i}{2} = 1 \pm i$

59. First write the equation as $t^3 - 13t^2 + 65t - 125 = 0$. Using long division to find the other factor:

$$\begin{array}{r}
t^2 - 8t + 25 \\
t-5 \overline{\smash{\big)} t^3 - 13t^2 + 65t - 125} \\
\underline{t^3 - 5t^2} \\
-8t^2 + 65t \\
\underline{-8t^2 + 40t} \\
25t - 125 \\
\underline{25t - 125} \\
0
\end{array}$$

So $t^3 - 13t^2 + 65t - 125 = (t-5)(t^2 - 8t + 25)$. One solution is 5, and the others can be found using the quadratic

formula: $t = \dfrac{8 \pm \sqrt{(-8)^2 - 4(1)(25)}}{2(1)} = \dfrac{8 \pm \sqrt{64-100}}{2} = \dfrac{8 \pm \sqrt{-36}}{2} = \dfrac{8 \pm 6i}{2} = 4 \pm 3i$

61. Simplifying: $(x+3)^2 - 2(x+3) - 8 = x^2 + 6x + 9 - 2x - 6 - 8 = x^2 + 4x - 5$

63. Simplifying: $(2a-3)^2 - 9(2a-3) + 20 = 4a^2 - 12a + 9 - 18a + 27 + 20 = 4a^2 - 30a + 56$

65. Simplifying:
$$2(4a+2)^2 - 3(4a+2) - 20 = 2(16a^2 + 16a + 4) - 3(4a+2) - 20$$
$$= 32a^2 + 32a + 8 - 12a - 6 - 20$$
$$= 32a^2 + 20a - 18$$

67. Solving the equation:
$$x^2 = \frac{1}{4}$$
$$x = \pm\sqrt{\frac{1}{4}} = \pm\frac{1}{2}$$

69. Since $\sqrt{x} \geq 0$, this equation has no solution.

71. Solving the equation:
$$x + 3 = 4$$
$$x = 1$$

73. Solving the equation:
$$y^2 - 2y - 8 = 0$$
$$(y+2)(y-4) = 0$$
$$y = -2, 4$$

75. Solving the equation:
$$4y^2 + 7y - 2 = 0$$
$$(4y-1)(y+2) = 0$$
$$y = -2, \frac{1}{4}$$

9.4 More Equations

1. Solving the equation:
$$(x-3)^2 + 3(x-3) + 2 = 0$$
$$(x-3+2)(x-3+1) = 0$$
$$(x-1)(x-2) = 0$$
$$x = 1, 2$$

3. Solving the equation:
$$2(x+4)^2 + 5(x+4) - 12 = 0$$
$$[2(x+4) - 3][(x+4) + 4] = 0$$
$$(2x+5)(x+8) = 0$$
$$x = -8, -\frac{5}{2}$$

5. Solving the equation:
$$x^4 - 6x^2 - 27 = 0$$
$$(x^2 - 9)(x^2 + 3) = 0$$
$$x^2 = 9, -3$$
$$x = \pm 3, \pm i\sqrt{3}$$

7. Solving the equation:
$$x^4 + 9x^2 = -20$$
$$x^4 + 9x^2 + 20 = 0$$
$$(x^2 + 4)(x^2 + 5) = 0$$
$$x^2 = -4, -5$$
$$x = \pm 2i, \pm i\sqrt{5}$$

9. Solving the equation:
$$(2a-3)^2 - 9(2a-3) = -20$$
$$(2a-3)^2 - 9(2a-3) + 20 = 0$$
$$(2a-3-4)(2a-3-5) = 0$$
$$(2a-7)(2a-8) = 0$$
$$a = \frac{7}{2}, 4$$

11. Solving the equation:
$$2(4a+2)^2 = 3(4a+2) + 20$$
$$2(4a+2)^2 - 3(4a+2) - 20 = 0$$
$$[2(4a+2) + 5][(4a+2) - 4] = 0$$
$$(8a+9)(4a-2) = 0$$
$$a = -\frac{9}{8}, \frac{1}{2}$$

13. Solving the equation:
$$6t^4 = -t^2 + 5$$
$$6t^4 + t^2 - 5 = 0$$
$$(6t^2 - 5)(t^2 + 1) = 0$$
$$t^2 = \frac{5}{6}, -1$$
$$t = \pm\sqrt{\frac{5}{6}} = \pm\frac{\sqrt{30}}{6}, \pm i$$

15. Solving the equation:
$$9x^4 - 49 = 0$$
$$(3x^2 - 7)(3x^2 + 7) = 0$$
$$x^2 = \frac{7}{3}, -\frac{7}{3}$$
$$x = \pm\sqrt{\frac{7}{3}}, \pm\sqrt{-\frac{7}{3}}$$
$$t = \pm\frac{\sqrt{21}}{3}, \pm\frac{i\sqrt{21}}{3}$$

17. Solving the equation:
$$x - 7\sqrt{x} + 10 = 0$$
$$(\sqrt{x} - 5)(\sqrt{x} - 2) = 0$$
$$\sqrt{x} = 2, 5$$
$$x = 4, 25$$
Both values check in the original equation.

19. Solving the equation:
$$t - 2\sqrt{t} - 15 = 0$$
$$(\sqrt{t} - 5)(\sqrt{t} + 3) = 0$$
$$\sqrt{t} = -3, 5$$
$$t = 9, 25$$
Only $t = 25$ checks in the original equation.

21. Solving the equation:
$$6x + 11\sqrt{x} = 35$$
$$6x + 11\sqrt{x} - 35 = 0$$
$$(3\sqrt{x} - 5)(2\sqrt{x} + 7) = 0$$
$$\sqrt{x} = \frac{5}{3}, -\frac{7}{2}$$
$$x = \frac{25}{9}, \frac{49}{4}$$

Only $x = \frac{25}{9}$ checks in the original equation.

23. Solving the equation:
$$(a - 2) - 11\sqrt{a - 2} + 30 = 0$$
$$(\sqrt{a - 2} - 6)(\sqrt{a - 2} - 5) = 0$$
$$\sqrt{a - 2} = 5, 6$$
$$a - 2 = 25, 36$$
$$a = 27, 38$$

25. Solving the equation:
$$(2x + 1) - 8\sqrt{2x + 1} + 15 = 0$$
$$(\sqrt{2x + 1} - 3)(\sqrt{2x + 1} - 5) = 0$$
$$\sqrt{2x + 1} = 3, 5$$
$$2x + 1 = 9, 25$$
$$2x = 8, 24$$
$$x = 4, 12$$

27. Solving for t:
$$16t^2 - vt - h = 0$$
$$t = \frac{v \pm \sqrt{v^2 - 4(16)(-h)}}{32} = \frac{v \pm \sqrt{v^2 + 64h}}{32}$$

29. Solving for x:
$$kx^2 + 8x + 4 = 0$$
$$x = \frac{-8 \pm \sqrt{64 - 16k}}{2k} = \frac{-8 \pm 4\sqrt{4 - k}}{2k} = \frac{-4 \pm 2\sqrt{4 - k}}{k}$$

31. Solving for x:
$$x^2 + 2xy + y^2 = 0$$
$$x = \frac{-2y \pm \sqrt{4y^2 - 4y^2}}{2} = \frac{-2y}{2} = -y$$

33. Solving for t (note that $t > 0$):
$$16t^2 - 8t - h = 0$$
$$t = \frac{8 + \sqrt{64 + 64h}}{32} = \frac{8 + 8\sqrt{1 + h}}{32} = \frac{1 + \sqrt{1 + h}}{4}$$

35. a. Sketching the graph:

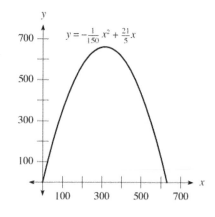

b. Finding the x-intercepts:
$$-\frac{1}{150}x^2 + \frac{21}{5}x = 0$$
$$-\frac{1}{150}x(x-630) = 0$$
$$x = 0, 630$$

The width is 630 feet.

37. a. The equation is $l + 2w = 160$.

b. The area is: $A = w \cdot l = w(160 - 2w) = -2w^2 + 160w$

c. Completing the table:

w	0	10	20	30	40	50	60	70	80
A	0	1,400	2,400	3,000	3,200	3,000	2,400	1,400	0

d. The maximum area is 3,200 square yards.

39. Evaluating when $x = 1$: $y = 3(1)^2 - 6(1) + 1 = 3 - 6 + 1 = -2$

41. Evaluating: $P(135) = -0.1(135)^2 + 27(135) - 500 = -1,822.5 + 3,645 - 500 = 1,322.5$

43. Solving the equation:
$$0 = a(80)^2 + 70$$
$$0 = 6400a + 70$$
$$6400a = -70$$
$$a = -\frac{7}{640}$$

45. Solving the equation:
$$x^2 - 6x + 5 = 0$$
$$(x-1)(x-5) = 0$$
$$x = 1, 5$$

47. Solving the equation:
$$-x^2 - 2x + 3 = 0$$
$$x^2 + 2x - 3 = 0$$
$$(x+3)(x-1) = 0$$
$$x = -3, 1$$

49. Solving the equation:
$$2x^2 - 6x + 5 = 0$$
$$x = \frac{6 \pm \sqrt{(-6)^2 - 4(2)(5)}}{2(2)} = \frac{6 \pm \sqrt{36 - 40}}{4} = \frac{6 \pm \sqrt{-4}}{4} = \frac{6 \pm 2i}{4} = \frac{3 \pm i}{2} = \frac{3}{2} \pm \frac{1}{2}i$$

51. Completing the square: $x^2 - 6x + 9 = (x-3)^2$

53. Completing the square: $y^2 + 2y + 1 = (y+1)^2$

9.5 Graphing Parabolas

1. First complete the square: $y = x^2 + 2x - 3 = (x^2 + 2x + 1) - 1 - 3 = (x+1)^2 - 4$

   The x-intercepts are $-3, 1$ and the vertex is $(-1, -4)$. Graphing the parabola:

   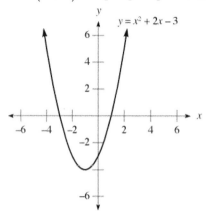

3. First complete the square: $y = -x^2 - 4x + 5 = -(x^2 + 4x + 4) + 4 + 5 = -(x+2)^2 + 9$

   The x-intercepts are $-5, 1$ and the vertex is $(-2, 9)$. Graphing the parabola:

   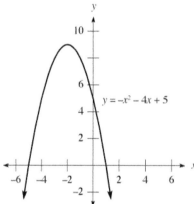

5. The x-intercepts are $-1, 1$ and the vertex is $(0, -1)$. Graphing the parabola:

   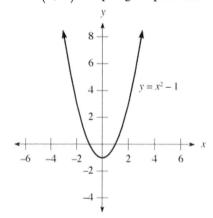

7. The x-intercepts are –3,3 and the vertex is $(0,9)$. Graphing the parabola:

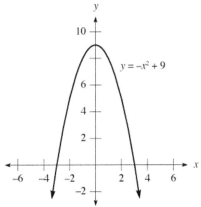

9. First complete the square: $y = 2x^2 - 4x - 6 = 2(x^2 - 2x + 1) - 2 - 6 = 2(x-1)^2 - 8$

   The x-intercepts are –1,3 and the vertex is $(1,-8)$. Graphing the parabola:

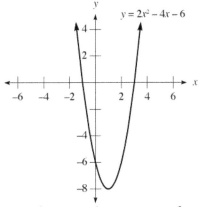

11. First complete the square: $y = x^2 - 2x - 4 = (x^2 - 2x + 1) - 1 - 4 = (x-1)^2 - 5$

    The x-intercepts are $1 \pm \sqrt{5}$ and the vertex is $(1,-5)$. Graphing the parabola:

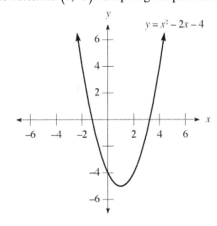

13. The vertex is $(1,3)$ and the y-intercept is 5. Graphing the parabola:

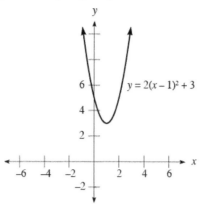

15. The vertex is $(-2,4)$ and the x-intercepts are –4 and 0. Graphing the parabola:

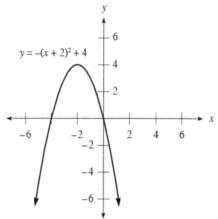

17. The vertex is $(2,-4)$ and the y-intercept is –2. Graphing the parabola:

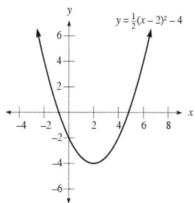

19. The vertex is $(4,-1)$ and the y-intercept is -33. Graphing the parabola:

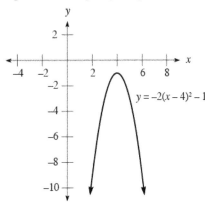

21. First complete the square: $y = x^2 - 4x - 4 = \left(x^2 - 4x + 4\right) - 4 - 4 = (x-2)^2 - 8$

    The vertex is $(2,-8)$. Graphing the parabola:

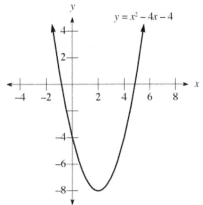

23. First complete the square: $y = -x^2 + 2x - 5 = -\left(x^2 - 2x + 1\right) + 1 - 5 = -(x-1)^2 - 4$

    The vertex is $(1,-4)$. Graphing the parabola:

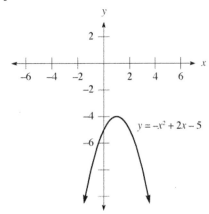

25. The vertex is $(0,1)$. Graphing the parabola:

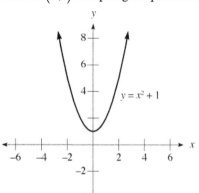

27. The vertex is $(0,-3)$. Graphing the parabola:

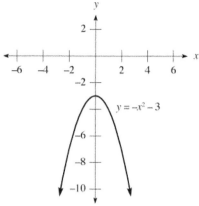

29. First complete the square: $g(x) = 3x^2 + 4x + 1 = 3\left(x^2 + \frac{4}{3}x + \frac{4}{9}\right) + 1 - \frac{4}{3} = 3\left(x + \frac{2}{3}\right)^2 - \frac{1}{3}$

    The vertex is $\left(-\frac{2}{3}, -\frac{1}{3}\right)$. Graphing the parabola:

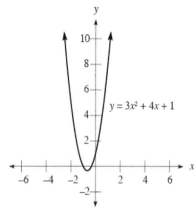

31. Completing the square: $y = x^2 - 6x + 5 = \left(x^2 - 6x + 9\right) - 9 + 5 = (x-3)^2 - 4$

    The vertex is $(3,-4)$, which is the lowest point on the graph.

33. Completing the square: $y = -x^2 + 2x + 8 = -\left(x^2 - 2x + 1\right) + 1 + 8 = -(x-1)^2 + 9$

    The vertex is $(1,9)$, which is the highest point on the graph.

35. Completing the square: $y = -x^2 + 4x + 12 = -\left(x^2 - 4x + 4\right) + 4 + 12 = -(x-2)^2 + 16$

    The vertex is $(2,16)$, which is the highest point on the graph.

37. Completing the square: $y = -x^2 - 8x = -\left(x^2 + 8x + 16\right) + 16 = -(x+4)^2 + 16$

    The vertex is $(-4,16)$, which is the highest point on the graph.

39. First complete the square:
    $P(x) = -0.002x^2 + 3.5x - 800 = -0.002\left(x^2 - 1750x + 765{,}625\right) + 1{,}531.25 - 800 = -0.002(x-875)^2 + 731.25$

    It must sell 875 patterns to obtain a maximum profit of \$731.25.

41. The ball is in her hand at times 0 sec and 2 sec.

    Completing the square: $h(t) = -16t^2 + 32t = -16\left(t^2 - 2t + 1\right) + 16 = -16(t-1)^2 + 16$

    The maximum height of the ball is 16 feet.

43. Completing the square: $R = xp = 1{,}200p - 100p^2 = -100\left(p^2 - 12p + 36\right) + 3{,}600 = -100(p-6)^2 + 3{,}600$

The price is $6.00 and the maximum revenue is $3,600. Sketching the graph:

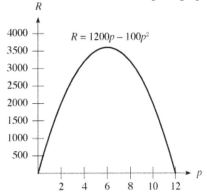

45. Completing the square: $R = xp = 1{,}700p - 100p^2 = -100\left(p^2 - 17p + 72.25\right) + 7{,}225 = -100(p-8.5)^2 + 7{,}225$

The price is $8.50 and the maximum revenue is $7,225. Sketching the graph:

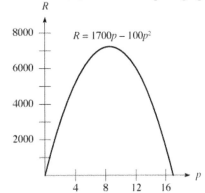

47. The equation is given on the graph:

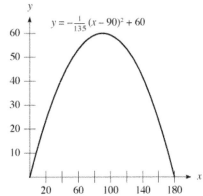

49. Solving the equation:

$x^2 - 2x - 8 = 0$
$(x-4)(x+2) = 0$
$x = -2, 4$

51. Solving the equation:

$6x^2 - x = 2$
$6x^2 - x - 2 = 0$
$(2x+1)(3x-2) = 0$
$x = -\dfrac{1}{2}, \dfrac{2}{3}$

53. Solving the equation:
$$x^2 - 6x + 9 = 0$$
$$(x-3)^2 = 0$$
$$x = 3$$

9.6 Quadratic Inequalities

1. Factoring the inequality:
$$x^2 + x - 6 > 0$$
$$(x+3)(x-2) > 0$$
Forming the sign chart:

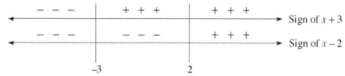

The solution set is $x < -3$ or $x > 2$. Graphing the solution set:

3. Factoring the inequality:
$$x^2 - x - 12 \le 0$$
$$(x+3)(x-4) \le 0$$
Forming the sign chart:

The solution set is $-3 \le x \le 4$. Graphing the solution set:

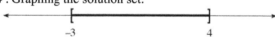

5. Factoring the inequality:
$$x^2 + 5x \ge -6$$
$$x^2 + 5x + 6 \ge 0$$
$$(x+2)(x+3) \ge 0$$
Forming the sign chart:

The solution set is $x \le -3$ or $x \ge -2$. Graphing the solution set:

7. Factoring the inequality:
$$6x^2 < 5x - 1$$
$$6x^2 - 5x + 1 < 0$$
$$(3x-1)(2x-1) < 0$$
Forming the sign chart:

The solution set is $\frac{1}{3} < x < \frac{1}{2}$. Graphing the solution set:

9. Factoring the inequality:
$$x^2 - 9 < 0$$
$$(x+3)(x-3) < 0$$
Forming the sign chart:

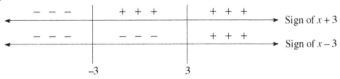

The solution set is $-3 < x < 3$. Graphing the solution set:

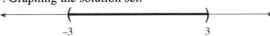

11. Factoring the inequality:
$$4x^2 - 9 \geq 0$$
$$(2x+3)(2x-3) \geq 0$$
Forming the sign chart:

The solution set is $x \leq -\frac{3}{2}$ or $x \geq \frac{3}{2}$. Graphing the solution set:

13. Factoring the inequality:
$$2x^2 - x - 3 < 0$$
$$(2x-3)(x+1) < 0$$
Forming the sign chart:

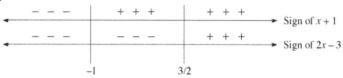

The solution set is $-1 < x < \dfrac{3}{2}$. Graphing the solution set:

15. Factoring the inequality:
$$x^2 - 4x + 4 \geq 0$$
$$(x-2)^2 \geq 0$$
Since this inequality is always true, the solution set is all real numbers. Graphing the solution set:

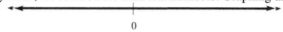

17. Factoring the inequality:
$$x^2 - 10x + 25 < 0$$
$$(x-5)^2 < 0$$
Since this inequality is never true, there is no solution.

19. Forming the sign chart:

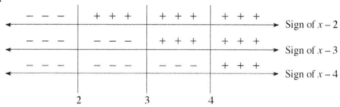

The solution set is $2 < x < 3$ or $x > 4$. Graphing the solution set:

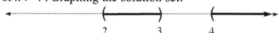

21. Forming the sign chart:

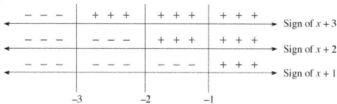

The solution set is $x \leq -3$ or $-2 \leq x \leq -1$. Graphing the solution set:

23. Forming the sign chart:

The solution set is $-4 < x \leq 1$. Graphing the solution set:

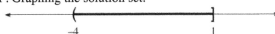

25. Write the inequality as $\dfrac{3x-8}{x+6} < 0$. Forming the sign chart:

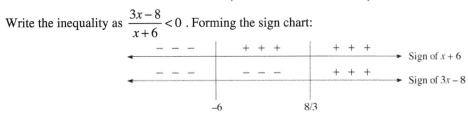

The solution set is $-6 < x < \dfrac{8}{3}$. Graphing the solution set:

27. Write the inequality as $\dfrac{4+x-6}{x-6} < 0$, or $\dfrac{x-2}{x-6} < 0$. Forming the sign chart:

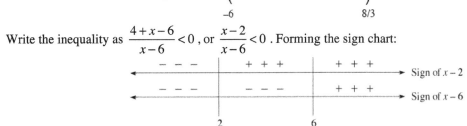

The solution set is $x < 2$ or $x > 6$. Graphing the solution set:

29. Forming the sign chart:

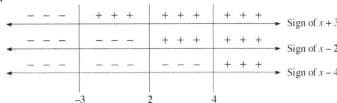

The solution set is $x < -3$ or $2 < x < 4$. Graphing the solution set:

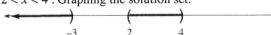

31. Simplify the inequality:

$$\frac{2}{x-4} - \frac{1}{x-3} < 0$$

$$\frac{2(x-3) - 1(x-4)}{(x-4)(x-3)} < 0$$

$$\frac{2x - 6 - x + 4}{(x-4)(x-3)} < 0$$

$$\frac{x-2}{(x-4)(x-3)} < 0$$

Forming the sign chart:

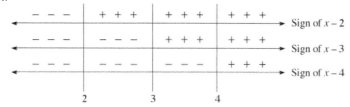

The solution set is $2 < x < 3$ or $x > 4$. Graphing the solution set:

33. Simplify the inequality:

$$\frac{x+7}{2x+12} + \frac{6}{x^2-36} \leq 0$$

$$\frac{x+7}{2(x+6)} + \frac{6}{(x+6)(x-6)} \leq 0$$

$$\frac{(x+7)(x-6) + 6 \cdot 2}{2(x+6)(x-6)} \leq 0$$

$$\frac{x^2 + x - 42 + 12}{2(x+6)(x-6)} \leq 0$$

$$\frac{x^2 + x - 30}{2(x+6)(x-6)} \leq 0$$

$$\frac{(x+6)(x-5)}{2(x+6)(x-6)} \leq 0$$

$$\frac{x-5}{2(x-6)} \leq 0$$

Forming the sign chart:

The solution set is $5 \leq x < 6$. Graphing the solution set:

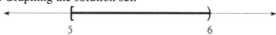

35. a. The solution set is $-2 < x < 2$. In interval notation, this is $(-2, 2)$.

b. The solution set is $x < -2$ or $x > 2$. In interval notation, this is $(-\infty, -2) \cup (2, \infty)$.

c. The solution set is $x = -2, 2$.

37. **a.** The solution set is $-2 < x < 5$. In interval notation, this is $(-2, 5)$.

b. The solution set is $x < -2$ or $x > 5$. In interval notation, this is $(-\infty, -2) \cup (5, \infty)$.

c. The solution set is $x = -2, 5$.

39. **a.** The solution set is $x < -1$ or $1 < x < 3$. In interval notation, this is $(-\infty, -1) \cup (1, 3)$.

b. The solution set is $-1 < x < 1$ or $x > 3$. In interval notation, this is $(-1, 1) \cup (3, \infty)$.

c. The solution set is $x = -1, 1, 3$.

41. Let w represent the width and $2w + 3$ represent the length. Using the area formula:
$$w(2w+3) \geq 44$$
$$2w^2 + 3w \geq 44$$
$$2w^2 + 3w - 44 \geq 0$$
$$(2w+11)(w-4) \geq 0$$

Forming the sign chart:

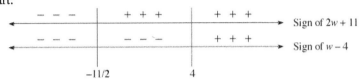

The width is at least 4 inches.

43. Solving the inequality:
$$1300p - 100p^2 \geq 4000$$
$$-100p^2 + 1300p - 4000 \geq 0$$
$$p^2 - 13p + 40 \leq 0$$
$$(p-8)(p-5) \leq 0$$

Forming the sign chart:

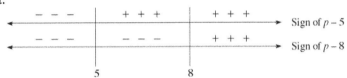

She should charge at least $5 but no more than $8 per radio.

45. Completing the square on the income:
$$y = (10{,}000 - 200x)(100 + 10x)$$
$$= 2{,}000(50-x)(10+x)$$
$$= 2{,}000\left(-x^2 + 40x + 500\right)$$
$$= -2{,}000\left(x^2 - 40x\right) + 1{,}000{,}000$$
$$= -2{,}000\left(x^2 - 40x + 400\right) + 800{,}000 + 1{,}000{,}000$$
$$= -2{,}000(x-20)^2 + 1{,}800{,}000$$

The union should have 20 increases of $10, so their new dues should be $100 + 20($10) = $300, and their income will be $1,800,000.

47. Let x represent the number of $2 increases in price. Completing the square on the income:
$$\begin{aligned} y &= (40-2x)(20+2x) \\ &= -4(x-20)(x+10) \\ &= -4(x^2 - 10x - 200) \\ &= -4(x^2 - 10x) + 800 \\ &= -4(x^2 - 10x + 25) + 100 + 800 \\ &= -4(x-5)^2 + 900 \end{aligned}$$

The business should have 5 increases of $2, so they should charge $20 + 5($2) = $30, and their income will be $900.

49. Using a calculator: $\dfrac{50{,}000}{32{,}000} = 1.5625$

51. Using a calculator: $\dfrac{1}{2}\left(\dfrac{4.5926}{1.3876} - 2\right) \approx 0.6549$

53. Solving the equation:
$$\begin{aligned} 2\sqrt{3t-1} &= 2 \\ \sqrt{3t-1} &= 1 \\ \left(\sqrt{3t-1}\right)^2 &= (1)^2 \\ 3t-1 &= 1 \\ 3t &= 2 \\ t &= \dfrac{2}{3} \end{aligned}$$
The solution is $\dfrac{2}{3}$.

55. Solving the equation:
$$\begin{aligned} \sqrt{x+3} &= x-3 \\ \left(\sqrt{x+3}\right)^2 &= (x-3)^2 \\ x+3 &= x^2 - 6x + 9 \\ 0 &= x^2 - 7x + 6 \\ 0 &= (x-6)(x-1) \\ x &= 1, 6 \quad (x=1 \text{ does not check}) \end{aligned}$$
The solution is 6.

57. Graphing the equation:

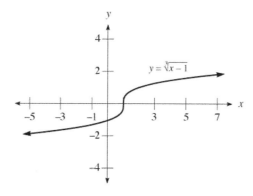

Chapter 9 Test

1. Solving the equation:
$$(2x+4)^2 = 25$$
$$2x+4 = \pm 5$$
$$2x+4 = -5, 5$$
$$2x = -9, 1$$
$$x = -\frac{9}{2}, \frac{1}{2}$$

2. Solving the equation:
$$(2x-6)^2 = -8$$
$$2x-6 = \pm\sqrt{-8}$$
$$2x-6 = \pm 2i\sqrt{2}$$
$$2x = 6 \pm 2i\sqrt{2}$$
$$x = 3 \pm i\sqrt{2}$$

3. Solving the equation:
$$y^2 - 10y + 25 = -4$$
$$(y-5)^2 = -4$$
$$y-5 = \pm\sqrt{-4}$$
$$y-5 = \pm 2i$$
$$y = 5 \pm 2i$$

4. Solving the equation:
$$(y+1)(y-3) = -6$$
$$y^2 - 2y - 3 = -6$$
$$y^2 - 2y + 3 = 0$$
$$y = \frac{2 \pm \sqrt{4-12}}{2} = \frac{2 \pm \sqrt{-8}}{2} = \frac{2 \pm 2i\sqrt{2}}{2} = 1 \pm i\sqrt{2}$$

5. Solving the equation:
$$8t^3 - 125 = 0$$
$$(2t-5)(4t^2 + 10t + 25) = 0$$
$$t = \frac{5}{2}, \frac{-10 \pm \sqrt{100-400}}{8} = \frac{-10 \pm \sqrt{-300}}{8} = \frac{-10 \pm 10i\sqrt{3}}{8} = -\frac{5}{4} \pm \frac{5i\sqrt{3}}{4}$$

6. Solving the equation:
$$\frac{1}{a+2} - \frac{1}{3} = \frac{1}{a}$$
$$3a(a+2)\left(\frac{1}{a+2} - \frac{1}{3}\right) = 3a(a+2)\left(\frac{1}{a}\right)$$
$$3a - a(a+2) = 3(a+2)$$
$$3a - a^2 - 2a = 3a + 6$$
$$-a^2 - 2a - 6 = 0$$
$$a^2 + 2a + 6 = 0$$
$$a = \frac{-2 \pm \sqrt{4-24}}{2} = \frac{-2 \pm \sqrt{-20}}{2} = \frac{-2 \pm 2i\sqrt{5}}{2} = -1 \pm i\sqrt{5}$$

7. Solving for r:
$$64(1+r)^2 = A$$
$$(1+r)^2 = \frac{A}{64}$$
$$1+r = \pm\frac{\sqrt{A}}{8}$$
$$r = -1 \pm \frac{\sqrt{A}}{8}$$

8. Solving by completing the square:
$$x^2 - 4x = -2$$
$$x^2 - 4x + 4 = -2 + 4$$
$$(x-2)^2 = 2$$
$$x-2 = \pm\sqrt{2}$$
$$x = 2 \pm \sqrt{2}$$

9. Solving the equation:
$$32t - 16t^2 = 12$$
$$-16t^2 + 32t - 12 = 0$$
$$4t^2 - 8t + 3 = 0$$
$$(2t-1)(2t-3) = 0$$
$$t = \frac{1}{2}, \frac{3}{2}$$
The object will be 12 feet above the ground after $\frac{1}{2}$ sec or $\frac{3}{2}$ sec.

10. Finding when the profit is equal to $200:
$$(25x - 0.2x^2) - (2x + 100) = 200$$
$$25x - 0.2x^2 - 2x - 100 = 200$$
$$-0.2x^2 + 23x - 300 = 0$$
$$-0.2(x^2 - 115x + 1,500) = 0$$
$$-0.2(x-15)(x-100) = 0$$
$$x = 15, 100$$
The company must sell 15 or 100 coffee cups to make a profit of $200.

11. First write the equation as $kx^2 - 12x + 4 = 0$. Setting the discriminant equal to 0:
$$(-12)^2 - 4(k)(4) = 0$$
$$144 - 16k = 0$$
$$-16k = -144$$
$$k = 9$$

12. First write the equation as $2x^2 - 5x - 7 = 0$. Finding the discriminant: $D = (-5)^2 - 4(2)(-7) = 25 + 56 = 81$
Since the discriminant is a perfect square $(9^2 = 81)$, the equation has two rational solutions.

13. Finding the equation:
$$(x-5)(3x+2) = 0$$
$$3x^2 - 13x - 10 = 0$$

14. Finding the equation:
$$(x+2)(x-2)(x-7) = 0$$
$$(x^2 - 4)(x-7) = 0$$
$$x^3 - 7x^2 - 4x + 28 = 0$$

15. Solving the equation:
$$4x^4 - 7x^2 - 2 = 0$$
$$(x^2 - 2)(4x^2 + 1) = 0$$
$$x^2 = 2, -\frac{1}{4}$$
$$x = \pm\sqrt{2}, \pm\frac{1}{2}i$$

16. Solving the equation:
$$(2t+1)^2 - 5(2t+1) + 6 = 0$$
$$4t^2 + 4t + 1 - 10t - 5 + 6 = 0$$
$$4t^2 - 6t + 2 = 0$$
$$2t^2 - 3t + 1 = 0$$
$$(2t-1)(t-1) = 0$$
$$t = \frac{1}{2}, 1$$

17. Solving the equation:
$$2t - 7\sqrt{t} + 3 = 0$$
$$(2\sqrt{t} - 1)(\sqrt{t} - 3) = 0$$
$$\sqrt{t} = \frac{1}{2}, 3$$
$$t = \frac{1}{4}, 9$$
Both values check in the original equation.

18. Solving for t:
$$16t^2 - 14t - h = 0$$
$$t = \frac{14 + \sqrt{196 - 4(16)(-h)}}{32} = \frac{14 + \sqrt{196 + 64h}}{32} = \frac{14 + 2\sqrt{49 + 16h}}{32} = \frac{7 + \sqrt{49 + 16h}}{16}$$
Note that only the positive answer was given here since t represents time, thus $t > 0$.

19. Completing the square: $y = x^2 - 2x - 3 = (x^2 - 2x + 1) - 1 - 3 = (x-1)^2 - 4$

    The vertex is $(1, -4)$, the x-intercepts are -1 and 3, and the y-intercept is -3. Graphing the parabola:

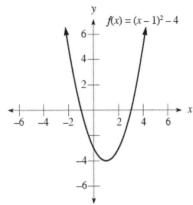

20. Completing the square: $y = -x^2 + 2x + 8 = -(x^2 - 2x + 1) + 1 + 8 = -(x-1)^2 + 9$

    The vertex is $(1, 9)$, the x-intercepts are -2 and 4, and the y-intercept is 8. Graphing the parabola:

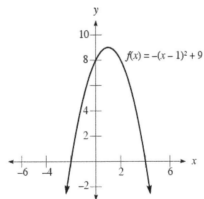

21. Factoring the inequality:
$$x^2 - x - 6 \leq 0$$
$$(x+2)(x-3) \leq 0$$
Forming a sign chart:

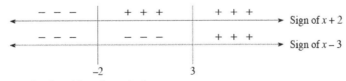

The solution set is $-2 \leq x \leq 3$. Graphing the solution set:

22. Factoring the inequality:
$$2x^2 + 5x > 3$$
$$2x^2 + 5x - 3 > 0$$
$$(2x-1)(x+3) > 0$$
Forming a sign chart:

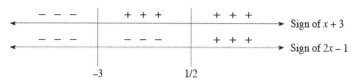

The solution set is $x < -3$ or $x > \dfrac{1}{2}$. Graphing the solution set:

23. Finding the profit function: $P(x) = R(x) - C(x) = (25x - 0.1x^2) - (5x + 100) = -0.1x^2 + 20x - 100$

Completing the square: $P(x) = -0.1x^2 + 20x - 100 = -0.1(x^2 - 200x + 10{,}000) + 1{,}000 - 100 = -0.1(x - 100)^2 + 900$

The maximum weekly profit is $900, obtained by selling 100 items per week.

Chapter 10
Exponential and Logarithmic Functions

10.1 Exponential Functions

1. Evaluating: $g(0) = \left(\dfrac{1}{2}\right)^0 = 1$

3. Evaluating: $g(-1) = \left(\dfrac{1}{2}\right)^{-1} = 2$

5. Evaluating: $f(-3) = 3^{-3} = \dfrac{1}{27}$

7. Evaluating: $f(2) + g(-2) = 3^2 + \left(\dfrac{1}{2}\right)^{-2} = 9 + 4 = 13$

9. Evaluating: $f(-1) + g(1) = 4^{-1} + \left(\dfrac{1}{3}\right)^1 = \dfrac{1}{4} + \dfrac{1}{3} = \dfrac{3}{12} + \dfrac{4}{12} = \dfrac{7}{12}$

11. Evaluating: $\dfrac{f(-2)}{g(1)} = \dfrac{4^{-2}}{\left(\dfrac{1}{3}\right)^1} = \dfrac{1}{16} \div \dfrac{1}{3} = \dfrac{3}{16}$

13. Graphing the function:

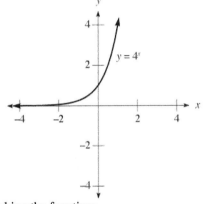

15. Graphing the function:

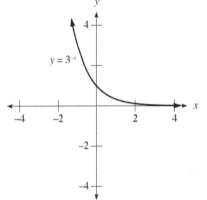

17. Graphing the function:

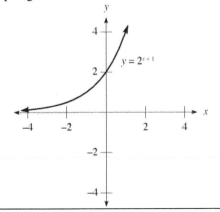

19. Graphing the function:

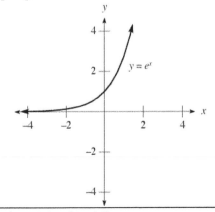

21. Graphing the function:

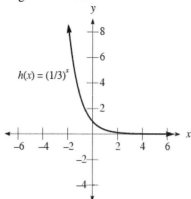

23. Graphing the function:

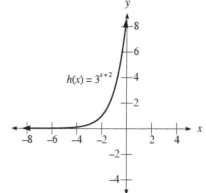

25. Graphing the functions:

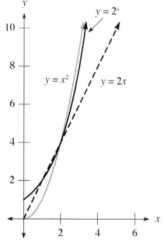

27. Graphing the functions:

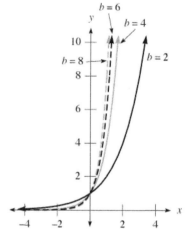

29. The equation is: $h(n) = 6\left(\dfrac{2}{3}\right)^n$. Substituting $n = 5$: $h(5) = 6\left(\dfrac{2}{3}\right)^5 \approx 0.79$ feet

31. From a graph, $Q(t) = 0.5$ when $t \approx 4.27$ days.

33.
    a. The equation is $A(t) = 1{,}200\left(1 + \dfrac{0.06}{4}\right)^{4t}$.

    b. Substitute $t = 8$: $A(8) = 1{,}200\left(1 + \dfrac{0.06}{4}\right)^{32} \approx \$1{,}932.39$

    c. Substitute $t = 8$ into the compound interest formula: $A(8) = 1{,}200 e^{0.06 \times 8} \approx \$1{,}939.29$

35. a. Substitute $t = 3.5$: $V(5) = 450{,}000(1 - 0.30)^5 \approx \$129{,}138.48$
    b. The domain is $\{t \mid 0 \le t \le 6\}$.
    c. Sketching the graph:

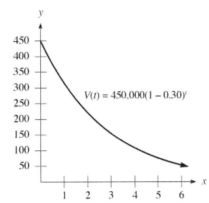

    d. The range is $\{V(t) \mid 52{,}942.05 \le V(t) \le 450{,}000\}$.
    e. From the graph, the crane will be worth $85,000 after approximately 4.7 years, or 4 years 8 months.

37. Finding the function values:
    $f(1) = 50 \cdot 4^1 = 200$ bacteria
    $f(2) = 50 \cdot 4^2 = 800$ bacteria
    $f(3) = 50 \cdot 4^3 = 3{,}200$ bacteria

39. Graphing the function:

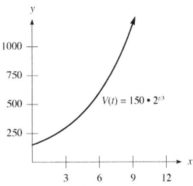

41. a. Evaluating when $t = 25$: $C(25) = 0.10 e^{0.0576(25)} \approx \$0.42$
    b. Evaluating when $t = 40$: $C(40) = 0.10 e^{0.0576(40)} \approx \$1.00$
    c. Evaluating when $t = 50$: $C(50) = 0.10 e^{0.0576(50)} \approx \$1.78$
    d. Evaluating when $t = 90$: $C(90) = 0.10 e^{0.0576(90)} \approx \$17.84$

43. Using the model: $B(3) = 0.798 \cdot 1.164^3 \approx 1{,}258{,}525$ bankruptcies
    The predicted model is 58,474 less bankruptcies less than the actual amount.

45. a. Evaluating when $t = 5$: $A(5) = 5{,}000{,}000 e^{-0.598(5)} \approx 231{,}437$ cells
    b. Evaluating when $t = 10$: $A(10) = 5{,}000{,}000 e^{-0.598(10)} \approx 12{,}644$ cells
    c. Evaluating when $t = 20$: $A(10) = 5{,}000{,}000 e^{-0.598(20)} \approx 32$ cells

47. Solving for y:
$$x = 2y - 3$$
$$2y = x + 3$$
$$y = \frac{x+3}{2}$$

49. Solving for y:
$$x = y^2 - 3$$
$$y^2 = x + 3$$
$$y = \pm\sqrt{x+3}$$

51. Solving for y:
$$x = \frac{y-4}{y-2}$$
$$x(y-2) = y - 4$$
$$xy - 2x = y - 4$$
$$xy - y = 2x - 4$$
$$y(x-1) = 2x - 4$$
$$y = \frac{2x-4}{x-1}$$

53. Solving for y:
$$x = \sqrt{y-3}$$
$$x^2 = y - 3$$
$$y = x^2 + 3$$

10.2 The Inverse of a Function

1. Let $y = f(x)$. Switch x and y and solve for y:
$$3y - 1 = x$$
$$3y = x + 1$$
$$y = \frac{x+1}{3}$$
The inverse is $f^{-1}(x) = \frac{x+1}{3}$.

3. Let $y = f(x)$. Switch x and y and solve for y:
$$y^3 = x$$
$$y = \sqrt[3]{x}$$
The inverse is $f^{-1}(x) = \sqrt[3]{x}$.

5. Let $y = f(x)$. Switch x and y and solve for y:
$$\frac{y-3}{y-1} = x$$
$$y - 3 = xy - x$$
$$y - xy = 3 - x$$
$$y(1-x) = 3 - x$$
$$y = \frac{3-x}{1-x} = \frac{x-3}{x-1}$$
The inverse is $f^{-1}(x) = \frac{x-3}{x-1}$.

7. Let $y = f(x)$. Switch x and y and solve for y:
$$\frac{y-3}{4} = x$$
$$y - 3 = 4x$$
$$y = 4x + 3$$
The inverse is $f^{-1}(x) = 4x + 3$.

9. Let $y = f(x)$. Switch x and y and solve for y:
$$\frac{1}{2}y - 3 = x$$
$$y - 6 = 2x$$
$$y = 2x + 6$$
The inverse is $f^{-1}(x) = 2x + 6$.

11. Let $y = f(x)$. Switch x and y and solve for y:
$$\frac{2}{3}y - 3 = x$$
$$2y - 9 = 3x$$
$$2y = 3x + 9$$
$$y = \frac{3}{2}x + \frac{9}{2}$$
The inverse is $f^{-1}(x) = \frac{3}{2}x + \frac{9}{2}$.

13. Let $y = f(x)$. Switch x and y and solve for y:

$$y^3 - 4 = x$$
$$y^3 = x + 4$$
$$y = \sqrt[3]{x+4}$$

The inverse is $f^{-1}(x) = \sqrt[3]{x+4}$.

15. Let $y = f(x)$. Switch x and y and solve for y:

$$\frac{4y-3}{2y+1} = x$$
$$4y - 3 = 2xy + x$$
$$4y - 2xy = x + 3$$
$$y(4 - 2x) = x + 3$$
$$y = \frac{x+3}{4-2x}$$

The inverse is $f^{-1}(x) = \frac{x+3}{4-2x}$.

17. Let $y = f(x)$. Switch x and y and solve for y:

$$\frac{2y+1}{3y+1} = x$$
$$2y + 1 = 3xy + x$$
$$2y - 3xy = x - 1$$
$$y(2 - 3x) = x - 1$$
$$y = \frac{x-1}{2-3x} = \frac{1-x}{3x-2}$$

The inverse is $f^{-1}(x) = \frac{1-x}{3x-2}$.

19. Finding the inverse:

$$2y - 1 = x$$
$$2y = x + 1$$
$$y = \frac{x+1}{2}$$

The inverse is $y = \frac{x+1}{2}$. Graphing each curve:

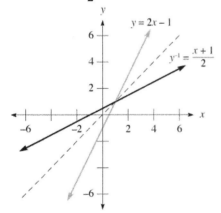

21. Finding the inverse:

$$y^2 - 3 = x$$
$$y^2 = x + 3$$
$$y = \pm\sqrt{x+3}$$

The inverse is $y = \pm\sqrt{x+3}$. Graphing each curve:

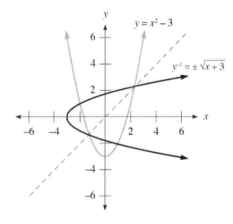

23. Finding the inverse:
$$y^2 - 2y - 3 = x$$
$$y^2 - 2y + 1 = x + 3 + 1$$
$$(y-1)^2 = x + 4$$
$$y - 1 = \pm\sqrt{x+4}$$
$$y = 1 \pm \sqrt{x+4}$$

The inverse is $y = 1 \pm \sqrt{x+4}$. Graphing each curve:

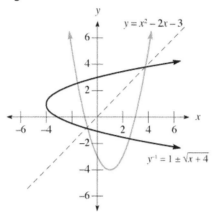

25. The inverse is $x = 3^y$. Graphing each curve:

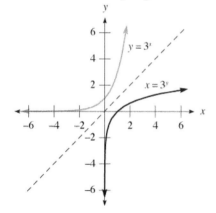

27. The inverse is $x = 4$. Graphing each curve:

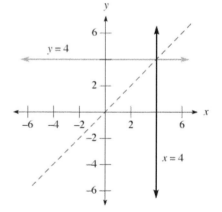

29. Finding the inverse:
$$\frac{1}{2}y^3 = x$$
$$y^3 = 2x$$
$$y = \sqrt[3]{2x}$$

The inverse is $y = \sqrt[3]{2x}$. Graphing each curve:

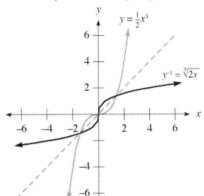

31. Finding the inverse:
$$\frac{1}{2}y + 2 = x$$
$$y + 4 = 2x$$
$$y = 2x - 4$$

The inverse is $y = 2x - 4$. Graphing each curve:

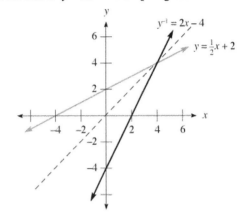

33. Finding the inverse:
$$\sqrt{y+2} = x$$
$$y + 2 = x^2$$
$$y = x^2 - 2$$

The inverse is $y = x^2 - 2, x \geq 0$. Graphing each curve:

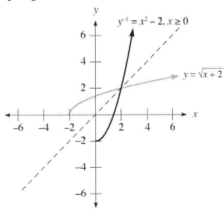

35.
   a. Yes, this function is one-to-one.
   b. No, this function is not one-to-one.
   c. Yes, this function is one-to-one.

37.
   a. Evaluating the function: $f(2) = 3(2) - 2 = 6 - 2 = 4$

   b. Evaluating the function: $f^{-1}(2) = \dfrac{2+2}{3} = \dfrac{4}{3}$

   c. Evaluating the function: $f(f^{-1}(2)) = f\left(\dfrac{4}{3}\right) = 3\left(\dfrac{4}{3}\right) - 2 = 4 - 2 = 2$

   d. Evaluating the function: $f^{-1}(f(2)) = f^{-1}(4) = \dfrac{4+2}{3} = \dfrac{6}{3} = 2$

39. Let $y = f(x)$. Switch x and y and solve for y:
$$\frac{1}{y} = x$$
$$y = \frac{1}{x}$$
The inverse is $f^{-1}(x) = \frac{1}{x}$.

41.
a. $3x$
b. $3x + 2$
c. $x - 2$
d. $\frac{x-2}{3}$

The inverse is $f^{-1}(x) = 7(x+2) = 7x + 14$.

43.
a. The value is –3.
b. The value is –6.
c. The value is 2.
d. The value is 3.
e. The value is –2.
f. The value is 3.
g. They are inverses of each other.

45.
a. Substituting $t = 25$: $s(15) = 16(15) + 249.4 = \$489.40$ billion
b. Finding the inverse:
$$s = 16t + 249.4$$
$$16t = s - 249.4$$
$$t(s) = \frac{s - 249.4}{16}$$
c. Substitute $s = 507$: $t(507) = \frac{507 - 249.4}{16} \approx 16$.
The payments will reach $507 billion in the year 2006.

47.
a. Substituting $m = 4520$: $f = \frac{22(4520)}{15} \approx 6{,}629$ feet per second
b. Finding the inverse:
$$f = \frac{22m}{15}$$
$$15f = 22m$$
$$m(f) = \frac{15f}{22}$$
c. Substituting $f = 2$: $m(2) = \frac{15(2)}{22} \approx 1.36$ mph

49. Simplifying: $3^{-2} = \frac{1}{3^2} = \frac{1}{9}$

51. Solving the equation:
$$2 = 3x$$
$$x = \frac{2}{3}$$

53. Solving the equation:
$$4 = x^3$$
$$x = \sqrt[3]{4}$$

55. Completing the statement: $8 = 2^3$

57. Completing the statement: $10{,}000 = 10^4$

59. Completing the statement: $81 = 3^4$

61. Completing the statement: $6 = 6^1$

10.3 Logarithms are Exponents

1. Writing in logarithmic form: $\log_2 16 = 4$
3. Writing in logarithmic form: $\log_5 125 = 3$
5. Writing in logarithmic form: $\log_{10} 0.01 = -2$
7. Writing in logarithmic form: $\log_2 \frac{1}{32} = -5$
9. Writing in logarithmic form: $\log_{1/2} 8 = -3$
11. Writing in logarithmic form: $\log_3 27 = 3$
13. Writing in exponential form: $10^2 = 100$
15. Writing in exponential form: $2^6 = 64$
17. Writing in exponential form: $8^0 = 1$
19. Writing in exponential form: $10^{-3} = 0.001$
21. Writing in exponential form: $6^2 = 36$
23. Writing in exponential form: $5^{-2} = \frac{1}{25}$

25. Solving the equation:
$$\log_3 x = 2$$
$$x = 3^2 = 9$$

27. Solving the equation:
$$\log_5 x = -3$$
$$x = 5^{-3} = \frac{1}{125}$$

29. Solving the equation:
$$\log_2 16 = x$$
$$2^x = 16$$
$$x = 4$$

31. Solving the equation:
$$\log_8 2 = x$$
$$8^x = 2$$
$$x = \frac{1}{3}$$

33. Solving the equation:
$$\log_x 4 = 2$$
$$x^2 = 4$$
$$x = 2$$

35. Solving the equation:
$$\log_x 5 = 3$$
$$x^3 = 5$$
$$x = \sqrt[3]{5}$$

37. Solving the equation:
$$\log_5 25 = x$$
$$5^x = 25$$
$$x = 2$$

39. Solving the equation:
$$\log_x 36 = 2$$
$$x^2 = 36$$
$$x = 6$$

41. Solving the equation:
$$\log_8 4 = x$$
$$8^x = 4$$
$$2^{3x} = 2^2$$
$$3x = 2$$
$$x = \frac{2}{3}$$

43. Solving the equation:
$$\log_9 \frac{1}{3} = x$$
$$9^x = \frac{1}{3}$$
$$3^{2x} = 3^{-1}$$
$$2x = -1$$
$$x = -\frac{1}{2}$$

45. Solving the equation:
$$\log_8 x = -2$$
$$x = 8^{-2} = \frac{1}{64}$$

47. Sketching the graph:

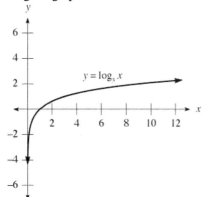

49. Sketching the graph:

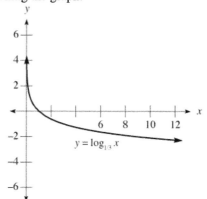

51. Sketching the graph:

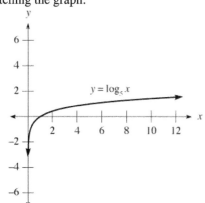

53. Sketching the graph:

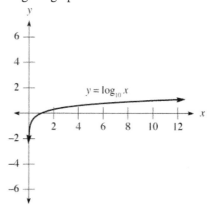

55. The equation is $y = 3^x$.

57. The equation is $y = \log_{1/3} x$.

59. Simplifying the logarithm:
$$x = \log_2 16$$
$$2^x = 16$$
$$x = 4$$

61. Simplifying the logarithm:
$$x = \log_{25} 125$$
$$25^x = 125$$
$$5^{2x} = 5^3$$
$$2x = 3$$
$$x = \frac{3}{2}$$

63. Simplifying the logarithm:
$$x = \log_{10} 1000$$
$$10^x = 1000$$
$$x = 3$$

65. Simplifying the logarithm:
$$x = \log_3 3$$
$$3^x = 3$$
$$x = 1$$

67. Simplifying the logarithm:
$$x = \log_5 1$$
$$5^x = 1$$
$$x = 0$$

69. Simplifying the logarithm:
$$x = \log_{17} 1$$
$$17^x = 1$$
$$x = 0$$

71. Simplifying the logarithm:
$x = \log_{16} 4$
$16^x = 4$
$4^{2x} = 4^1$
$2x = 1$
$x = \dfrac{1}{2}$

73. Simplifying the logarithm:
$x = \log_{100} 1000$
$100^x = 1000$
$10^{2x} = 10^3$
$2x = 3$
$x = \dfrac{3}{2}$

75. First find $\log_2 8$:
$x = \log_2 8$
$2^x = 8$
$x = 3$
Now find $\log_3 3$:
$x = \log_3 3$
$3^x = 3$
$x = 1$

77. First find $\log_3 81$:
$x = \log_3 81$
$3^x = 81$
$x = 4$
Now find $\log_{1/2} 4$:
$x = \log_{1/2} 4$
$\left(\dfrac{1}{2}\right)^x = 4$
$2^{-x} = 2^2$
$x = -2$
Now find $\log_3 1$:
$x = \log_3 1$
$3^x = 1$
$x = 0$
Now find $\log_2 4$:
$x = \log_2 4$
$2^x = 4$
$x = 2$

79. First find $\log_6 6$:
$x = \log_6 6$
$6^x = 6$
$x = 1$

81. First find $\log_2 16$:
$x = \log_2 16$
$2^x = 16$
$x = 4$
Now find $\log_4 2$:
$x = \log_4 2$
$4^x = 2$
$2^{2x} = 2$
$2x = 1$
$x = \dfrac{1}{2}$

83. Completing the table:

Prefix	Multiplying Factor	$\log_{10}$ (Multiplying Factor)
Nano	0.000000001	−9
Micro	0.000001	−6
Deci	0.1	−1
Giga	1,000,000,000	9
Peta	1,000,000,000,000,000	15

85. Using the relationship $M = \log_{10} T$:
$M = \log_{10} 100$
$10^M = 100$
$M = 2$

87. It is 10^8 times as large.

89. Since $M = 6$, there are 120 earthquakes.

91. Simplifying: $8^{2/3} = \left(8^{1/3}\right)^2 = \left(\sqrt[3]{8}\right)^2 = 2^2 = 4$

93. Solving the equation:
$$(x+2)(x) = 2^3$$
$$x^2 + 2x = 8$$
$$x^2 + 2x - 8 = 0$$
$$(x+4)(x-2) = 0$$
$$x = -4, 2$$

95. Solving the equation:
$$\frac{x-2}{x+1} = 9$$
$$x - 2 = 9(x+1)$$
$$x - 2 = 9x + 9$$
$$-8x = 11$$
$$x = -\frac{11}{8}$$

97. Writing in exponential form: $2^3 = (x+2)(x)$

99. Writing in exponential form: $3^4 = \frac{x-2}{x+1}$

10.4 Properties of Logarithms

1. Using properties of logarithms: $\log_3 4x = \log_3 4 + \log_3 x$

3. Using properties of logarithms: $\log_6 \frac{5}{x} = \log_6 5 - \log_6 x$

5. Using properties of logarithms: $\log_2 y^5 = 5\log_2 y$

7. Using properties of logarithms: $\log_9 \sqrt[3]{z} = \log_9 z^{1/3} = \frac{1}{3}\log_9 z$

9. Using properties of logarithms: $\log_6 x^2 y^4 = \log_6 x^2 + \log_6 y^4 = 2\log_6 x + 4\log_6 y$

11. Using properties of logarithms: $\log_5 \left(\sqrt{x} \cdot y^4\right) = \log_5 x^{1/2} + \log_5 y^4 = \frac{1}{2}\log_5 x + 4\log_5 y$

13. Using properties of logarithms: $\log_b \frac{xy}{z} = \log_b xy - \log_b z = \log_b x + \log_b y - \log_b z$

15. Using properties of logarithms: $\log_{10} \frac{4}{xy} = \log_{10} 4 - \log_{10} xy = \log_{10} 4 - \log_{10} x - \log_{10} y$

17. Using properties of logarithms: $\log_{10} \frac{x^2 y}{\sqrt{z}} = \log_{10} x^2 + \log_{10} y - \log_{10} z^{1/2} = 2\log_{10} x + \log_{10} y - \frac{1}{2}\log_{10} z$

19. Using properties of logarithms: $\log_{10} \frac{x^3 \sqrt{y}}{z^4} = \log_{10} x^3 + \log_{10} y^{1/2} - \log_{10} z^4 = 3\log_{10} x + \frac{1}{2}\log_{10} y - 4\log_{10} z$

21. Using properties of logarithms: $\log_b \sqrt[3]{\frac{x^2 y}{z^4}} = \log_b \frac{x^{2/3} y^{1/3}}{z^{4/3}} = \log_b x^{2/3} + \log_b y^{1/3} - \log_b z^{4/3} = \frac{2}{3}\log_b x + \frac{1}{3}\log_b y - \frac{4}{3}\log_b z$

23. Using properties of logarithms: $\log_3 \sqrt[3]{\frac{x^2 y}{z^6}} = \log_3 \frac{x^{2/3} y^{1/3}}{z^2} = \log_3 x^{2/3} + \log_3 y^{1/3} - \log_3 z^2 = \frac{2}{3}\log_3 x + \frac{1}{3}\log_3 y - 2\log_3 z$

25. Using properties of logarithms:
$$\log_a \frac{4x^5}{9a^2} = \log_a 4x^5 - \log_a 9a^2 = \log_a 2^2 + \log_a x^5 - \log_a 3^2 - \log_a a^2 = 2\log_a 2 + 5\log_a x - 2\log_a 3 - 2$$

27. Writing as a single logarithm: $\log_b x + \log_b z = \log_b (xz)$

29. Writing as a single logarithm: $2\log_3 x - 3\log_3 y = \log_3 x^2 - \log_3 y^3 = \log_3 \left(\frac{x^2}{y^3}\right)$

31. Writing as a single logarithm: $\frac{1}{2}\log_{10} x + \frac{1}{3}\log_{10} y = \log_{10} x^{1/2} + \log_{10} y^{1/3} = \log_{10} \left(\sqrt{x}\sqrt[3]{y}\right)$

33. Writing as a single logarithm: $3\log_2 x + \frac{1}{2}\log_2 y - \log_2 z = \log_2 x^3 + \log_2 y^{1/2} - \log_2 z = \log_2\left(\frac{x^3\sqrt{y}}{z}\right)$

35. Writing as a single logarithm: $\frac{1}{2}\log_2 x - 3\log_2 y - 4\log_2 z = \log_2 x^{1/2} - \log_2 y^3 - \log_2 z^4 = \log_2\left(\frac{\sqrt{x}}{y^3 z^4}\right)$

37. Writing as a single logarithm: $\frac{3}{2}\log_{10} x - \frac{3}{4}\log_{10} y - \frac{4}{5}\log_{10} z = \log_{10} x^{3/2} - \log_{10} y^{3/4} - \log_{10} z^{4/5} = \log_{10}\left(\frac{x^{3/2}}{y^{3/4} z^{4/5}}\right)$

39. Writing as a single logarithm: $\frac{1}{2}\log_5 x + \frac{2}{3}\log_5 y - 4\log_5 z = \log_5 x^{1/2} + \log_5 y^{2/3} - \log_5 z^4 = \log_5\left(\frac{\sqrt{x}\cdot\sqrt[3]{y^2}}{z^4}\right)$

41. Writing as a single logarithm:
$\log_3(x^2-16) - 2\log_3(x+4) = \log_3(x^2-16) - \log_3(x+4)^2 = \log_3\frac{(x+4)(x-4)}{(x+4)^2} = \log_3\left(\frac{x-4}{x+4}\right)$

43. Solving the equation:

$\log_2 x + \log_2 3 = 1$
$\log_2 3x = 1$
$3x = 2^1$
$3x = 2$
$x = \frac{2}{3}$

45. Solving the equation:

$\log_3 x - \log_3 2 = 2$
$\log_3 \frac{x}{2} = 2$
$\frac{x}{2} = 3^2$
$\frac{x}{2} = 9$
$x = 18$

47. Solving the equation:

$\log_3 x + \log_3(x-2) = 1$
$\log_3(x^2 - 2x) = 1$
$x^2 - 2x = 3^1$
$x^2 - 2x - 3 = 0$
$(x-3)(x+1) = 0$
$x = 3, -1$

The solution is 3 (–1 does not check).

49. Solving the equation:

$\log_3(x+3) - \log_3(x-1) = 1$
$\log_3\frac{x+3}{x-1} = 1$
$\frac{x+3}{x-1} = 3^1$
$x+3 = 3x - 3$
$-2x = -6$
$x = 3$

51. Solving the equation:

$\log_2 x + \log_2(x-2) = 3$
$\log_2(x^2 - 2x) = 3$
$x^2 - 2x = 2^3$
$x^2 - 2x - 8 = 0$
$(x-4)(x+2) = 0$
$x = 4, -2$

The solution is 4 (–2 does not check).

53. Solving the equation:

$\log_8 x + \log_8(x-3) = \frac{2}{3}$
$\log_8(x^2 - 3x) = \frac{2}{3}$
$x^2 - 3x = 8^{2/3}$
$x^2 - 3x - 4 = 0$
$(x-4)(x+1) = 0$
$x = 4, -1$

The solution is 4 (–1 does not check).

55. Solving the equation:
$$\log_3(x+2) - \log_3 x = 1$$
$$\log_3 \frac{x+2}{x} = 1$$
$$\frac{x+2}{x} = 3^1$$
$$x+2 = 3x$$
$$2x = 2$$
$$x = 1$$

57. Solving the equation:
$$\log_2(x+1) + \log_2(x+2) = 1$$
$$\log_2(x^2 + 3x + 2) = 1$$
$$x^2 + 3x + 2 = 2^1$$
$$x^2 + 3x = 0$$
$$x(x+3) = 0$$
$$x = 0, -3$$

The solution is 0 (–3 does not check).

59. Solving the equation:
$$\log_9 \sqrt{x} + \log_9 \sqrt{2x+3} = \frac{1}{2}$$
$$\log_9 \sqrt{2x^2 + 3x} = \frac{1}{2}$$
$$\sqrt{2x^2 + 3x} = 9^{1/2}$$
$$2x^2 + 3x = 9$$
$$2x^2 + 3x - 9 = 0$$
$$(2x-3)(x+3) = 0$$
$$x = \frac{3}{2}, -3$$

The solution is $\frac{3}{2}$ (–3 does not check).

61. Solving the equation:
$$4\log_3 x - \log_3 x^2 = 6$$
$$4\log_3 x - 2\log_3 x = 6$$
$$2\log_3 x = 6$$
$$\log_3 x = 3$$
$$x = 3^3$$
$$x = 27$$

63. Solving the equation:
$$\log_5 \sqrt{x} + \log_5 \sqrt{6x+5} = 1$$
$$\log_5 \sqrt{6x^2 + 5x} = 1$$
$$\frac{1}{2}\log_5(6x^2 + 5x) = 1$$
$$\log_5(6x^2 + 5x) = 2$$
$$6x^2 + 5x = 5^2$$
$$6x^2 + 5x - 25 = 0$$
$$(3x-5)(2x+5) = 0$$
$$x = \frac{5}{3}, -\frac{5}{2}$$

The solution is $\frac{5}{3}$ ($-\frac{5}{2}$ does not check).

65. Rewriting the formula:
$$D = 10\log_{10}\left(\frac{I}{I_0}\right)$$
$$D = 10(\log_{10} I - \log_{10} I_0)$$

67. a. Finding the value: $\log_{10} 40 = \log_{10}(8 \cdot 5) = \log_{10} 8 + \log_{10} 5 = 0.903 + 0.699 = 1.602$
    b. Finding the value: $\log_{10} 320 = \log_{10}(8^2 \cdot 5) = \log_{10} 8^2 + \log_{10} 5 = 2\log_{10} 8 + \log_{10} 5 = 2(0.903) + 0.699 = 2.505$
    c. Finding the value:
       $\log_{10} 1600 = \log_{10}(8^2 \cdot 5^2) = \log_{10} 8^2 + \log_{10} 5^2 = 2\log_{10} 8 + 2\log_{10} 5 = 2(0.903) + 2(0.699) = 3.204$

69. Rewriting the equation: $pH = 6.1 + \log_{10}\left(\dfrac{x}{y}\right) = 6.1 + \log_{10} x - \log_{10} y$

71. Solving for M: $M = 0.21 \log_{10} \dfrac{1}{10^{-12}} = 0.21 \log_{10} 10^{12} = 0.21(12) = 2.52$

73. Simplifying: $5^0 = 1$

75. Simplifying: $\log_3 3 = \log_3 3^1 = 1$

77. Simplifying: $\log_b b^4 = 4$

79. Using a calculator: $10^{-5.6} \approx 2.5 \times 10^{-6}$

81. Using a calculator: $\dfrac{2.00 \times 10^8}{3.96 \times 10^6} \approx 51$

10.5 Common Logarithms and Natural Logarithms

1. Evaluating the logarithm: $\log 378 \approx 2.5775$
3. Evaluating the logarithm: $\log 37.8 \approx 1.5775$
5. Evaluating the logarithm: $\log 3{,}780 \approx 3.5775$
7. Evaluating the logarithm: $\log 0.0378 \approx -1.4225$
9. Evaluating the logarithm: $\log 37{,}800 \approx 4.5775$
11. Evaluating the logarithm: $\log 600 \approx 2.7782$
13. Evaluating the logarithm: $\log 2{,}010 \approx 3.3032$
15. Evaluating the logarithm: $\log 0.00971 \approx -2.0128$
17. Evaluating the logarithm: $\log 0.0314 \approx -1.5031$
19. Evaluating the logarithm: $\log 0.399 \approx -0.3990$

21. Solving for x:
    $\log x = 2.8802$
    $x = 10^{2.8802} \approx 759$

23. Solving for x:
    $\log x = -2.1198$
    $x = 10^{-2.1198} \approx 0.00759$

25. Solving for x:
    $\log x = 3.1553$
    $x = 10^{3.1553} \approx 1{,}430$

27. Solving for x:
    $\log x = -5.3497$
    $x = 10^{-5.3497} \approx 0.00000447$

29. Solving for x:
    $\log x = -7.0372$
    $x = 10^{-7.0372} \approx 0.0000000918$

31. Solving for x:
    $\log x = 10$
    $x = 10^{10}$

33. Solving for x:
    $\log x = -10$
    $x = 10^{-10}$

35. Solving for x:
    $\log x = 20$
    $x = 10^{20}$

37. Solving for x:
    $\log x = -2$
    $x = 10^{-2} = \dfrac{1}{100}$

39. Solving for x:
    $\log x = \log_2 8$
    $\log x = 3$
    $x = 10^3 = 1{,}000$

41. Solving for x:
    $\ln x = 1$
    $x = e^{-1} = \dfrac{1}{e}$

43. Solving for x:
    $\log x = 2 \log 5$
    $\log x = \log 5^2$
    $x = 25$

45. Solving for x:
    $\ln x = -3 \ln 2$
    $\ln x = \ln 2^{-3}$
    $x = \dfrac{1}{8}$

47. Simplifying the logarithm: $\ln e = \ln e^1 = 1$

49. Simplifying the logarithm: $\ln e^5 = 5$

51. Simplifying the logarithm: $\ln e^x = x$

53. Simplifying the logarithm: $\log 10{,}000 = \log 10^4 = 4$

55. Simplifying the logarithm: $\ln \dfrac{1}{e^3} = \log e^{-3} = -3$

57. Simplifying the logarithm: $\log \sqrt{1000} = \log 10^{3/2} = \dfrac{3}{2}$

59. Using properties of logarithms: $\ln 10e^{3t} = \ln 10 + \ln e^{3t} = 3t + \ln 10$

61. Using properties of logarithms: $\ln Ae^{-2t} = \ln A + \ln e^{-2t} = -2t + \ln A$

63. Using properties of logarithms: $\log\left[100(1.01)^{3t}\right] = \log 10^2 + \log 1.01^{3t} = 2 + 3t \log 1.01$

65. Using properties of logarithms: $\ln(Pe^{rt}) = \ln P + \ln e^{rt} = \ln P + rt$

67. Using properties of logarithms: $-\log(4.2 \times 10^{-3}) = -\log 4.2 - \log 10^{-3} = 3 - \log 4.2$

69. Evaluating the logarithm: $\ln 15 = \ln(3 \cdot 5) = \ln 3 + \ln 5 = 1.0986 + 1.6094 = 2.7080$

71. Evaluating the logarithm: $\ln \dfrac{1}{3} = \ln 3^{-1} = -\ln 3 = -1.0986$

73. Evaluating the logarithm: $\ln 9 = \ln 3^2 = 2 \ln 3 = 2(1.0986) = 2.1972$

75. Evaluating the logarithm: $\ln 16 = \ln 2^4 = 4 \ln 2 = 4(0.6931) = 2.7724$

77. For the 1906 earthquake:
$\log T = 8.3$
$T = 10^{8.3} = 1.995 \times 10^8$
For the atomic bomb test:
$\log T = 5.0$
$T = 10^{5.0} = 1 \times 10^5$
Dividing the two values: $\dfrac{1.995 \times 10^8}{1 \times 10^5} \approx 2000$. San Francisco earthquake was approximately 2,000 times greater.

79. Completing the table:

x	$(1+x)^{1/x}$
1	2
0.5	2.25
0.1	2.5937
0.01	2.7048
0.001	2.7169
0.0001	2.7181
0.00001	2.7183

It appears to approach e.

81. Substituting $s = 15$:
$5 \ln x = 15$
$\ln x = 3$
$x = e^3 \approx 20$
Approximately 15% of students enrolled are in the age range in the year 2009.

83. Computing the pH: $\text{pH} = -\log(6.50 \times 10^{-4}) \approx 3.19$

85. Finding the concentration:
$4.75 = -\log[H^+]$
$-4.75 = \log[H^+]$
$[H^+] = 10^{-4.75} \approx 1.78 \times 10^{-5}$

87. Finding the magnitude:
$$5.5 = \log T$$
$$T = 10^{5.5} \approx 3.16 \times 10^5$$

89. Finding the magnitude:
$$8.3 = \log T$$
$$T = 10^{8.3} \approx 2.00 \times 10^8$$

91. Completing the table:

Location	Date	Magnitude (M)	Shockwave (T)
Moresby Island	January 23	4.0	1.00×10^4
Vancouver Island	April 30	5.3	1.99×10^5
Quebec City	June 29	3.2	1.58×10^3
Mould Bay	November 13	5.2	1.58×10^5
St. Lawrence	December 14	3.7	5.01×10^3

93. Finding the rate of depreciation:
$$\log(1-r) = \frac{1}{5}\log\frac{4500}{9000}$$
$$\log(1-r) \approx -0.0602$$
$$1 - r \approx 10^{-0.0602}$$
$$r = 1 - 10^{-0.0602}$$
$$r \approx 0.129 = 12.9\%$$

95. Finding the rate of depreciation:
$$\log(1-r) = \frac{1}{5}\log\frac{5750}{7550}$$
$$\log(1-r) \approx -0.0237$$
$$1 - r \approx 10^{-0.0237}$$
$$r = 1 - 10^{-0.0237}$$
$$r \approx 0.053 = 5.3\%$$

97. Solving the equation:
$$5(2x+1) = 12$$
$$10x + 5 = 12$$
$$10x = 7$$
$$x = \frac{7}{10}$$

99. Using a calculator: $\frac{100{,}000}{32{,}000} = 3.1250$

101. Using a calculator: $\frac{1}{2}\left(\frac{-0.6931}{1.4289} + 3\right) \approx 1.2575$

103. Rewriting the logarithm: $\log 1.05^t = t\log 1.05$

105. Simplifying: $\ln e^{0.05t} = 0.05t$

10.6 Exponential Equations and Change of Base

1. Solving the equation:
$$3^x = 5$$
$$\ln 3^x = \ln 5$$
$$x \ln 3 = \ln 5$$
$$x = \frac{\ln 5}{\ln 3} \approx 1.4650$$

3. Solving the equation:
$$5^x = 3$$
$$\ln 5^x = \ln 3$$
$$x \ln 5 = \ln 3$$
$$x = \frac{\ln 3}{\ln 5} \approx 0.6826$$

5. Solving the equation:
$$5^{-x} = 12$$
$$\ln 5^{-x} = \ln 12$$
$$-x \ln 5 = \ln 12$$
$$x = -\frac{\ln 12}{\ln 5} \approx -1.5440$$

7. Solving the equation:
$$12^{-x} = 5$$
$$\ln 12^{-x} = \ln 5$$
$$-x \ln 12 = \ln 5$$
$$x = -\frac{\ln 5}{\ln 12} \approx -0.6477$$

9. Solving the equation:
$$8^{x+1} = 4$$
$$2^{3x+3} = 2^2$$
$$3x + 3 = 2$$
$$3x = -1$$
$$x = -\frac{1}{3} \approx -0.3333$$

11. Solving the equation:
$$4^{x-1} = 4$$
$$4^{x-1} = 4^1$$
$$x - 1 = 1$$
$$x = 2$$

13. Solving the equation:
$$3^{2x+1} = 2$$
$$\ln 3^{2x+1} = \ln 2$$
$$(2x+1)\ln 3 = \ln 2$$
$$2x + 1 = \frac{\ln 2}{\ln 3}$$
$$2x = \frac{\ln 2}{\ln 3} - 1$$
$$x = \frac{1}{2}\left(\frac{\ln 2}{\ln 3} - 1\right) \approx -0.1845$$

15. Solving the equation:
$$3^{1-2x} = 2$$
$$\ln 3^{1-2x} = \ln 2$$
$$(1-2x)\ln 3 = \ln 2$$
$$1 - 2x = \frac{\ln 2}{\ln 3}$$
$$-2x = \frac{\ln 2}{\ln 3} - 1$$
$$x = \frac{1}{2}\left(1 - \frac{\ln 2}{\ln 3}\right) \approx 0.1845$$

17. Solving the equation:
$$15^{3x-4} = 10$$
$$\ln 15^{3x-4} = \ln 10$$
$$(3x-4)\ln 15 = \ln 10$$
$$3x - 4 = \frac{\ln 10}{\ln 15}$$
$$3x = \frac{\ln 10}{\ln 15} + 4$$
$$x = \frac{1}{3}\left(\frac{\ln 10}{\ln 15} + 4\right) \approx 1.6168$$

19. Solving the equation:
$$6^{5-2x} = 4$$
$$\ln 6^{5-2x} = \ln 4$$
$$(5-2x)\ln 6 = \ln 4$$
$$5 - 2x = \frac{\ln 4}{\ln 6}$$
$$-2x = \frac{\ln 4}{\ln 6} - 5$$
$$x = \frac{1}{2}\left(5 - \frac{\ln 4}{\ln 6}\right) \approx 2.1131$$

21. Solving the equation:
$$3^{-4x} = 81$$
$$3^{-4x} = 3^4$$
$$-4x = 4$$
$$x = -1$$

23. Solving the equation:
$$5^{3x-2} = 15$$
$$\ln 5^{3x-2} = \ln 15$$
$$(3x-2)\ln 5 = \ln 15$$
$$3x - 2 = \frac{\ln 15}{\ln 5}$$
$$3x = \frac{\ln 15}{\ln 5} + 2$$
$$x = \frac{1}{3}\left(\frac{\ln 15}{\ln 5} + 2\right) \approx 1.2275$$

25. Solving the equation:

$$100e^{3t} = 250$$
$$e^{3t} = \frac{5}{2}$$
$$3t = \ln\frac{5}{2}$$
$$t = \frac{1}{3}\ln\frac{5}{2} \approx 0.3054$$

27. Solving the equation:

$$1200\left(1+\frac{0.072}{4}\right)^{4t} = 25000$$
$$\left(1+\frac{0.072}{4}\right)^{4t} = \frac{125}{6}$$
$$\ln\left(1+\frac{0.072}{4}\right)^{4t} = \ln\frac{125}{6}$$
$$4t\ln\left(1+\frac{0.072}{4}\right) = \ln\frac{125}{6}$$
$$t = \frac{\ln\frac{125}{6}}{4\ln\left(1+\frac{0.072}{4}\right)} \approx 42.5528$$

29. Solving the equation:

$$50e^{-0.0742t} = 32$$
$$e^{-0.0742t} = \frac{16}{25}$$
$$-0.0742t = \ln\frac{16}{25}$$
$$t = \frac{\ln\frac{16}{25}}{-0.0742} \approx 6.0147$$

31. Evaluating the logarithm: $\log_8 16 = \dfrac{\log 16}{\log 8} = \dfrac{\log 2^4}{\log 2^3} = \dfrac{4}{3} \approx 1.3333$

33. Evaluating the logarithm: $\log_{16} 8 = \dfrac{\log 8}{\log 16} = \dfrac{\log 2^3}{\log 2^4} = \dfrac{3}{4} = 0.7500$

35. Evaluating the logarithm: $\log_7 15 = \dfrac{\log 15}{\log 7} \approx 1.3917$

37. Evaluating the logarithm: $\log_{15} 7 = \dfrac{\log 7}{\log 15} \approx 0.7186$

39. Evaluating the logarithm: $\log_8 240 = \dfrac{\log 240}{\log 8} \approx 2.6356$

41. Evaluating the logarithm: $\log_4 321 = \dfrac{\log 321}{\log 4} \approx 4.1632$

43. Evaluating the logarithm: $\ln 345 \approx 5.8435$

45. Evaluating the logarithm: $\ln 0.345 \approx -1.0642$

47. Evaluating the logarithm: $\ln 10 \approx 2.3026$

49. Evaluating the logarithm: $\ln 45,000 \approx 10.7144$

51. Using the compound interest formula:
$$500\left(1+\frac{0.06}{2}\right)^{2t} = 1000$$
$$\left(1+\frac{0.06}{2}\right)^{2t} = 2$$
$$\ln\left(1+\frac{0.06}{2}\right)^{2t} = \ln 2$$
$$2t\ln\left(1+\frac{0.06}{2}\right) = \ln 2$$
$$t = \frac{\ln 2}{2\ln\left(1+\frac{0.06}{2}\right)} \approx 11.72$$

It will take 11.72 years.

53. Using the compound interest formula:
$$1000\left(1+\frac{0.12}{6}\right)^{6t} = 3000$$
$$\left(1+\frac{0.12}{6}\right)^{6t} = 3$$
$$\ln\left(1+\frac{0.12}{6}\right)^{6t} = \ln 3$$
$$6t\ln\left(1+\frac{0.12}{6}\right) = \ln 3$$
$$t = \frac{\ln 3}{6\ln\left(1+\frac{0.12}{6}\right)} \approx 9.25$$

It will take 9.25 years.

55. Using the compound interest formula:
$$P\left(1+\frac{0.08}{4}\right)^{4t} = 2P$$
$$\left(1+\frac{0.08}{4}\right)^{4t} = 2$$
$$\ln\left(1+\frac{0.08}{4}\right)^{4t} = \ln 2$$
$$4t\ln\left(1+\frac{0.08}{4}\right) = \ln 2$$
$$t = \frac{\ln 2}{4\ln\left(1+\frac{0.08}{4}\right)} \approx 8.75$$

It will take 8.75 years.

57. Using the compound interest formula:
$$25\left(1+\frac{0.06}{2}\right)^{2t} = 75$$
$$\left(1+\frac{0.06}{2}\right)^{2t} = 3$$
$$\ln\left(1+\frac{0.06}{2}\right)^{2t} = \ln 3$$
$$2t\ln\left(1+\frac{0.06}{2}\right) = \ln 3$$
$$t = \frac{\ln 3}{2\ln\left(1+\frac{0.06}{2}\right)} \approx 18.58$$

It was invested 18.58 years ago.

59. Using the continuous interest formula:
$$500e^{0.06t} = 1000$$
$$e^{0.06t} = 2$$
$$0.06t = \ln 2$$
$$t = \frac{\ln 2}{0.06} \approx 11.55$$

It will take 11.55 years.

61. Using the continuous interest formula:
$$500e^{0.06t} = 1500$$
$$e^{0.06t} = 3$$
$$0.06t = \ln 3$$
$$t = \frac{\ln 3}{0.06} \approx 18.31$$

It will take 18.31 years.

63. Using the continuous interest formula:
$$1000e^{0.08t} = 2500$$
$$e^{0.08t} = 2.5$$
$$0.08t = \ln 2.5$$
$$t = \frac{\ln 2.5}{0.08} \approx 11.45$$

It will take 11.45 years.

65. Using the population model:
$$32{,}000 e^{0.05t} = 64{,}000$$
$$e^{0.05t} = 2$$
$$0.05t = \ln 2$$
$$t = \frac{\ln 2}{0.05} \approx 13.9$$

The city will reach 64,000 toward the end of the year 2018 (October).

67. Using the exponential model:
$$466 \cdot 1.035^t = 900$$
$$1.035^t = \frac{900}{466}$$
$$\ln 1.035^t = \ln \frac{900}{466}$$
$$t \ln 1.035 = \ln \frac{900}{466}$$
$$t = \frac{\ln \frac{900}{466}}{\ln 1.035} \approx 19$$

In the year 2009 it is predicted that 900 million passengers will travel by airline.

69. Using the exponential model:
$$78.16(1.11)^t = 800$$
$$1.11^t = \frac{800}{78.16}$$
$$\ln 1.11^t = \ln \frac{800}{78.16}$$
$$t \ln 1.11 = \ln \frac{800}{78.16}$$
$$t = \frac{\ln \frac{800}{78.16}}{\ln 1.11} \approx 22$$

In the year 1992 it was estimated that $800 billion will be spent on health care expenditures.

71. Using the compound interest formula:
$$20{,}972 \left(1 + \frac{0.07}{2}\right)^{2t} = 41{,}944$$
$$\left(1 + \frac{0.07}{2}\right)^{2t} = 2$$
$$\ln \left(1 + \frac{0.07}{2}\right)^{2t} = \ln 2$$
$$2t \ln \left(1 + \frac{0.07}{2}\right) = \ln 2$$
$$t = \frac{\ln 2}{2 \ln \left(1 + \frac{0.07}{2}\right)} \approx 10.07$$

It will take 10.07 years for the money to double.

73. Using the exponential formula:
$$0.10 e^{0.0576t} = 1.00$$
$$e^{0.0576t} = 10$$
$$0.0576t = \ln 10$$
$$t = \frac{\ln 10}{0.0576} \approx 40$$

A Coca Cola will cost $1.00 in the year 2000.

75. Completing the square: $y = 2x^2 + 8x - 15 = 2(x^2 + 4x + 4) - 8 - 15 = 2(x+2)^2 - 23$. The lowest point is $(-2, -23)$.

77. Completing the square: $y = 12x - 4x^2 = -4\left(x^2 - 3x + \dfrac{9}{4}\right) + 9 = -4\left(x - \dfrac{3}{2}\right)^2 + 9$. The highest point is $\left(\dfrac{3}{2}, 9\right)$.

79. Completing the square: $y = 64t - 16t^2 = -16(t^2 - 4t + 4) + 64 = -16(t-2)^2 + 64$

    The object reaches a maximum height after 2 seconds, and the maximum height is 64 feet.

Chapter 10 Test

1. Graphing the function:

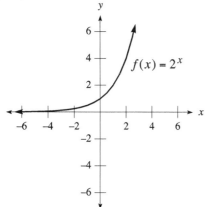

2. Graphing the function:

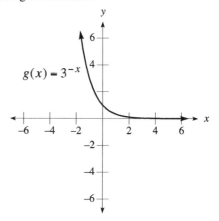

3. Finding the inverse:
$$2y - 3 = x$$
$$2y = x + 3$$
$$y = \dfrac{x+3}{2}$$

The inverse is $f^{-1}(x) = \dfrac{x+3}{2}$. Sketching the graph:

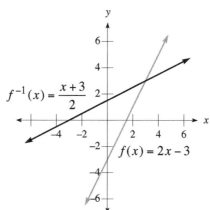

4. Graphing the function and its inverse:

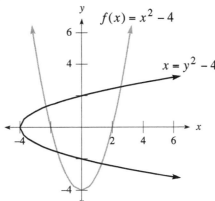

5. Solving for x:
$$\log_4 x = 3$$
$$x = 4^3 = 64$$

6. Solving for x:
$$\log_x 5 = 2$$
$$x^2 = 5$$
$$x = \sqrt{5}$$

7. Graphing the function:

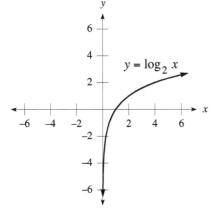

8. Graphing the function:

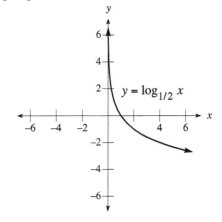

9. Evaluating the logarithm:
$$x = \log_8 4$$
$$8^x = 4$$
$$2^{3x} = 2^2$$
$$3x = 2$$
$$x = \frac{2}{3}$$

10. Evaluating the logarithm: $\log_7 21 = \dfrac{\ln 21}{\ln 7} \approx 1.5646$

11. Evaluating the logarithm: $\log 23{,}400 \approx 4.3692$

12. Evaluating the logarithm: $\log 0.0123 \approx -1.9101$

13. Evaluating the logarithm: $\ln 46.2 \approx 3.8330$

14. Evaluating the logarithm: $\ln 0.0462 \approx -3.0748$

15. Expanding the logarithm: $\log_2\left(\dfrac{8x^2}{y}\right) = \log_2 2^3 + \log_2 x^2 - \log_2 y = 3 + 2\log_2 x - \log_2 y$

16. Expanding the logarithm: $\log\left(\dfrac{\sqrt{x}}{y^4 \sqrt[5]{z}}\right) = \log\left(\dfrac{x^{1/2}}{y^4 z^{1/5}}\right) = \log x^{1/2} - \log y^4 - \log z^{1/5} = \dfrac{1}{2}\log x - 4\log y - \dfrac{1}{5}\log z$

17. Writing as a single logarithm: $2\log_3 x - \frac{1}{2}\log_3 y = \log_3 x^2 - \log_3 y^{1/2} = \log_3\left(\dfrac{x^2}{\sqrt{y}}\right)$

18. Writing as a single logarithm: $\frac{1}{3}\log x - \log y - 2\log z = \log x^{1/3} - \log y - \log z^2 = \log\left(\dfrac{\sqrt[3]{x}}{yz^2}\right)$

19. Solving for x:
    $\log x = 4.8476$
    $x = 10^{4.8476} \approx 70{,}404$

20. Solving for x:
    $\log x = -2.6478$
    $x = 10^{-2.6478} \approx 0.00225$

21. Solving for x:
    $3^x = 5$
    $\ln 3^x = \ln 5$
    $x\ln 3 = \ln 5$
    $x = \dfrac{\ln 5}{\ln 3} \approx 1.4650$

22. Solving for x:
    $4^{2x-1} = 8$
    $2^{4x-2} = 2^3$
    $4x - 2 = 3$
    $4x = 5$
    $x = \dfrac{5}{4}$

23. Solving for x:
    $\log_5 x - \log_5 3 = 1$
    $\log_5 \dfrac{x}{3} = 1$
    $\dfrac{x}{3} = 5^1$
    $x = 15$

24. Solving for x:
    $\log_2 x + \log_2(x-7) = 3$
    $\log_2(x^2 - 7x) = 3$
    $x^2 - 7x = 2^3$
    $x^2 - 7x - 8 = 0$
    $(x-8)(x+1) = 0$
    $x = 8, -1$
    The solution is 8 (−1 does not check).

25. Finding the pH: $\text{pH} = -\log(6.6 \times 10^{-7}) \approx 6.18$

26. Using the compound interest formula: $A = 400\left(1 + \dfrac{0.10}{2}\right)^{2 \cdot 5} = 400(1.05)^{10} \approx 651.56$

    There will be $651.56 in the account after 5 years.

27. Using the compound interest formula:
    $600\left(1 + \dfrac{0.08}{4}\right)^{4t} = 1800$
    $\left(1 + \dfrac{0.08}{4}\right)^{4t} = 3$
    $\ln(1.02)^{4t} = \ln 3$
    $4t\ln 1.02 = \ln 3$
    $t = \dfrac{\ln 3}{4\ln 1.02} \approx 13.87$

    It will take 13.87 years for the account to reach $1,800.

28. Using the depreciation formula: $V(4) = 18{,}000(1-0.20)^4 = 18{,}000(0.8)^4 \approx \$7{,}372.80$
    The car will be worth approximately $7,373 after 4 years.

Appendix A
Piecewise Defined Functions

Problem Set A.1

1. Graphing the function:

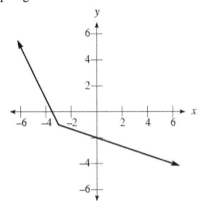

3. Graphing the function:

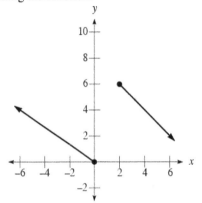

5. For problem 1, the domain is all real numbers, or $(-\infty,\infty)$.
    For problem 2, the domain is all real numbers, or $(-\infty,\infty)$.
    For problem 3, the domain is $(-\infty,0] \cup [2,\infty)$.

7. For the left branch $(x < -5)$, use the point $(-5,-3)$ and slope $= -1$ in the point-slope formula:
    $$y - (-3) = -1(x - (-5))$$
    $$y + 3 = -x - 5$$
    $$y = -x - 8$$
    For the right branch $(x \geq 1)$, use the point $(1,-3)$ and slope $= -2$ in the point-slope formula:
    $$y - (-3) = -2(x - 1)$$
    $$y + 3 = -2x + 2$$
    $$y = -2x - 1$$
    The piecewise function is $f(x) = \begin{cases} -x - 8 & \text{if } x < -5 \\ -2x - 1 & \text{if } x \geq 1 \end{cases}$.

9. Writing in piecewise form: $|x+3| = \begin{cases} -(x+3) & \text{if } x \le -3 \\ x+3 & \text{if } x > -3 \end{cases} = \begin{cases} -x-3 & \text{if } x \le -3 \\ x+3 & \text{if } x > -3 \end{cases}$

Sketching the graph:

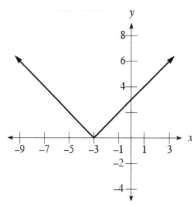